Power Systems

Yong-Hua Song · Xi-Fan Wang (Eds.)

Springer
London
Berlin
Heidelberg
New York
Hong Kong
Milan
Paris
Tokyo

Yong-Hua Song · Xi-Fan Wang (Eds.)

Operation of Market-oriented Power Systems

With 181 Figures and 80 Tables

Springer

Professor Yong-Hua Song
Department of Electronic and Computer Engineering
Brunel Institute of Power Systems
Brunel University
Uxbridge
Middlesex, UB8 3PH, UK

Professor Xi-Fan Wang
Department of Electrical Engineering
Xi'an, Jiaotong University
Xi'an
China

British Library Cataloguing in Publication Data
Operation of market-oriented power systems. – (Power Systems)
1.Electric utilities – Management 2.Electric utilities –
Rates 3.Electric power systems – Simulation methods
I.Song, Yong-Hua II.Wang, Xi-Fan
333.7'932

Library of Congress Cataloging-in-Publication Data
Operation of market-oriented power systems / Yong-Hua Song and Xi-Fan Wang (eds.)
p. cm. – (power systems)
Includes bibliographical references and index.

1. Electric power systems--Management. 2. Electric power systems--Load dispatching.
3. Electric utilities--Cost effectiveness. I. Song, Yong-Hua. II. Wang, Xi-Fan III. Series.
TK1001.064 2003
621.31--dc21 2003042774

Power Systems
ISBN 978-1-84996-893-5

http://www.springer.co.uk

69/3830-543210 Printed on acid-free paper

Preface

World-wide unprecedented reform and restructuring of the electric power industry has imposed tremendous challenges on the operation of power systems under this new environment. Regardless of the market structures that may emerge in various parts of the world, system security, reliability and quality of supply must be maintained. Faced by an increasingly complicated co-existence of technical and economical considerations, new computational tools and software systems are in great demand by generators, system operators, retailers, and other market participants to meet operating, scheduling, planning, and financial requirements.

In recent years there have been many books published on deregulation of the power industry but most of them placed emphasis on the market structure and policy issues. From an engineering point of view, how to develop effective computational tools for efficiently operating restructured power systems is still a big challenge. During the past several years, with funding from both research council and industry, we have been working on different computational models and methods for operation and control of market-oriented power systems. This book, resulting from these successful projects, covers all the major operational issues, such as scheduling and dispatch, congestion management, available transfer capability calculation, price forecasting and optimal bidding strategies. In addition, a comprehensive review of international research and world-wide industry practice is presented in each chapter before describing our methods, so as to give readers a broader state-of-the-art in this exciting field. Thus this book should be a useful reference for professional managers and engineers involved in the operation and control of market-oriented power systems. It would also be of considerable value to postgraduate researchers.

We are very grateful to various sponsors for their generous funding of our research. We thank our former/current PhD students, research follows and colleagues for their cooperation and contributions. We wish to thank Oliver Jackson of Springer for his assistance in the preparation of the book. We would also like to thank Angelo Centonza for re-setting the style of the whole book by overcoming incompatible word processing formats.

Yong-Hua Song, Brunel University, UK

Xi-Fan Wang, Xi'an Jiaotong University, China

Acknowledgements

The authors gratefully acknowledge various sponsors for their generous funding to the research reported in the book, in particular, the National Grid Company (UK), the Royal Academy of Engineering, the EPSRC, the British Energy, the ORS (UK), The Royal Society, Brunel University, the National Energy Policy Office (Thailand), the Chinese National Key Basic Research Fund (G19980310), Chinese Natural Science Foundation for Overseas Young Scientist Cooperation (No. 59928705), Cheung Kong Scholar Professorship, Chinese Natural Science Foundation Key Research Project (No. 59937150), and Chinese Academy of Sciences for Outstanding Overseas Scientists.

Table of Contents

List of Contributors

M. Eremia
Department of Power Engineering
University Polytechnic of Bucharest
Bucharest, Romania

Y. He
Office of Electricity Regulation
Beijing, China

M.R. Irving
Brunel Institute of Power Systems
Brunel University
Uxbridge, UK

H. Javid
Department of Electrical Engineering
Ferdowsi University
Mashhad, Iran

J.Z. Liu
North China Electric Power University
Beijing, China

Q. Lu
Department of Electrical Engineering
Tsinghua University
Beijing, China

J. Ma
Department of Electrical Engineering
Tsinghua University
Beijing, China

A. Maiorano
Sell Energy Ltd
London, UK

S. Phichaisawat
Department of Electrical Power Engineering
Chulalongkorn University
Bangkok, Thailand.

M. Rashidinejad
Department of Electrical Engineering
Shahid Bahonar University
Kerman, Iran

Y.H. Song,
Brunel Institute of Power Systems
Brunel University
Uxbridge, UK

Y.Z. Sun
Department of Electrical Engineering
Tsinghua University
Beijing, China

M. Tan
Institute of Automation
Chinese Academy of Sciences
Beijing, China

G. Taylor
Brunel Institute of Power Systems
Brunel University, Uxbridge, UK

M. Trovato
Department of Electrical Engineering
Politechico Di Bari
Bari, Italy

X. Wang
Alstom ESCA Corporation
Bellevue, USA

X.F. Wang
Department of Electrical Engineering
Xi'an Jiaotong University
Xi'an, China

Y. Xiao
Alstom ESCA Corporation
Bellevue, USA

K. Xie
ABB
Santa Clara, USA

I.K. Yu
College of Engineering
Changwon National University
Changwon, Korea

1. Operation of Restructured Power Systems

Y.H. Song, X. Wang and J.Z. Liu

There has been a world-wide trend towards restructuring and deregulation of the power industry over the last decade. The competition in the wholesale generation market and the retail market together with the open access to the transmission network can bring many benefits to the end consumers, such as lower electricity prices and better services. However, this competition also brings many new technical issues and challenges to the operation of restructured power systems. In recent years there have been many publications [1–4] devoted to the regulation and policy issues of establishing markets for electricity. This book will focus on the development of computational tools for effectively and efficiently operating such restructured systems.

1.1 System Operation in a Competitive Environment

Regardless of the market structures that may emerge in various parts of the world, one fact that seems always to be true is that transmission and generation services will be unbundled from one another. The generation market will become fully competitive, with many market participants who will be able to sell their energy services (or demand side management). On the other hand, the operation of a transmission system is expected to remain a regulated monopoly whose function is to allow open, non-discriminatory and comparable access to all suppliers and consumers of electrical energy. This function can be implemented by an entity called the Independent System Operator (ISO) [5–13].

Although electricity markets may have many different ISO designs and approaches all over the world, there are nonetheless elements that are necessary to all types of ISOs in order to meet their common basic requirements. Basically, the ISO has responsibility for the reliability functions in its region of operation and for assuring that all participants have open and nondiscriminatory access to transmission services through its planning and operation of the power transmission system. The ISO should conduct all of its functions in an impartial manner so that all participants are treated equitably. The main functions of the ISO can be categorized into reliability-related functions and market-related functions.

1.1.1 Reliability-related Functions

The reliability-related functions include two aspects:

- System operation and coordination. The ISO should perform system security monitoring functions and redispatch generation as necessary to eliminate real-time transmission congestion and to maintain system reliability, including taking all necessary emergency actions to maintain the security of the system in both normal and abnormal operating conditions.
- Transmission planning and construction. The ISO should carry out reliability studies and planning activities in coordination with the transmission owners and other market participants to assure the adequacy of the transmission system. The ISO should publish data, studies and plans relating to the adequacy of the transmission system. Data might include locational congestion prices and planning studies that identify options for actions that might be taken to remedy reliability problems on the grid and cost data for some of these actions.

1.1.2 Market-related Functions

First of all, an ISO must be a market enabler with no commercial interest in the competitive generation market. The market-related functions of an ISO must be carried out according to transparent, understandable rules and protocols. The following operational functions are necessary to enable a competitive generation market:

- Determine Available Transmission Capability (ATC) for all paths of interest within the ISO region.
- Receive and process all requests for transmission service within and through the ISO region from all participants, including transmission owners.
- Schedule all transactions it has approved.
- Operate or participate in an Open Access Same Time Information System (OASIS) for information publishing.
- Establish a clear ranking of transmission rights of all the participants on the ISO transmission system. Facilitate trading of transmission rights on its grid among participants.
- Manage transmission congestion in accordance with established rules and procedures for generation redispatch and its cost allocation.
- Assure the provision of ancillary services required to support all scheduled delivery transactions.
- Market settlement and billing functions.

The minimum functions of the ISO should include the operation and coordination of the power system to ensure security. In this case, a separate market operator (for example, the Power Exchange in California) is needed to perform the market-related functions. On the other hand, the maximum functions of the ISO will include all the reliability-related and market-related functions mentioned above and in addition the ISO is the transmission owner (for example, the National Grid Company of UK). The functions of the ISO at various sizes and time scales are shown in Figure 1.1.

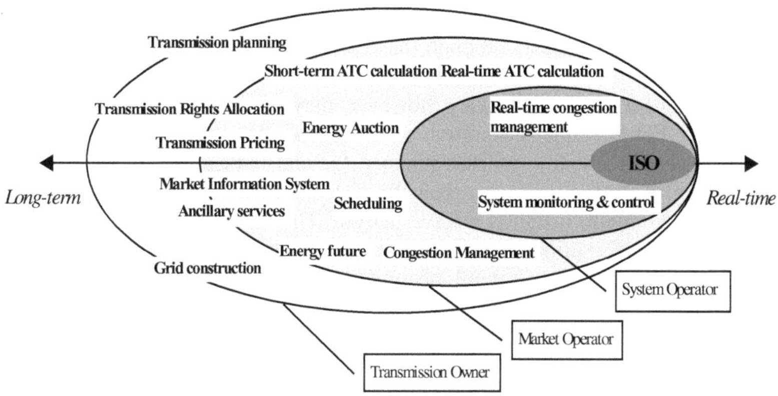

Figure 1.1. Functions of the ISO at various sizes and time scales

1.2 Effects of Industry Restructuring on System Reliability

Maintaining reliability involves two sets of operations: normal operations and emergency operations. Markets can do much to maintain reliability and prevent outages (by preparing resources for use in emergencies) during normal operations. Markets alone may be much less effective during actual emergencies [11].

Response time is the key factor that will determine whether the independent actions of participants in competitive markets can perform some reliability functions or whether technical standards and direct control will be required. Roughly speaking, competition is likely to work well for actions that occur half an hour or more in the future. Given this lead time, buyers and sellers can find the price level for each service that will balance supply and demand. For shorter time periods, however, system control is still likely to be required. Technical standards may be needed to specify the amount of each service that is required and to establish metrics for judging the adequacy of service delivery; markets can then determine the least-cost ways to deliver the required services. Disturbance response and generation planning provide useful examples of the two ends of the temporal spectrum.

The system operator must have the ultimate authority to compel actions needed to maintain reliability in real time and to restore the system quickly and safely after an outage occurs, although after-the-fact disputes may occur over who pays for what. If the system operator deemed it necessary to reduce flows on a particular transmission line, to take a line out of service, to reduce output at a particular generator, or to increase output at another generator, the operators of those pieces of equipment would be required to comply with the orders from the system operator. Such real-time operating authority is necessary to ensure system security in the future, as in the past, although these services may be obtained in a market-based means.

Providing for system adequacy, however, may be different in the future from in the past. For example, generation planning will be entirely different from its past practice. Historically, utilities planned for and built power plants to meet a predetermined reserve criterion, typically a 1-day-in-10-years loss-of-load probability or a minimum installed reserve margin. The regulator then determined the extent to which the utility would recover the costs of these generators through rates charged to the utility's retail customers. In addition, these costs were generally reflected in embedded-cost rates that did not vary from hour to hour. In the future, in a market-based model for providing adequate generation resources, decisions on retirement or repowering of existing generators and the construction of new units are likely to be made by investors with much less regulatory involvement. Of course, governments will still oversee the siting and environmental consequences of these decisions. But with retail choice of generation suppliers, markets (investors and consumers), rather than economic regulators, will decide which supplies are needed and economical.

These decisions will be made on the basis of trends in market prices and projected revenues from the sale of electricity relative to the construction and operating costs of the unit in question. Generators will be built when projected market prices of electricity are high enough to yield a profit. Prices in the future are likely to vary from hour to hour throughout the year, based on the units in operation each hour and the balance between unconstrained demand and supply online. When demand begins to exhaust the available supply, prices will rise, sometimes sharply, which in turn will suppress demand and induce investment in new supply. Spot prices will stop rising only when constrained demand is brought down, supply is increased, or both. Although these spot prices are likely to be quite low for most hours, they may be very high for a few hours each year. It is the level, frequency, and duration of these high prices that will signal markets to build more generating capacity, rather than the decisions of planners in vertically integrated utilities. This price volatility will also signal customers on the benefits of managing their loads in real time.

In electricity markets, customer response to real-time pricing signals could also help to improve reliability. High prices will encourage the construction of new generating units and the prompt restoration to service of existing units that are off-line. Similarly, with real-time price information, consumers can decide whether they want to conserve or reduce their usage at times of high prices. Together, these supply and demand responses to price will reduce the need to maintain expensive generating capacity that is only rarely used. Thus, economics can substitute for engineering to maintain real-time reliability when demand would otherwise exceed supply. The challenge of restructuring the electricity industry is to find an appropriate mix of economic incentives and performance standards that maintain reliability at the lowest reasonable cost.

1.3 New Requirement for Computation Tools and Software Systems in Electricity Markets

New computational tools and software systems are needed for generators, retailers, the ISO and other market participants to meet the operating, scheduling, planning, and financial requirements in the emerging competitive market environment. For example, generation companies may need new bidding systems to decide their bidding strategies and to communicate their bidding information with the market operator; the retailers and distribution companies may need new billing systems and new load management system to meet the time varying spot prices.

The most complex requirement on software systems will come from the ISO, who is in charge of the secure operation of the power system and may even run a few markets for energy auction, ancillary services procurement, and transmission rights auction, etc. Historically, the main software system in the control centre of the power system is the well-known Energy Management System (EMS), which consists of four major elements [11–13]:

- Supervisory Control and Data Acquisition (SCADA), including data acquisition, control, alarm processing, online topology processor, etc.
- Generation scheduling and control applications, including Automatic Generation Control (AGC), Economic Dispatch (ED), Unit Commitment (UC), hydrothermal coordination, short term load forecast, interchange scheduling, etc.
- Network analysis application, including topology processor, state estimator, power flow, contingency analysis, Optimal Power Flow (OPF), security enhancement, voltage and reactive power optimisation, stability analysis, etc.
- Dispatch Training Simulator (DTS), including all the three above components but in a separate off-line environment.

The EMS is still needed by the ISO in the electricity market, but some of its functions will change to meet the new requirement. For example, some generation scheduling applications might be removed or redesigned to be something like energy market trading applications while some other network analysis application, like OPF, should be extended to be able to perform new functions. DTS is also facing significant changes. It must include all the market applications and power system applications.

Besides EMS, some new software systems will be needed in the ISO. These new systems may include:

- Market long-term planning subsystem, including applications like a plan for future transmission expansion, long-term ATC determination, maintenance of transmission facilities, etc. This subsystem needs coordination between the ISO and transmission owners.
- Market trading subsystem, including all the possible functions associated with market administration roles of the ISO or a separate market operator. These functions could be a day-ahead energy auction to match supply offers and demand bids (a spot market), electricity futures trading, ancillary services procurement, transmission rights auction, etc.
- Market operation planning subsystem, including power system scheduling function, short-term ATC determination, short-run transmission-related services pricing, congestion management, etc.
- Market real-time dispatching subsystem, including power system dispatch function, system balancing, real-time ATC determination, real-time congestion management, etc.
- Market settlement and billing subsystem, determining deviations from the schedules and bilateral contracts, determining payments to suppliers and ancillary services providers, determining payments to financial instrument holders.
- Market information subsystem. All ISOs are expected to provide a system of open communication for information related to power system operations. In the US, some of this information will be published on the FERC mandated OASIS. The Information that would assist with the efficiency and security of system operation should include: system information on transmission congestion, locational market clearing prices, need and bid for ancillary services and their prices, and all applicable ATCs, etc.

These new software subsystems are linked tightly with each other and must coordinate with the existing systems in the control room to support the implementation of electricity market. Therefore, besides the development of new applications, there is still enormous work on software system integration to be done. An overview of possible software systems in the competitive market environment and the relationship between them are given in Figure 1.2.

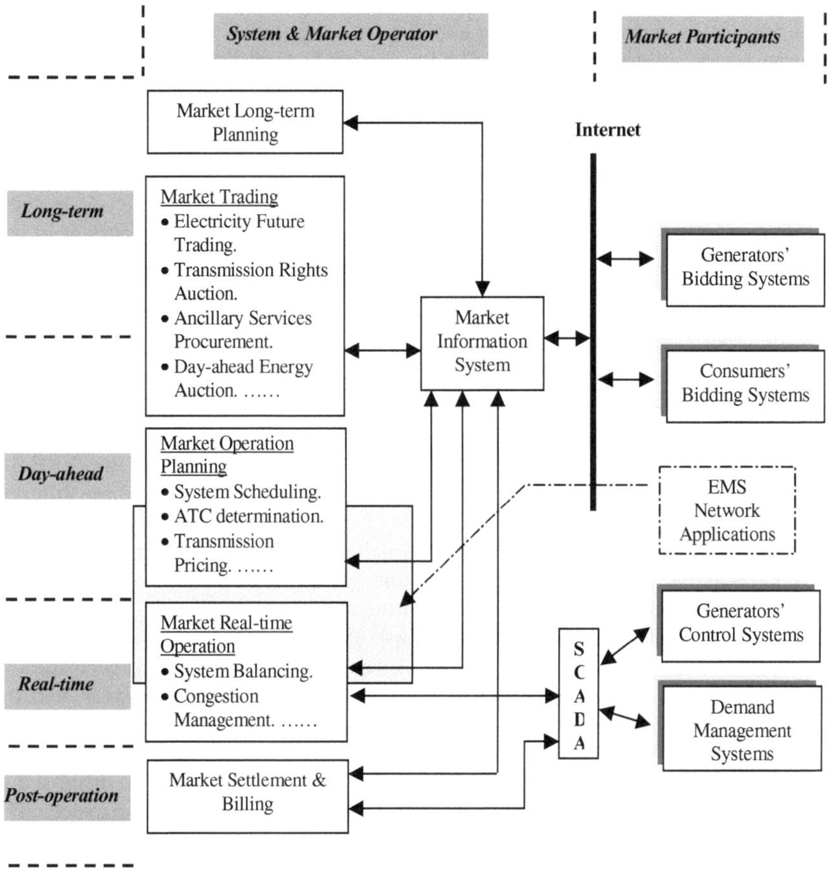

Figure 1.2. Overview of software systems in the competitive electricity market

1.4 Outline of the Book

In recent years, there have been many books published on deregulation of the power industry but most of them placed emphasis on the market structure and policy issues. From an engineering point of view, how to develop effective computational tools for efficiently operating restructured power systems is a big challenge. Several books [5–7] have been published recently on this topic. During the past several years, with funding from both research council and industry, we have been working on different computational models and methods for operation and control of such systems. The book is based on our recent PhD theses and over 30 papers published in IEEE and IEE journals/conferences. It covers all the major operational issues, such as scheduling and dispatch, congestion management, available transfer capability calculation, price forecasting and optimal bidding strategies. In addition, a review of international research and world-wide industry practice is covered in each chapter before presenting our methods so as to give the readers a broader state-of-the-art in this exciting field. The contents of the book are described in more detail in this section.

The main issue addressed in Chapter 2 is the investigation of the oligopolistic aspects of an electricity market as market structure is critically important in any industry. In this respect, a mathematical model is proposed for a wholesale electricity spot market, which takes into account the oligopolistic aspect of the market. The model is developed on the basis of the Cournot paradigm for the analysis of oligopolistic markets and allows the presence of private contracts (bilateral agreements) to be easily taken into account to assess their influence on profits and on the market-clearing price. The proposed methodology allows several features, typical of oligopolistic competition in the hourly market of electricity, to be analysed. The effectiveness of the proposed methodology is tested on the IEEE 30-bus system, assuming that the relative marketplace is composed of three different generating companies.

In the emerging electricity market, which relies on price-based competition, an unambiguous, transparent and predictable pricing framework of electricity for both active and reactive power is one of the major issues. Chapter 3 describes the development of an Integrated Optimal Spot Pricing (IOSP) model after reviewing various existing models. The proposed model includes the detailed derivation of optimal nodal specific real-time prices for active and reactive powers, and the method to decompose them into different components corresponding to generation, loss and many selected ancillary services such as spinning reserve, voltage control and security control. The features of the proposed model are discussed in relationship to existing pricing models and classical economic dispatch. The model is then implemented by extending the IPOPF method developed in the previous chapters. Insight is given by the case studies on a 5-bus, IEEE 30-bus and IEEE 118-bus systems.

It is well known that the generation and consumption of electricity must occur essentially at the same time. Therefore, real-time (minute-to-minute) operations and the associated markets and pricing approaches are crucial to design and implement a successful competitive wholesale electricity market. In Chapter 4, real-time dispatch methodologies for deregulated power systems are reviewed. An efficient dispatch framework for unbundled electricity markets through a real-time balancing mechanism is then presented. This framework is able to meet the real-time energy imbalance caused by unexpected contingencies and unpredictable load fluctuation with all the available adjustment bids in the real-time balancing market. Test results demonstrate that the proposed coordinated dispatch method implemented with a modified OPF can deal with system imbalance and network congestion simultaneously and successfully.

With the recent advent of world-wide power industry deregulation, measuring and modelling transmission system transfer capability is assuming greater importance. In Chapter 5, first of all, to give the reader some insight into the terms, history of standard terms of transfer capability of transmission networks is reviewed, together with the definitions and analysis of their weaknesses. To get accurate and dependable available transfer capability (ATC) measures, major uncertain factors associated with ATC estimation are identified and analysed. For short-term operational planning purposes, this chapter focuses on the development of a practical approach to evaluate the steady-state ATC of interconnected systems. To reflect the physical realities of the transmission network, a stochastic model is formulated for ATC calculation, where the key uncertainties affecting the level of ATC are modeled explicitly. To deal with the complex problem with both discrete random variables and continuous random variables involved efficiently, the solution methodology rests on a novel hybrid stochastic programming formulation. The results on the IEEE 118-bus system are encouraging, and clearly illustrate the effectiveness and efficiency of the proposed methodology.

Chapter 6 deals with congestion management. To find out what is going on in the real world, the congestion management approaches of five typical electricity markets in the world are investigated. A new scheme for an augmented LR-based regionally decomposed OPF is presented in this chapter. Applying the regionally decomposed OPF algorithm to active power congestion management across interconnected regions through the RBM, presented in Chapter 4, provides an efficient redispatch method to relieve the inter/intra-regional congestion without exchanging too much information between regional ISOs. The proposed method is of particular interest to a multi-utility or a multi-national interconnected system, such as the USA and Europe, where the independence of regional dispatching should be retained. Case studies based on the three-region IEEE RTS-96 are presented to illustrate the proposed method.

Chapter 7 analyses the dynamic security issue in the restructuring power market. A general framework is proposed to manage this issue in combination with the market mechanism and the power system intrinsic characteristics. Under this framework, the Security Management Market has been specially set up and categorized into several markets based on the market nature of the participants. By bidding into the respective market with their offers, different market participants provide their own control measures for utilization under emergency. Based on the available resources in this market, ISO would secure the system in the most economic way. Thus, power system security can be ensured under the market mechanism. Mathematically, the problem is formulated as an optimal control problem.

In Chapters 4 and 6, congestion management in the real-time operation of the electricity markets is discussed. A framework was proposed and implemented, in which the Independent System Operator (ISO) can balance its system and relieve transmission congestion efficiently through a real-time balancing market. This chapter is about how to avoid and manage transmission congestion during short-term (day-ahead to hour-ahead) scheduling of electricity markets. In this chapter, the concepts of two typical financial instruments, CfDs and FTRs, and how they have been used to hedge against price risks, are introduced. the basic model of optimal dispatch in the spot market and the fundamentals of LMP theory are presented. In particular, the impact of limits of bus generation and load on nodal prices are emphasised from the analysis of different forms of nodal price. After that, on the basis of the typical spot market dispatch model, some new individual revenue adequacy constraints are added to produce a more reasonable result for bilateral contract delivery in transmission congestion situations. An iterative procedure is employed to solve the formulated complex problem with dual variables in constraints. A 5-bus system and the IEEE 30-bus system are analysed to illustrate the proposed approach.

Ancillary services, a new terminology, are those services necessary to provide security and reliability issues that can be applied to dispatching and scheduling problem. Furthermore, ancillary services are required to maintain system reliability and to effect commercial transactions. Chapter 9 introduces the concept of ancillary services and the various ancillary service markets. In particular, we discuss the procurement and pricing of reserves in electric power markets. A joint dispatch model for energy and reserve with regards to real-time and day-ahead markets is presented. Finally, development of option pricing model is described.

Chapter 10 focuses on voltage security as a function of ancillary services that can be described in the context of reactive power markets and associated pricing mechanisms. First, an introduction to voltage security and reactive power management is presented and the consequences of recent electricity industry restructuring in the UK are highlighted. Second, an overview of recently developed reactive power markets is presented and several examples are discussed in detail. Finally, several recently developed algorithmic techniques are presented. The techniques introduce the concepts of transition-optimisation and congestion management in the context of a complete framework for the monitoring, assessment and control of voltage security.

Short-term load forecasting (STLF) plays an important role in the operational planning and the security functions of an energy management system. The STLF is aimed at predicting electric loads for a period of minutes, hours, days or weeks for the purpose of providing fundamental load profiles to the system. On the other hand, the electricity supply industry is undergoing unprecedented restructuring world-wide and there is a growing interest in the prediction of system marginal price (SMP) under the competitive market structure of deregulated power systems. The prediction of SMP improves the financial performance of an independent power producer bidding in the day-ahead market. Chapter 11 reviews conventional forecasting techniques in brief followed by novel load and price forecasting techniques via wavelet transform and neural networks with case studies to aid comprehension. The wavelet transform is a recently developed mathematical tool for signal analysis. A novel composite technique for short-term load forecasting using the Kohonen neural network and wavelet transform is described in this chapter. Practical examples on the South Korea system are reported. Then similar techniques are applied to the prediction of system marginal prices.

Chapters 12 and 13 are devoted to the generating sector. In Chapter 12, a methodology to simulate the strategic behaviour of generating companies in an oligopolistic electricity market, by using strategic supply functions, is proposed. In particular, electricity producers are supposed to bid in a pool-based electricity market. In order to simulate strategic competition among producers in the electricity market, the bidding process is expressed using linear supply functions. Another important aspect to be taken into account when analysing electricity market, as already mentioned in the previous chapter, is the presence of private contracts between generating companies and customers. In the developed tween generating companies and customers. In the developed methodology, the general case of an asymmetric oligopoly is analysed and the strategic behaviour of producers is investigated, assuming linear supply functions. The results of the analysis are expressed in terms of market clearing price, profit-maximizing value of the supply functions' slope, and hence real power output sold and profit made, of each producer sharing the market. Moreover, the proposed methodology allows the presence of private contracts, set up as Contracts for Difference (CfDs), to be taken into account, evaluating their effects on the market equilibrium conditions.

Chapter 13 presents the research work on bidding strategies of generators in electricity markets. Under deregulation, it is a common practice that generators produce and submit the bids to system operator/market operator based on market-oriented self-scheduling with the objective function of maximizing their own profits while considering the whole market situations such as system demands, network constraints, competitors' bidding behaviors and market prices. Four sets of bids sensitivities are defined and derived based on IPOPF model assuming each generator submits a linear bidding curve. They are bids-output sensitivity, bids-market price sensitivity, bids-profit sensitivity, and bids-line power flower sensitivity, which are valuable indexes and information in a pool-based electricity market. A single-period pool-based bidding model is then proposed based on bids sensitivities with the assumption that multi-round bidding and discriminatory pricing scheme are accepted as basic bidding rules. This model takes interactions among generators into consideration. The optimal bids (the coefficients of linear bidding curves) are obtained when the system reaches Nash equilibrium. Generators temporal constraints and demand-side bidding are not considered in this single-period bidding problem. Furthermore, the model is extended to take coalition into account. It allows the considered generator to make any potential coalition with other generators (grand coalition is not allowed). The algorithm is presented to find the optimal bids for the coalition subgroup.

Chapter 14 describes the aspect of transmission service improvement. Generally speaking, transmission services should be provided in a way to support all the required functions of transmission systems reliably. The advent of Flexible AC Transmission Systems (FACTS) technology has coincided with the major restructuring of the electric power industry. By the use of power electronics-based controllable components to control line impedance, magnitude and phase angle of nodal voltage individually and simultaneously, FACTS can provide benefits in increasing system transmission capacity and power flow control flexibility and rapidity. Therefore, it is able to play an important role in transmission services improvement. Taking the two important applications of FACTS control, ATC enhancement and FTR auction, as examples, this chapter addresses this issue with the intention of giving deeper insights into the ability of FACTS technology to facilitate electricity market operation and transaction.

1.5 References

[1] F.C. Schweppe, M.C. Caramanis, R.D. Tabors, and R.E. Bohn: "Spot Pricing of Electricity" (Kluwer Academic Publishers, 1988)
[2] M. Ilic, F. Galiana, and L. Fink: "Power System Restructuring: Engineering and Economics" (Kluwer Academic Publishers, 1998)
[3] M. Einhorn, and R. Siddiqui: "Electricity Transmission Pricing and Technology" (Kluwer Academic Publishers, 1996)
[4] S. Stoft: "Power Systems Economics", IEEE and Wiley-Interscience, 2002
[5] K. Bhattacharya, M. Bollen, and J.E. Daalder: "Operation of restructured Power Systems", Kluwer Academic Publishers, 2001
[6] M. Shahidiehpour, H. Yamin, and Z. Li: "Market Operations in Electric Power Systems", IEEE and Wiley-Interscience, 2002
[7] Y.H. Song: "Modern Optimisation Techniques in Power Systems"(Kluwer Academic Publishers, 1999)
[8] T.J. Overbye: "Reengineering the Electric Grid", American Scientist, May-June 2000, pp. 220-229
[9] S. Hunt, G. Shuttleworth: "Unlocking the Grid", IEEE Spectrum, July 1996, pp.20-25
[10] P.G. Harris: "Impacts of Deregulation on the Electric Power Industry", IEEE Power Engineering Review, October 2000, pp.4-6
[11] Secretary of Energy Advisory Board, U.S. Department of Energy: "Maintain Reliability in a Competitive U.S. Electricity Industry", September 29, 1998
[12] F.A. Rahimi, A. Vojdani: "Meet the Emerging Transmission Market Segments", IEEE Computer Application in Power, January 1999, pp. 26-32
[13] R. Hakvoort: "Technology and Restructuring the Electricity Market", Proceedings of the International Conference on Electric Utility Deregulation

2. Modelling and Analysis of Electricity Markets

A. Mariorano, Y.H. Song and M. Trovato

Around the globe, the electricity industry is undergoing a restructuring process moving from the corporate environment towards privatisation and re-regulation. The expectation is that a market-driven structure of the electricity industry can encourage competition in generation and supply, with open transmission and distribution access to enable it, and that this process can finally lead to efficiency, savings and reduced prices [1–6].

The idea behind the introduction of competition is that electric energy can be separated commercially as a product from transmission as a service. When considered as a commodity, electric energy can be traded in a free market, regulated only by consumer demands and supply bids. In particular, a *commercial market* and an *operational market* are usually identified [7, 8]. In the commercial market, the electricity is traded in a financial framework. In the operational market actual generation schedules are produced, which are re-optimised at regular time intervals, during the system operation [9–11].

As a consequence, while in the case of an integrated or nationalised utility the generation resources are centrally controlled and the objective function is to minimise the total cost made up of capital charges and operating costs, in a deregulated environment emphasis has to be necessarily given to the maximisation of the profit of several actors, such as generators, distribution companies and customers.

However, for each player the possibility of improving its own benefit depends on the market structure and on the type of competition really established. The market structure affects the freedom of action of each player and the nature of the competition determines how much the behaviour of a single player can condition the action of other players.

The main components of the electricity industry are generators, transmission owners, dispatching entities and distribution companies. However, depending on the adopted regulatory framework, some of these components can be aggregated or unbundled and have different levels of interaction [12–17].

The main issue addressed in this chapter is the investigation of the oligopolistic aspects of an electricity market, which can influence the competition level. The privatisation and deregulation of the power sector should lead to gain in technical and economical efficiency, but some studies also indicate that market participants may have an incentive to substantially increase prices [18–30]. Moreover, even when market rules should allow a perfect competition among the participants, a generator exercising market power could influence the action of other generators [31,32]. These problems increase when the electricity market answers to the description of an *oligopoly*.

Strictly interpreted, oligopoly is a market situation where, due to the presence of few firms operating within the market, the actions of a firm have, quite invariably, a significant influence on the behaviour of the remaining firms and, consequently, on their profits [33].

The term '*few firms*' is not explicitly defined, but it is usually intended to cover the condition between two extremes: the presence of a unique producer (monopoly) or of a large enough number of producers to allow perfect competition. Depending on the market rules, it has been noted that in oligopolistic markets prices and incomes may be raised in conjunction without fear of loss of market share, and profits would then be much higher.

2.1 Types of Markets

The task of integrated utilities was to predict future energy and demand needs and assess the need for new generation and transmission to maintain economic and secure operation to published standards. Therefore, they determined both the location and the type of generation that best met the overall needs of the system to maintain optimum performance. They planned the timely closure of older generating units and maintained and integrated planning processes for the development of the complete system.

In the new regime, individual generators and transmission companies have to make their investment decisions independently and with little knowledge of the commercial plans of their competitors. Therefore, it becomes of major concern to understand the new market mechanism and its rules. For this purpose, the electricity marketplace features are described in detail. In particular, since all the different approaches to deregulate the electricity industry involve the presence of brokerage systems and auction markets, electricity auctions are analysed, focusing on their structures, the different ways of bundling demand into lots for auctions and the different pricing and sequencing rules applied in the electricity sector.

As mentioned before, the main concept behind the deregulation of the electricity industry is that electric energy as a commodity can be analysed in a financial framework. For this reason, different markets, within the same brokerage system, can be set up. In the electricity marketplace the most common type of markets are: cash, futures, options, swap and planning markets, each of them having a different time frame and involving the use of different financial derivatives. Finally, the features of these financial derivatives are analysed, in terms of contract components, basic contract types and how they are carried out in electricity markets.

2.1.1 Fundamental Market Structure and Mechanism

Although the common theme of electricity industry deregulation is promoting competition, there is a wide spectrum of opinions on the best approach to be implemented. An overview of alternative market structures and mechanisms is provided, particularly for enabling generation competition. Also, the restructuring schemes adopted for the electricity industry in different countries of the world are discussed.

In a *vertically* or *fully integrated utility* (often a state utility), all the generation is owned by a unique company, together with the transmission and sometimes distribution. In this case, there is a monopoly condition at all levels and any form of competition is unfeasible. This electricity industry structure has been the paradigm for a century and its main advantage is to enable integrated generation and transmission planning. On the contrary, the absence of competition may lead to inefficiencies and over-investment.

The *mixed generation scheme* is arrangement of the electricity industry where an elementary form of energy market can be implemented and competition developed. In this case, a vertically integrated utility is retained but is required to enable access into the market of non-utility generators (NUGs). However, the level of competition is low since the utility invariably favours its own generation.

In the *Single Buyer* (SB) model, a nominated purchasing authority is allowed to buy energy and services from generators and Independent Power Producers (IPPs) on behalf of all registered consumers. Although, it enables the development of generation and transmission in a coordinated manner, the merits of this arrangement are quite debatable in that the authority represents a monopoly, which is not itself subject to any market force. Moreover, the buyer should not own generating plants since its main objective is promoting full competition at the generation level. A further disadvantage is that the single buyer scheme usually requires a minimum number of long-term contracts between the purchasing agency and producers, to insure them against market risk. Consequently, this risk is passed, through the purchasing agency, to captive consumers.

In recent years, greater emphasis has been placed on realising more efficient electricity industry restructuring models both in the USA [1] and in Europe [2,3]. The US Federal Energy Regulatory Commission (FERC) made it mandatory for transmission network owners to post rates for open access and use of their network. The result is an open market structure that includes, in general, generating companies (gencos), distribution companies (discos), transmission companies (transcos), and an Independent System Operator (ISO). The ISO operates the transmission network and provides transmission services to all transmission customers. Depending on the adopted deregulation policy, the ISO may or may not have the responsibility for running an energy market or Power Exchange (PX) [4].

The PX may simply provide a forum for bilateral (or multilateral) trades (unregulated PX) or to act as a pool for energy supply and demand bids, and establish a market-clearing price (regulated PX). The former case involves most of the energy being traded directly between producers and consumers, with minimal regulatory intervention. Since generation availability and actual demand level are not accurately predictable, energy transactions have only to be coordinated in order to clear any residual energy or not traded demand.

When a poolco-based model is adopted for the PX, all energy, or most of the energy, is traded through the PX. Depending on the specific market operating rules, different markets can coexist. Usually, a day-ahead market is implemented for trading energy one day before each operating day. In other situations, the day-ahead market can be preceded by a long-term market, and integrated by an hour-ahead market, which allows energy trading up to one or two hours before the operating hour. Then, based on the above mentioned market alternatives and ISO responsibilities, various ISO/PX structures can be implemented.

In Europe, electricity industry restructuring moved from a different situation where the primary need was to split up formerly integrated, government utility organisations. This involved starting unbundling processes for the separation of the original utility into separate and independent private organisations owned by shareholders. One of the issues of the EC-Directive, which was enacted in December 1996, is the organisation of access to the grid. Two institutional, not exclusive, arrangements are considered: the negotiated third-party access (NTPA) and the Single Buyer model. In the NTPA scheme, all energy producers and suppliers and all eligible customers can sign supply contracts with each other. The modalities of grid access have to be negotiated between the parties to the contract and the network operator. It can be observed that, although each country of the European Union has enacted state laws conforming to the EC-Directive, the actual restructuring process is, at the moment, still in progress in several countries. However, the England and Wales system and the Nord Pool system, which covers Norway and Sweden, seem to constitute well-established electricity industry structures. These two cases will be further analysed since each one represents different form of the electricity industry deregulation.

2.2 Commodity Markets

Electric energy as a commodity can be analysed in a financial framework [13,14]. A commodity is defined as anything useful or valuable that can be turned into a commercial or other advantage. Commodity markets provide the setting for trading commodities. Commodity market prices generally have a maximum allowable price move in a given day to provide price stability. It is the element of risk that commodity markets uniquely address. If no market participant can impose a position through a change in actions, then a state of general equilibrium exists. A stable market process is one where revised bids and offers lead to a uniform price with uniform quantities offered and demanded at that price.

It is assumed that there are five markets commonly operating: cash market, futures market, option market, swap market and planning market.

2.2.1 Cash Market

The two contracts normally traded in cash markets are spot and forward. The spot contracts reflect the current price of transactions. The forward contracts reflect future systems conditions. Transactions are executed immediately for spot contracts. In the forward market, prices are determined at the time of the contract but the transactions occur at some time in the future. The auction mechanism for the cash market is developed in [14].

2.2.2 Futures Market

A futures market is a place where buyers and sellers can meet readily. A futures market creates competition because it unifies diverse and scattered local markets and stabilises prices. The contracts in futures markets are risky because price movements over time can result in large gains or losses. There is a link between cash market and futures market which allows price volatility. The reduction of price volatility could be accomplished by increasing market friction through an increase in transactions costs, an increase in margins, limiting arbitrage or banning trading in futures. The components of a futures contract include trading unit, trading hours, trading months, price quotation, minimum price fluctuation, last trading day, exercise of options, option strike prices, delivery, delivery period, alternate delivery procedure, exchanged of futures for, or in connection with physical transactions, quality specifications and customer margin requirements [15].

2.2.3 Options Market

Options allow the purchaser flexibility in exercising the right to activate a contract or cancel it. Claims to buy are "call" options. Claims to sell are "put" options. Options contracts redistribute the risk from those wishing to avoid it to those willing to accept it. They contribute to price stability in that they increase supply when prices rise and increase demand when prices fall. Options can be traded on the cash or futures markets. Thus options, as well as futures, are financial vehicles to provide risk management.

2.2.4 Swap Market

In the swap market, contract position can be closed with an exchange of physical or financial substitutions. The trader can find another trader who will accept delivery and ends the trader's delivery obligation. The acceptor of the obligation is compensated through a price discount or a premium relative to the going price.

2.2.5 Planning Market

The growth of the transmission grid requires contracts of expected usage for transmission companies to finance such projects. The planning market would underwrite the use of equipment subject to the long-term commitments which the generation and distribution companies are bound by the rules of network expansion to maintain a fair marketplace. The network expansion should be built to maximise the use of the transmission grid for all players.

In Figure 2.1, all the different markets and their time frame and interactions are shown.

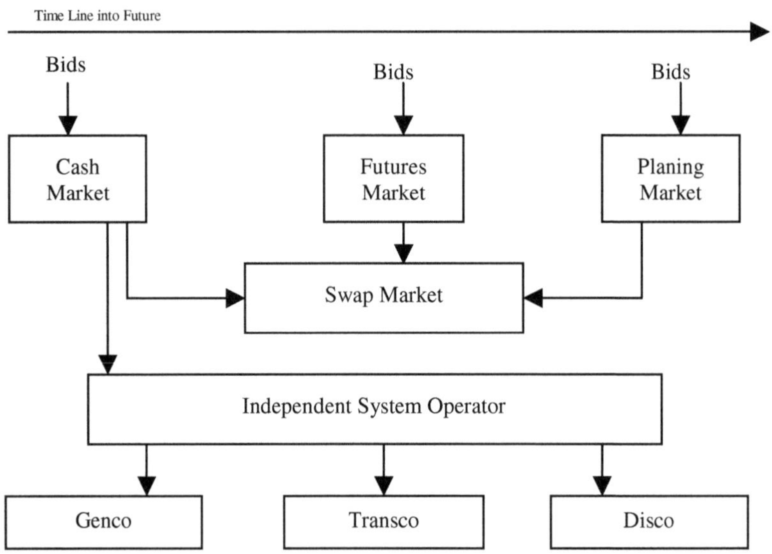

Figure 2.1. Electricity market

2.3 Perfect Competition and Oligopolistic Market

The aim of this section is to give a better understanding of economic markets, in terms of different models, different rules and different "power" of the participants according to the particular situation.

In particular, the first section concerns the general market rule of supply and demand: how supply and demand curves can represent the general market condition for a particular good and how their intersection can reveal the possible equilibrium conditions. Moreover, a brief introduction to the elasticity of demand and supply respect to the price, and versus other goods, in terms of substitutes and/or complements, is illustrated.

According to the market configuration, the number of firms sharing the same industry, the relative size of each firm compared to the total industry output, physical constraints and so on, particular conditions can take place entitling each participant to a different "market power". The first configuration analysed is the perfect competitive market model. This is an ideal condition and it was the market paradigm towards which the electricity marketplace was supposed to tend when liberalising the electricity industry. In this kind of situation each participant's output is supposed to represent such a small part of the total amount produced by the whole industry that the only possible way of maximising its own profit is to bid at marginal costs.

Unfortunately, as will be analysed in detail in the next several sections, because of the relative size of competitors and/or the presence of transmission constraints, set by weak transmission lines, the electricity market cannot be considered as a perfect competitive market structure. Therefore, different theories, based on imperfect competition, called *Oligopoly*, can be better implemented to represent this situation.

Because inter-firm interactions in imperfect markets take many forms, oligopoly theory lacks unambiguous results of the sort that can be obtained for perfect competition and monopoly. Instead, a variety of results can be developed, derived from different behavioural assumptions, with each model potentially relevant to certain real-world situations.

The Cournot model theory is analysed in detail, showing how it can be interpreted in terms of isoprofit curves and reaction functions, since it is the basis of the work proposed in this thesis. Nevertheless, two other oligopolistic models, the Bertrand model and the Stackelberg model, are briefly illustrated and the results are compared to the one obtained by applying the Cournot Model to the same market conditions.

2.3.1 Market Equilibrium: The Law of Supply and Demand

Supply–demand analysis is a fundamental and powerful tool that can be applied to a wide variety of interesting and important problems. In this section, a review of how supply and demand curves are used to describe the market mechanism is reported [16,17].

In fact, without government interventions, through the imposition of price controls or some other regulatory policy, supply and demand will come into equilibrium to determine the market price of a good and the total quantity produced. What that price and quantity will be depends on the particular characteristics of supply and demand curves relative to the particular market under consideration.

Let us start with a review of the basic supply–demand diagram as shown in Figure 2.2. The vertical axis shows the price of a good p, measured in a particular currency per unit. This is the price that sellers receive for a given quantity supplied and that buyers will pay for a given quantity demanded. The horizontal axis shows the total quantity demanded and supplied, X, measured in number of units per period. The *supply curve S* says how much producers are willing to sell for each price that they receive in the market. This can be expressed as follows:

$$X_S = X_S(p) \tag{2.1}$$

As shown in Figure 2.2, the supply curve slopes upward because the higher the price, the more firms are usually able and willing to produce and sell.

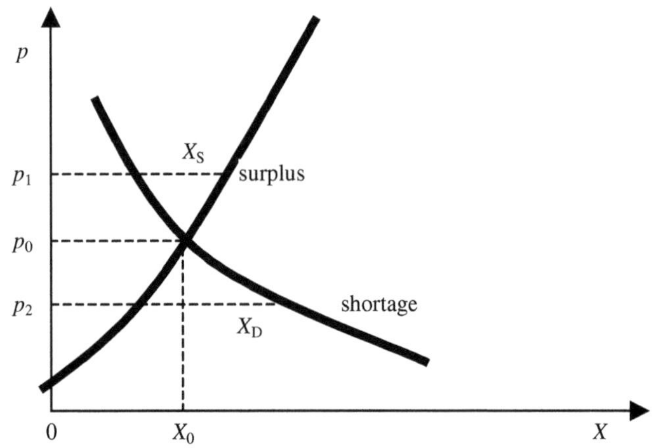

Figure 2.2. Supply and demand

The *demand curve D* says how much consumers are willing to buy for each price per unit that they must pay. It can be expressed as follows:

$$X_D = X_D(p) \tag{2.2}$$

As shown in the figure, it slopes downward because consumers are usually ready to buy more if the price is lower.

The two curves intersect at the *equilibrium*, or *market-clearing*, price and quantity. At this price p_0, the quantity supplied and the quantity demanded are just equal, to X_0. The market mechanism is the tendency in a free market for the price to change until the market clears, that means until quantity supplied and quantity demanded are equal. At this point there is neither shortage nor excess of supply, so there is no pressure for the price to change further. Supply and demand might not always be in equilibrium and some market might not clear quickly when conditions change suddenly, but the tendency is for the market to clear.

In fact, let us suppose that the price was initially above the market-clearing level, p_1 in Figure 2.2.

Then producers would try to produce more and sell more than consumers were willing to buy. A surplus would accumulate, and to sell this surplus, or at least prevent it from growing, producers begin to lower their prices. Eventually price would fall, quantity demanded would increase, and quantity supplied would decrease until the equilibrium price p_0 was reached.

The opposite would happen if the price were initially below p_0, p_2 in Figure 2.2. A shortage would develop because consumers would be unable to purchase all they would like at this price. This would put upward pressure on price as consumers tried to outbid one another for existing supplies and producers reacted by increasing and expanding their output. Again, the price would eventually reach p_0.

2.3.1.1 Elasticities of Supply and Demand

Supply and demand curves show how producers and consumers are willing to sell and buy as functions of the price they receive and pay. But supply and demand are also determined by other variables beside price. For example, the quantity that producers are willing to sell depends on their production costs, including wages, interest charges and cost of raw materials. Quantity demanded depends on the total disposable income available to consumers, the price of other products and other variables [21–23].

An *elasticity* is defined as *a measure of the sensitivity of one variable to another*. In this context, it is used to determine the percentage change that will occur in one variable in response to a 1 per cent change in another variable. In particular, the *price elasticity of demand* E_p measures the sensitivity of quantity demanded to price changes and it can be expressed as follows:

$$E_p = \frac{\% \Delta X_D}{\% \Delta p} \tag{2.3}$$

Since the percentage change in a variable is just the absolute change in the variable divided by the original level of the variable, Equation 2.3 can be rewritten as follows:

$$E_p = \frac{p}{X_D} \frac{\Delta X_D}{\Delta p} \tag{2.4}$$

The price elasticity of demand is usually a negative number. When the price of a good increases, the quantity demanded usually falls, so $\frac{\Delta X_D}{\Delta p}$ is negative and therefore E_p is negative.

When the price elasticity is greater than 1 in magnitude, the demand is said to be price elastic because the percentage decline in quantity demanded is greater than the percentage increment in price. If the price elasticity is less than 1 in magnitude, demand is said to be price inelastic.

In general, the elasticity of demand for a good depends on the availability of other goods that can be substituted for it. When there are close substitutes, a price increment will cause the consumer to buy less of the goods and more of the substitute. Demand will be then highly price elastic.

Equation 2.6 says that the price elasticity of demand is the change in quantity associated to a change in price times the ratio of price to quantity. But moving down the demand curve $\frac{\Delta X_D}{\Delta p}$ may change, and the price and quantity always change. Therefore, the price elasticity of demand must be measured at a particular point on the demand curve and will generally change when moving along the curve itself.

As mentioned before, the demand for some goods is also affected by the prices of other goods. A cross-price elasticity of demand refers to the percentage change in the quantity demanded for a good that results from a 1 per cent increment in the price of another goods. By indicating with X_{D1}, X_{D2}, p_{D1} and p_{D2} the quantities demanded and the price of each goods, respectively, the following expression can be obtained:

$$E_{X_{D1},p_{D2}} = \frac{p_{D2}}{X_{D1}} \frac{\Delta X_{D1}}{\Delta p_{D2}} \tag{2.5}$$

The cross-price elasticity will be positive if the goods are *substitutes*, they compete in the market. On the contrary, the cross-price elasticity is negative if they are *complements*, they tend to be used together so that an increase in the price of one tends to push down the consumption of the other.

Elasticities of supply are defined in a similar way. The price elasticity of supply is the percentage change in the quantity supplied resulting from a 1 percent increase in the price. This elasticity is usually positive because a higher price gives producers an incentive to increase output. Moreover, for most manufactured goods, the elasticity of supply with respect to the price of raw materials are negative. In fact, an increment in the price of raw material input means higher cost for the firm so, other things being equal, the quantity supplied will fall.

2.3.2 Perfect Competition

The theory of perfect competition analyses a situation where there are a large number of small firms producing an identical output for which there is a single market price which is not influenced by the actions of any individual firm. So the firms are price-takers, and they seek to maximise profits. There is no impediment to new firms entering the industry and each firm faces identical cost conditions which include average costs rising at a level of output which is small compared to the total market [20–22].

In the short run, the "capital" of the firm is fixed and this means that the number of firms is fixed. Thus firms outside the industry with zero capital in the industry cannot enter the industry and firms in the industry cannot leave it.

Then the firm faces the problem of maximising its profit over the short-run period for which it is making its decision. The expression for the profit is:

$$\pi = pX - C(X) \tag{2.6}$$

where p is the market price, X the level of output and C the relevant total cost.

The firm has only to decide the level of output. The maximisation of this profit function requires:

$$\frac{d\pi}{dX} = p - \frac{dC}{dX} = 0 \tag{2.7}$$

$$\frac{d^2\pi}{dX^2} = -\frac{d^2C}{dX^2} < 0 \tag{2.8}$$

so that the output is adjusted to bring price and marginal cost into equality, and marginal cost are rising rather than falling.

In the long run, firms can adjust their capital stock, thus enabling firms to enter or leave the industry. The free-entry assumption leads to a long-run equilibrium condition that is characterised by the fact that economic profits are zero; otherwise firms would be entering or leaving the industry.

Zero profits, where profits refer to excess of revenue over all costs including the cost of capital, yield to the long-run condition in which price equals average cost.

The model of perfect competition is widely used but there are a number of possible objections to it which need to be investigated.

The theory of perfect competition is essentially a theory of equilibrium, and in particular, with firms acting as price-takers. It does not incorporate any explanation of how price changes. In fact, within the model set above there is no way for prices to change.

The second assumption is that the theory of perfect competition has proceeded on the basis that all firms have the same cost curves and face the same demand conditions, that means the same price. The individual firm examined, then, can be thought of as the representative firm of the industry, with the experience of other firms paralleling that of the firm examined. An alternative interpretation is that the firm examined is the "marginal " firm in the industry in the sense that is the firm on the margin of leaving the industry since it is only just earning normal profits. The other, non-marginal, firms would be earning supernormal profits. However, their output decision rule would be the same if they wished to maximise their profits, equate marginal costs to price, but in the long-run equilibrium their profits would be above the normal level. For, by assumption, firms not in the industry would have cost curves if they entered the industry which would never yield even normal profits. In other words, their average cost curves lie everywhere above the prevailing price.

Finally, the third assumption is that of increasing costs setting in at a level of output which is small relative to the industry total. The presence of economies of scale would undermine perfect competition in at least two ways. First, whichever firm were larger would have lower costs and could expand either by investing their profit or by a departure from price-taking by setting a price lower than that which can be matched by smaller firms. Secondly, when there are economies of scale, the marginal cost curve lies below the average cost curve. Thus the rule of equating price and marginal cost would generate losses for the firm, since average costs are greater than marginal costs (price) for any level of output. Therefore, competitive conditions, in terms of a large number of small firms in an industry and the existence of unexploited economies of scale and/or of decreasing short-run costs facing the firm, would appear to be incompatible.

2.3.3 Classical Theories of Oligopoly

Models of enterprise decision-making in oligopoly derive their special character from the fact that firms in an oligopolistic industry are, and know they are, interdependent [18,19 22]. In competitive markets, there are enough other producers that each firm can safely ignore the reactions of rivals to its decisions when it makes its output choices. This is not necessarily the case where there are only a few producers; and two broad approaches can be taken to solve the problem created by this fact.

First the oligopolist can be thought of as making assumptions about the variables to which competitors will react and about the nature of their reactions. Models dealing with such behaviour are referred to as non-collusive because, though in equilibrium the expectations of each firm about the reaction of the rivals are realised, the two sides never communicate with each other directly about their likely reactions. Alternatively, firms can be thought of as agreeing to cooperate in setting price and quantity from the outset. Collusive models deal with such behaviour. In these models firms form a cartel, and output–price combinations equivalent to those that would emerge under monopoly are chosen for the industry as a whole. In this thesis, only non-collusive models are taken into account.

Because interfirm interactions in imperfect markets take many forms, oligopoly theory lacks unambiguous results of the sort that can be obtained for perfect competition and monopoly. Instead, a variety of results can be developed, derived from different behavioural assumptions.

2.3.3.1 The Cournot Model

The best known non-collusive model is the one derived by Augustin Cournot in 1838. In this formulation, the strategic variable to which the firm reacts is the output of its rival and the firm is assumed to take its rival's output choice as given.

In order to simplify the analysis that follows, it will be assumed that there are only two firms in the industry, a situation called duopoly. The results can easily be extended to the more general case of n firms.

If X_1 and X_2 are the production levels of the two duopolists, the aggregated demand function, expressed as price in function of quantity, can be formulated as follows:

$$p = F(X) = F(X_1 + X_2) \tag{2.9}$$

In a duopoly case, the revenue of each firm depends on the level of output of both firms because output price is given by the industry demand curve:

$$R_1 = X_1 F(X_1 + X_2) = R_1(X_1, X_2) \tag{2.10}$$
$$R_2 = X_2 F(X_1 + X_2) = R_2(X_1, X_2) \tag{2.11}$$

Consequently, the profits of the two firms are given by:

$$\pi_1 = R_1(X_1, X_2) - C_1(X_1) = pX_1 - C_1(X_1) \tag{2.12}$$
$$\pi_2 = R_2(X_1, X_2) - C_2(X_2) = pX_2 - C_2(X_2) \tag{2.13}$$

where $C(\bullet)$ represents each firm's cost function.

The Cournot assumption is that each firm chooses its output level taking the output of its rivals as given. Hence firm 1 chooses X_1 to maximise π_1 given X_2, and firm 2 chooses X_2 to maximise π_2 given X_1. This can be expressed as:

$$\begin{aligned}
&\left[\frac{\partial \pi_1}{\partial X_1}\right]_{X_2=cost} = \left[\frac{\partial R_1}{\partial X_1}\right]_{X_2=cost} - \frac{dC_1}{dX_1} = 0; \quad \left[\frac{\partial R_1}{\partial X_1}\right]_{q_2=cost} = \frac{dC_1}{dX_1} \\
&\left[\frac{\partial \pi_2}{\partial X_2}\right]_{X_1=cost} = \left[\frac{\partial R_2}{\partial X_2}\right]_{q_1=cost} - \frac{dC_2}{dX_2} = 0; \quad \left[\frac{\partial R_2}{\partial X_2}\right]_{X_1=cost} = \frac{dC_2}{dX_2}
\end{aligned} \tag{2.14}$$

Therefore, the first-order conditions require that each producer chooses its output level so that marginal revenues are equal to marginal costs. The marginal revenues for each firm are given by:

$$\left[\frac{\partial R_i(X_1, X_2)}{\partial X_i}\right]_{X_j=cost} = p + X_i \cdot \frac{\partial p}{\partial X} \cdot \frac{\partial X}{\partial X_i} = p + X_i \cdot \frac{\partial p}{\partial X} \qquad (i = 1,2) \tag{2.15}$$

The second-order condition for each producer requires (for i = 1,2) that:

$$\left[\frac{\partial^2 \pi_i}{\partial X_i^2}\right]_{X_j=cost} = \left[\frac{\partial^2 R_i}{\partial X_i^2}\right]_{X_j=cost} - \frac{d^2 C_i}{dX_i^2} < 0; \quad \left[\frac{\partial^2 R_i}{\partial X_i^2}\right]_{X_j=cost} < \frac{d^2 C_i}{dX_i^2} \tag{2.16}$$

that is that the marginal revenue has to increase less steeper than marginal costs.

The market is in an equilibrium condition at which each firm maximises its profit, given the rival's output, and none of them feels the need to change its output level. The equilibrium condition can then be revealed by solving the set of Equations (2.14).

2.3.3.2 Isoprofit Curves and Reaction Functions

The Cournot model can also be explained diagrammatically with reference to *Isoprofit Curves* and *Reaction Functions*. Let us assume that the oligopoly under consideration is, for simplicity, a duopoly.

An *Isoprofit Curve* for firm 1 is the locus of points in (X_1, X_2) space defined by different levels of output for both firm 1 and firm 2 that yield the same level of profit for firm 1. In duopoly, the profits of each firm depend on the output decision of its rival because the price obtained for the goods depends on industry output ($X_1 + X_2$).

Isoprofit curves are concave to the axis of the firm to which they relate, with the level of profits declining with the height of the isoprofit curve above the horizontal axis. A family of isoprofit curves is illustrated in Figure 2.3.

It should be noted that if firm 2 produces nothing, firm 1 is a monopolist, the output is M_1 and the isoprofit curve corresponding to monopoly profits is a point on the axis at M_1. Similarly, M_2 represents monopoly output for firm 2.

In order to analyse the Cournot equilibrium, the impact of a change of firm 2's output upon the quantity produced by firm 1, and vice versa, have to be considered. The relationships through which these interactions are summarised are called Reaction Functions: firm 1's reaction function shows how firm 1 will change its output in response to a change in X_2. Consequently, firm 2's reaction function shows how firm 2 will respond to a change in X_1. It is assumed that both firms seek to maximise their profit and that the industry faces a downward-sloping demand curve along which price varies with the quantity produced by both firms, that is $X_1 + X_2$.

With the Cournot behavioural assumption that firms take the output of their rival as given, an increase in X_2 lowers the price both firms receive from the market, lowers the marginal revenue obtained by firm 1 at each level of output and therefore, for given marginal costs, leads firm 1 to reduce its output. Similarly, this happens for firm 2, suggesting that the reaction functions for both firms are backward bending.

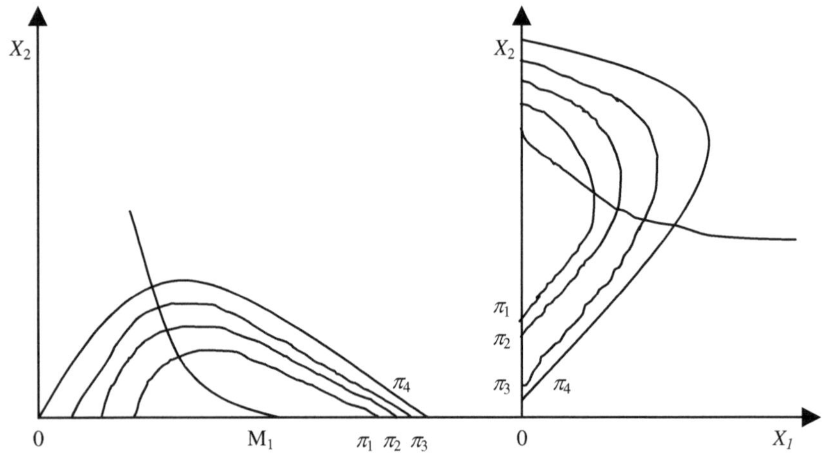

Figure 2.3. Isoprofit curves

This is illustrated in Figure 2.4. For a given level of output in firm 2, X_2^1, firm 1 seeks to maximise its profit, which entails choosing an output level X_1^1 which yields the lowest isoprofit curve consistent with X_2^1 at the tangency between $\pi 1^1$ and X_2^1. If firm 2 increases its output to level X_2^2, the highest level of attainable is at level X_1^2 at the tangency between π_2^1 and X_2^2. The locus of the tangencies between given values of X_2 and isoprofit lines plots out the reaction function for firm 1. This is the locus of maxima of the isoprofit curves. Each reaction function, therefore, plots the profit-maximising level of output for each firm given every conceivable expected value of production which could be chosen by its rival. The Cournot equilibrium is reached when the expectation of each firm about its rival's output choice proves to be correct. Clearly, this occurs at the intersection of the two reaction functions, where each firm is choosing to produce exactly the level of output which its rival expects. This mutual realisation of output expectations is illustrated at output levels (X_1^*, X_2^*) in Figure 2.5.

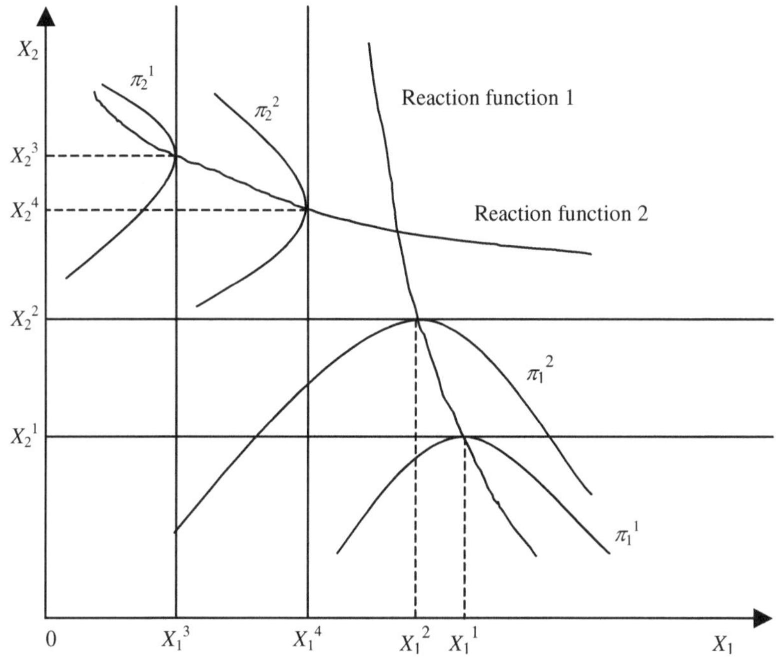

Figure 2.4. Adjustment process

It is also possible to illustrate the path of adjustment towards equilibrium implied by the Cournot behaviour. If it is arbitrarily assumed that firm 1 chooses X_1^1, profits are maximised for firm 2 at the point where X_1^1 intersects firm 2's reaction function at X_2^1. But this output pair, (X_1^1, X_2^1) is not a Cournot equilibrium because firm 1's expectation of firm 2's behaviour is not confirmed by firm 2. Firm 1 chose X_1^1 because it expected firm 2 to produce X_2^0, but in fact it produced X_2^1. The process of adjusting output in order to maximise profit in response to unrealised expectations about the output choice of rivals will stop when expectations are mutually consistent at (X_1^*, X_2^*) where the reaction functions cross.

The process described here is stable, in the sense that from the initial starting point on either firm 1 or firm 2's reaction function, it will lead to the Cournot equilibrium.

2.3.3.3 The Bertrand Model

The Bertrand model is analogous to the Cournot model and gives the same equilibrium condition even if it is based on different assumptions. In fact, the main assumption is that each producer chooses price rather than output, and makes that decision on the belief that its rival will keep its price constant.

Let us assume that the demand curve is linear and expressed by:

$$p = a - bX = a - b(X_1 + X_2) \tag{2.17}$$

Once again, profits can be expressed as follows:

$$\pi_1 = pX_1 - C_1(X_1) \tag{2.18}$$
$$\pi_2 = pX_2 - C_2(X_2) \tag{2.19}$$

The first-order condition for profit maximisation, derived respect to the price,

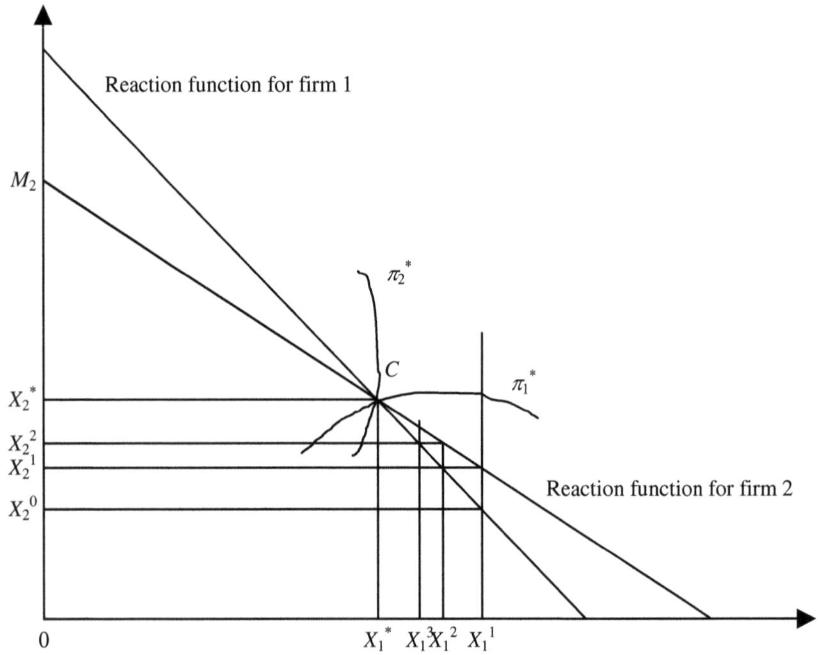

is given by:

Figure 2.5. Reaction functions and Cournot equilibrium

$$\begin{aligned}\left[\frac{\partial \pi_1}{\partial p}\right]_{X_2=\cos t} &= X_1 + p\cdot\left[\frac{\partial X_1}{\partial p}\right]_{X_2=\cos t} - \frac{dC_1}{dq_1}\cdot\left[\frac{\partial X_1}{\partial p}\right]_{X_2=\cos t} = 0\\ \left[\frac{\partial \pi_2}{\partial p}\right]_{X_1=\cos t} &= X_2 + p\cdot\left[\frac{\partial X_2}{\partial p}\right]_{X_1=\cos t} - \frac{dC_2}{dq_2}\cdot\left[\frac{\partial X_2}{\partial p}\right]_{X_1=\cos t} = 0\end{aligned} \tag{2.20}$$

then substituting the value of X_1 determined from Equation 2.17 gives:

$$\begin{aligned}X_1 + \left[\frac{\partial\left(\text{a/b} - p/b - X_2\right)}{\partial p}\right]_{X_2=\cos t}\cdot\left(p - \frac{dC_1}{dX_1}\right) = 0\\ X_2 + \left[\frac{\partial\left(\text{a/b} - p/b - X_1\right)}{\partial p}\right]_{X_1=\cos t}\cdot\left(p - \frac{dC_2}{dX_2}\right) = 0\end{aligned} \tag{2.21}$$

and then:

$$\begin{aligned}X_1 + (-1/b)\cdot\left(p - \frac{dC_1}{dX_1}\right) = 0\\ X_2 + (-1/b)\cdot\left(p - \frac{dC_2}{dX_2}\right) = 0\end{aligned} \tag{2.22}$$

from Equation 2.22 it is possible to determine the expression of the price as follows:

$$p = \frac{1}{3}a + \frac{1}{3}\left(C_1' + C_2'\right) \tag{2.23}$$

so that the optimal values for X_1 and X_2 are given by:

$$\begin{aligned}X_1 = (1/b)\cdot\left(\frac{1}{3}a + \frac{1}{3}\left(C_1' + C_2'\right) - C_1'\right) = \frac{a}{3b} + \frac{C_2'}{3b} + \frac{2C_1'}{3b}\\ X_2 = (1/b)\cdot\left(\frac{1}{3}a + \frac{1}{3}\left(C_1' + C_2'\right) - C_2'\right) = \frac{a}{3b} + \frac{C_1'}{3b} + \frac{2C_2'}{3b}\end{aligned} \tag{2.24}$$

The second-order condition, of the profit maximisation problem, can be obtained by imposing that the second derivative of the profit function, respect to the price, is negative for the output pair solving Equation 2.20.

It should be noted that the solution to the Bertrand model is the same as would be obtained by using the Cournot model, under the same market conditions.

2.3.3.4 The Stackelberg Equilibrium

The Cournot model is based on the idea that each firm takes its rival's output as given when choosing its level of production. Stackelberg analysed what would happen if one firm understood the structure of the Cournot model sufficiently well to work out how the other firm would react and then use this information to improve upon its position in the equilibrium. For this reason, the Stackelberg model is known as the leader/follower model. It can be developed as an extension of the Cournot framework. Firm 1, which is assumed to be the leader, seeks to maximise its profit in the knowledge that firm 2, the follower, will treat firm 1's output as given. Hence firm 2 will always make decisions along its reaction function, while firm 1 maximises its profit subject to firm 2's reaction function. The resulting Stackelberg equilibrium is at the tangency between firm 1's isoprofit curve and firm 2's reaction function.

Let us supposed that the demand curve and the cost functions for both producers are linear:

$$p = a - b(X_1 + X_2) \tag{2.25}$$

$$C(X_1) = F + cX_1 ; C(X_2) = F + cX_2 \tag{2.26}$$

Firm 1's profit is given by:

$$\pi_1 = aX_1 - bX_1^2 - bX_1X_2 - F - cX_1 \tag{2.27}$$

while firm 2's reaction function can be expressed as follows:

$$X_2 = \frac{a-c}{2b} - \frac{X_1}{2} \tag{2.28}$$

In the Stackelberg equilibrium, firm 1 chooses X_1 to maximise Equation (2.27) subject to the constraint that firm 2 chooses X_2 on its reaction function, expressed by Equation 2.28. Therefore, by substituting Equation 2.28 into Equation 2.27 and maximising the profit, we obtain:

$$\frac{\partial \pi_1}{\partial X_1} = a - 2b\,X_1 - \left(\frac{a-c}{2}\right) + b\,X_1 - c = 0 \tag{2.29}$$

$$X_1^* = \frac{a-c}{2b} \tag{2.30}$$

Finally, substituting the profit-maximising output level for firm 1 into firm 2's reaction function, Equation 2.28, results in:

$$X_2^* = \frac{a-c}{4b} \tag{2.31}$$

Under the same market conditions, in terms of demand curve and cost functions, the profit maximising level output in a Cournot model for both producers is given by:

$$X_1^* = X_2^* = \frac{a-c}{3b} \tag{2.32}$$

It can be then noted that the leader is producing more and the follower less than the Cournot equilibrium.

2.4 Oligopolistic Electricity Market

Throughout this section, the electricity market is modelled on the basis of the Cournot model for the analysis of an oligopolistic non-collusive market. The assumption for this choice is that although many attempts have been made all over the world to increase the degree of competition, especially at the wholesale level, the conditions for a perfect competitive market have not taken place yet. In fact, in any market structure in which the generators set their bids, those generators in areas constrained by weak transmission lines should see their market power boosted because they are isolated, by the constraints, from the competition of other generators. As shown in [24], locational market power will exist because of loop flow and transmission constraints that produce geographically and temporally relevant markets.

Moreover, as shown by the experience of deregulation in the British electric power market [25], global market power can also exist, transmission constrained set aside, because of the relative size of competitors. Therefore, it should not be assumed that the market is perfect, but rather acknowledge the potential existence of market power in any form of proposed deregulation and try to limit this power.

In this section, the effects of a non-perfect market structure, caused by the low number of companies sharing the market, are analysed.

For this purpose, the traditional assumptions of a perfect market have to be replaced with more realistic oligopolistic models. A mathematical model is proposed in this section for a wholesale electricity spot market, which takes into account the oligopolistic aspect of the market. The model is developed on the basis of the Cournot paradigm for the analysis of oligopolistic markets and allows the presence of private contracts (bilateral agreements) to be easily taken into account to assess their influence on profits and on the market-clearing price.

The proposed methodology is supposed to give a useful means to analyse the strategical behaviour carried out by producers and to investigate the market equilibrium conditions revealed.

In order to apply the Cournot model to an electricity market, some of its main features have to be investigated and appropriately taken into account. In particular, three main assumptions have been made:

- the analysis of the hourly electricity market is restricted only to a given time interval;
- the consumers' behaviour is considered from a global point of view, that is, the presence of a single equivalent consumer consuming the total load demand of the system, is assumed;
- the generators belonging to each company are dispatched following the *same marginal criterium.*

In this section, these three aspects are examined in detail, showing their impact on the development of the model itself.

The proposed methodology allows several features, typical of an oligopolistic kind of competition in the hourly market of electricity, to be analysed. In particular, after the number of participating generating companies and their operating characteristics are chosen, the following issues will be investigated:

- evaluation of market equilibrium conditions, in terms of market-clearing price, real power amount sold by each participant and its profit, for a given load demand level required for the time interval under consideration;
- comparison of such a market equilibrium condition with the one relative to a perfect competitive market, assuming the same operating conditions;
- evaluation of the behaviour of the market equilibrium condition, in terms of the same variables, when *two-way contracts for differences* are assumed to be set up;
- investigation of the opportunity for a generating company to set up bilateral contracts.

The effectiveness of the proposed methodology is tested on the IEEE 30-bus system, assuming that the relative marketplace is composed of three different generating companies.

2.4.1 Modelling of the Load Demand

The electricity market is very peculiar since, unlike any other commodities, energy cannot be stored in large quantities. In fact, electricity has demand and supply that must be carefully balanced, in the sense that instant demand requires instant generation. Therefore in the electricity spot market prices are determined in advance for each level of demand expected during the following day. In fact, the total real power demand P_{load} of any transmission system is subject to random factors, but its variation over time is much more significant. Moreover, for the purpose of the analysis developed here, it is very important to highlight the dependency of the load demand on the spot price of electricity ρ [28–32]. Therefore, the demand function can be expressed as follows:

$$P_{load}(\tau,\rho)=P^0_{load}(\tau)+P_{load}(\rho) \tag{2.33}$$

Finally, since the load demand varies within the day and the season, it cannot be expected that the market-clearing price of electricity would be constant. For this reason, the time of day is divided into several periods, typically of half hour, and in each of them the load demand is considered constant over time and dependent on the value of the price of electricity related to the particular time interval under consideration. This means that for each period into which the day is divided, a different electricity market has to be considered. Therefore, throughout this chapter, a particular interval of time during the day is chosen, $\tau=\overline{\tau}$, and that particular electricity market is analysed.

Consumers' behaviour is considered from a global point of view; that is the presence of a single equivalent consumer consuming the total load demand of the system, is taken into account.

The behaviour of the equivalent consumer is quantified by the use of a utility function, in analogy to the producers' cost function. The related operating curve is the monetary utility, $u_L(P_{load})$ [$], of consuming P_{load} [MW], quantity of real power and it is assumed to have the following expression:

$$u_L(P_{load})=-\frac{a}{2}P^2_{load}+bP_{load}+c \qquad a>0 \tag{2.34}$$

The consumer's objective function is to maximise its benefits β_L [$] by choosing acceptable optimal consumption level(s) P_{load} for a given market electricity price ρ.

The consumer's benefit function can be expressed as follows:

$$\beta_L=u_L(P_{load})-\rho P_{load} \tag{2.35}$$

The optimal consumption level is chosen by maximising the consumer's benefit as follows:

$$\frac{\partial\beta_L}{\partial P_{load}}=0 \tag{2.36}$$

$$\frac{\partial^2\beta_L}{\partial P^2_{load}}<0 \tag{2.37}$$

It should be noted that Equation 2.39 is always satisfied since a is greater than zero for all the loads present in the system. Then, by substituting Equations 2.34 and 2.35 into Equation 2.36:

$$P_{load}(\rho)=\frac{b}{a}-\frac{\rho}{a} \qquad a>0 \tag{2.38}$$

Finally, under these assumptions, Equation 2.33 can be modified as follows:

$$\overline{P}_{load}(\rho) - \overline{P}_{load}^{0} + m\rho \qquad (2.39)$$

where:

$$\begin{aligned} \overline{P}_{load}^{0} &= \overline{P}_{load}^{0}(\overline{\tau}) + \frac{b}{a} \\ m &= -\frac{1}{a}, \quad m < 0 \end{aligned} \qquad (2.40)$$

It should be noted that the coefficient m represents the elasticity of the total demand versus price. Equation (2.39) represents how the total load demand of the entire system varies according to changes in the electricity price, for any given time interval.

2.4.2 Evaluation of Company Marginal Cost Functions

Let us assume that the electricity market under consideration is composed of n_c generating companies sharing the n_g generating units. Moreover, let us assume a quadratic cost function and, consequently, linear marginal cost function for each generator:

$$C_i(P_{gi}) = \frac{c_i}{2} P_{gi}^2 + d_i P_{gi} + e_i \qquad i = 1, \ldots, n_g \qquad (2.41)$$

$$C_i'(P_{gi}) = c_i P_{gi} + d_i \qquad i = 1, \ldots, n_g \qquad (2.42)$$

with the following capacity constraints:

$$P_{\min_i} \leq P_{gi} \leq P_{\max_i} \qquad i = 1, \ldots, n_g \qquad (2.43)$$

Since each generating company has to maximise its profit-making appropriate output decision, it is important to derive a global marginal cost function, which takes into account the marginal cost function of each generating unit belonging to the company and combines them. We refer to this function as the *Company Cost Function* Cco_i of the ith company and it is expressed as a function of the total power generated by the company. The main assumption is that each company will bid every generating unit it owns, following the same marginal cost criterion.

Let us suppose that there are n_{gi} generating units belonging to the ith company, the total cost function of the company has the following expression:

$$Cco_i(P_{Ci}) = \sum_{k=1}^{n_{gi}} C_k(P_{gk}) \qquad (2.44)$$

where

$$P_{Ci} = \sum_{k=1}^{n_{gi}} P_{gk} \tag{2.45}$$

and

$$\frac{dC_k}{dP_{gk}} = \frac{dC_j}{dP_{gj}} = \lambda \qquad \forall\, k, j = 1, \ldots n_{gi} \tag{2.46}$$

A simple procedure to generate $Cco_i(P_{Ci})$ consists of adjusting λ from $\lambda^{\min}$ to $\lambda^{\max}$ in specified increments, where:

$$\begin{aligned} \lambda^{\min} &= \min\left(\frac{dC_k}{dP_{gk}}, \quad k = 1, \ldots n_{gi} \right) \\ \lambda^{\max} &= \max\left(\frac{dC_k}{dP_{gk}}, \quad k = 1, \ldots n_{gi} \right) \end{aligned} \tag{2.47}$$

At each step, the total production costs Cco_i^{α} and the total power output for all the units P_{Ci}^{α} is calculated. If one of the generating units hits a limit, its output is held constant equal to the limit reached. The applied procedure is illustrated in Figure 2.6. At the end of the process, all the points $P_{Ci}^{\alpha}, Cco_i^{\alpha}$ are fitted to a curve to obtain the company cost function for the given company.

In particular, a quadratic expression for this function is chosen, in analogy with the cost function of each generator:

$$Cco_i(P_{Ci}) = \frac{\tilde{c}_i}{2} P_{Ci}^2 + \tilde{d}_i\, P_{Ci} + \tilde{e}_i \qquad i = 1, \ldots, n_C \tag{2.48}$$

Consequently, a linear company marginal cost function is derived:

$$Cco_i'(P_{Ci}) = \tilde{c}_i\, P_{Ci} + \tilde{d}_i \qquad i = 1, \ldots, n_C \tag{2.49}$$

2.4.3 Presence of Transmission Losses

Let us suppose that in the transmission system, related to the electricity market under consideration, there are n_g generating units, and denote by P_{gi} the real power output of the ith unit. Demand and supply can be balanced using the following expression:

$$\sum_{i=1}^{n_g} P_{gi} = \overline{P}_{load} + P_{loss} \tag{2.50}$$

where P_{loss} represents the transmission losses throughout the system. The real power output that every generating company can produce depends on the relative amount produced by the other companies. Unfortunately, transmission losses are not constant but are a function of the structure and the topology of the transmission network and of the amount and location of the generated power. For a given transmission system, our concern is to evaluate how transmission losses vary according to changes in output level of the generating units present in the system. For this purpose, we can evaluate transmission losses using the B-matrix loss formula [33]:

$$P_{loss} = \mathbf{P}_g^T \, B \, \mathbf{P}_g + b_1^T \, \mathbf{P}_g + b_0 \tag{2.51}$$

where $\mathbf{P}_g = [P_{g1}, \ldots, P_{gi}, \ldots, P_{gng}]^T$ is the $(n_g{*}1)$ vector of all real generated power, B is a $(n_g{*}n_g)$ of the loss coefficient B_{ij}, b_1 is a vector of the same length as $\mathbf{P}_g$ of coefficient b_{1i} and b_0 is a real constant.

Then, using Equations 2.39, 2.50 and 2.51, it is possible to determine the *Inverse Demand Function* characteristic of the electricity market under consideration:

$$\rho = -\frac{1}{m}\left[\sum_{i=1}^{ng}\sum_{j=1}^{ng} P_{gi} \, B_{ij} \, P_{gj} + \sum_{i=1}^{ng}(b_{1i} - 1)P_{gi} + \left(\overline{P}_{load}^{0} + b_0\right)\right] \tag{2.52}$$

Equation 2.52 expresses the relation between the real power supplied by each generating unit present in the transmission system and the corresponding value of the spot price, related to the particular market under consideration, given a specified time interval.

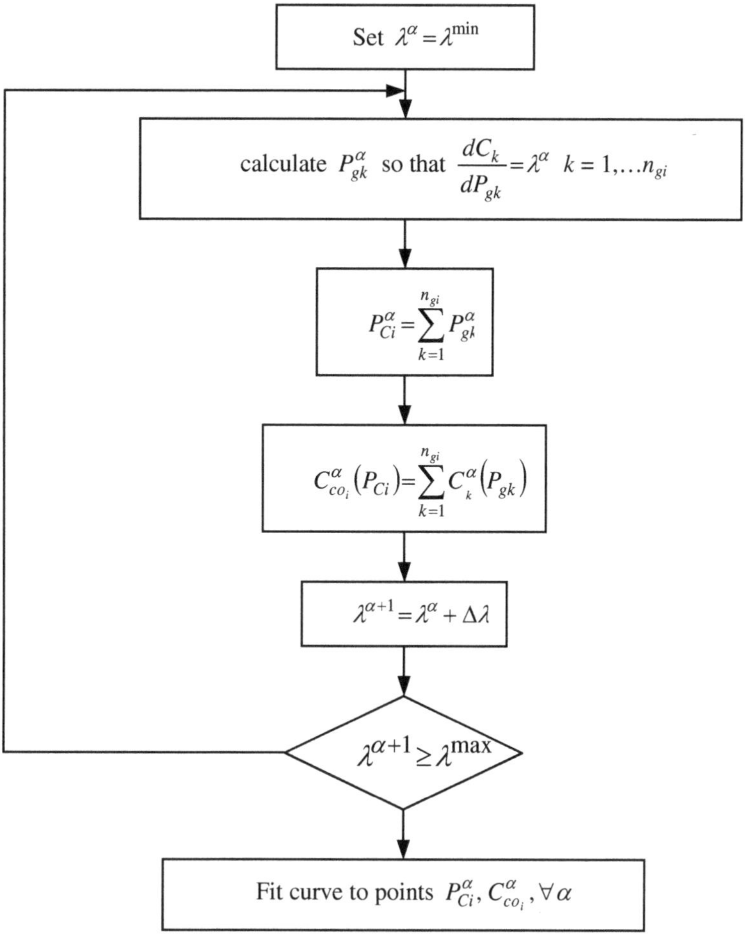

Figure 2.6. Company marginal cost function evaluation

2.4.4 Proposed Model

At this stage, it is possible to apply the model illustrated in Section 2.2.2 to an electricity market composed by few companies, which share all the generating units present in the related transmission system, and taking into account all the main features of such a type of market [26,27]. Following the same marginal cost criterium, as illustrated in Section 2.2.4, the real power output of the *k*th generator belonging to the *i*th company can be expressed as a function of the total real power output of the company, as follows:

$$P_{gk} = \alpha_i^k \, P_{Ci} \qquad k=1, \ldots, n_{gi} \tag{2.53}$$

where

$$\sum_{k=1}^{n_{gi}} \alpha_i^k = 1 \qquad i=1, \ldots, n_C \tag{2.54}$$

Under this assumption, the transmission losses can be determined as functions of the companies' outputs:

$$P_{loss} = \sum_{i=1}^{n_C}\sum_{j=1}^{n_C} P_{Ci} A_{ij} P_{Cj} + \sum_{i=1}^{n_C} a_i P_{Ci} + a_0 \tag{2.55}$$

where:

$$\begin{aligned} A_{ij} &= \sum_{r=1}^{n_{gi}}\sum_{s=1}^{n_{gj}} \alpha_i^r B_{ij}^{rs} \alpha_j^s \\ a_i &= \sum_{r=1}^{n_{gi}} \alpha_i^r b_{1i}^r \\ a_0 &= b_0 \end{aligned} \tag{2.56}$$

Equation 2.1, expressing the profit of a company in an oligopolistic market, can be detailed to highlight the peculiar aspects of the electricity market and then rewritten as follows:

$$\pi_{Ci} = P_{Ci}\left\{ -\frac{1}{m}\left[\sum_{i=1}^{nC}\sum_{j=1}^{nC} P_{Ci} A_{ij} P_{Cj} + \sum_{i=1}^{nC} (a_i - 1) P_{Ci} + \left(\overline{P}_{load}^0 + a_0\right)\right]\right\} - \left[\frac{\tilde{c}_i}{2} P_{Ci}^2 + \tilde{d}_i P_{Ci} + \tilde{e}_i\right] \tag{2.57}$$

It should be noted how the profit of the *i*th company is highly dependent not only on the total power output of the remaining companies sharing the market but also on the way this output is shared among the generating units.

In particular, the relationships through which the interactions among rivals are expressed, as mentioned in Section 3.3.1.1, can be evaluated in terms of *Reaction Functions*. For each generating company, the profit maximising level of real power output is plotted given every conceivable expected value of production that would be chosen by its rivals.

Then, the equilibrium conditions can be determined by the intersections of the reaction curves of all the generating companies sharing the market, as shown below:

$$\frac{\delta \pi_{Ci}}{\delta P_{Ci}} = \rho\left(P_{C1}, \ldots, P_{Ck}, \ldots, P_{Cn_C}\right) + \frac{\delta \rho}{\delta P_{Ci}} P_{Ci} - \tilde{c}_i P_{Ci} - \tilde{d}_i = 0 \qquad i=1, \ldots, n_C \tag{2.58}$$

Moreover, using the proposed model it is possible to determine *Isoprofit Curves*, another important feature of the behaviour of the generating companies sharing the market. An *Isoprofit Curve* for a particular supply company is the locus of points in the all suppliers' real power output space defined by different levels of outputs for the suppliers sharing the market, which yields to the same level of profit for the supplier under consideration.

2.4.5 Presence of Bilateral Contracts

An important aspect to take into account when analysing electricity markets is the presence of bilateral contracts stipulated between generating companies and customers. When participation in the wholesale market is mandatory, these agreements are private and assume the form of Contracts for Difference (CfDs), as shown in Section 2.2.5.

This section illustrates how they can be taken into account in the proposed model. The expression for the profit of the *i*th generating company, assuming that a private contract has been set up, can be formulated as follows:

$$\pi_{Ci} = P_{Ci}\rho(P_{Ci}, P_{-Ci}) - X_{Ci}(\rho(P_{Ci}, P_{-Ci}) - z_{Ci}) - C_{Ci}(P_{Ci}) \tag{2.59}$$

where z_{Ci} indicates the strike price of the contract itself and X_{Ci} the traded amount of real power. Equation 2.59 shows that when the value of the spot price ρ is greater than the strike price, the producer has to reimburse the consumer. For this reason its profit decreases by the quantity $(\rho - z_{Ci}) \cdot X_{Ci}$. In particular, if the amount of real power sold by the producers is not significant, the decrement can provoke an actual loss for the producer itself. On the contrary, if the value of the spot price of electricity is lower than the strike price, the generating company's profit increases by the amount $(z_{Ci} - \rho) \cdot X_{Ci}$, since it is reimbursed by the consumer.

For these reasons the presence of private contracts works as an incentive to stimulate producers to sell more real power, even for lower values of the spot prices. The test results show how an oligopolistic market in which private contracts have been set up as an equilibrium condition is closer to the one of a competitive market.

Finally, the expression for the profit of the *k*th generating company, taking into account all the features peculiar to the electricity market can be formulated as follows:

$$\begin{aligned}\pi_k = (P_{Ck} - X_{Ck})&\left[-\frac{1}{m}\left(\sum_{i=1}^{n_C}\sum_{j=1}^{n_C} P_{Ci}A_{ij}P_{Cj} + \sum_{i=1}^{n_C}\tilde{a}_i P_{Ci} + \tilde{a}_0\right)\right] \\ &+ X_{Ck}z_{Ck} - \left[\frac{\tilde{c}_k}{2}P_{Ck}^2 + \tilde{d}_k P_{Ck} + \tilde{e}_k\right]\end{aligned} \tag{2.60}$$

where the last term, in square brackets, indicates the company cost function C_{Ck}. It is assumed to have a quadratic expression and it is evaluated by following the same procedure illustrated in Section 2.2.4.

It should be noted how the profit of the *k*th company is highly dependent not only on the total power output of the remaining companies sharing the market but also on the way this output is shared among generating units.

The market equilibrium condition is revealed by determining the profit-maximising levels of real power output for each company. In particular, by deriving Equation (2.60) with respect to the real power output itself, following the Cournot model, one can obtain:

$$\begin{aligned} &3A_{kk}P_{Ck}^2 + 2P_{Ck}\sum_{i\neq k}(A_{ki}+A_{ik})P_{Ci} + (2\tilde{a}_k + a\tilde{c}_k)P_{Ck} - 2A_{kk}P_{Ck}X_{Ck} \\ &-X_{Ck}\left(\sum_{i\neq k}(A_{ki}+A_{ik})P_{Ci} + \tilde{a}_k\right) + \sum_{i\neq k}\sum_{j\neq k}P_{Ci}A_{ij}P_{Cj} + \sum_{i\neq k}\tilde{a}_i P_{Ci} + \tilde{a}_0 + \tilde{d}_k = 0 \end{aligned} \tag{2.61}$$

It should be noted how useful it could be for any generating company, taking part in the day-ahead spot market to take the most accurate estimate of the value of the spot price of electricity for the next day. In fact, this estimate will allow them to submit the most appropriate bids, which should than permitt them to take part in the operating phase maximising the company's profit.

2.5 Case Studies and Results Analysis

The proposed model has been applied to the electricity marketplace related to the IEEE 30-bus system, in terms of generating unit characteristics and network topology, while the expression for the inverse demand function has been separately defined. In Figure 2.7, the network topology of the IEEE 30-bus system is illustrated. In order to define a corresponding marketplace, it is supposed that there are three generating companies and the generating units present in the system are shared by the companies as shown in Table 2.1. The number of companies is set to three.

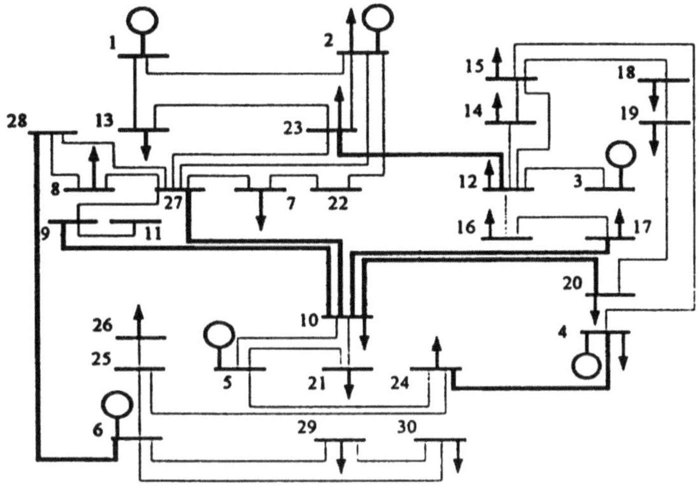

Figure 2.7. IEEE 30-bus system

Table 2.1 shows the real power constraints for each generator and, consequently, for each generating company.

Table 2.1. IEEE 30-bus system electricity market structure

Company no.	Generator no.	P_g^{min} [MW]	P_g^{max} [MW]	P_C^{min} [MW]	P_C^{max} [MW]
1	1	0	80	0	80
2	2	0	80	0	130
	3	0	50		
3	4	0	55	0	125
	5	0	30		
	6	0	40		

A quadratic expression has been chosen for each generator's cost function and in Table 2.2 the relative values of its parameters have been reported. The expression is:

$$C_i\left(P_{gi}\right)=\frac{c_i}{2}P_{gi}^2+d_i\,P_{gi}+e_i \qquad i=1,\ldots,n_g \qquad (2.62)$$

The first step is to apply the procedure illustrated in Section 4.4 in order to evaluate the Company Marginal Cost Functions. The values of the parameters are obtained by the fitting process of the data into a quadratic curve, for the Company

Cost Function C_{coi}, and then a linear expression is derived for the Company Marginal Cost Function. The values of these parameters are reported in Table 2.3.

Table 2.2. Generator cost function coefficients

Generator number	e_i [$]	d_i [$/MW]	c_i [$2/MW]
1	0	2	0.02
2	0	1.75	0.0175
3	0	3	0.025
4	0	1	0.0625
5	0	3	0.025
6	0	3.25	0.0083

A particular time interval during the day, $\tau = \overline{\tau}$, has been chosen so that the total real power demand of the system is equal to the one corresponding to the standard loading condition of the IEEE 30-bus system and it is supposed that a values of minus 10 [MW/$] has been chosen for the parameter *m*, which represents the elasticity of demand versus price. Under these assumptions, Equation (2.62) can be rewritten for this particular electricity market as follows:

$$\overline{P}_{load}(\rho) = 189.2 - 10\,\rho \tag{2.63}$$

Table 2.3. Company marginal cost function parameters

Company no.	$\tilde{c}_i$ [$/MW^2]	$\tilde{d}_i$ [$/MW]
1	0.04	2.000
2	3.194x10^{-2}	1.354
3	1.124x10^{-2}	3.087

In Table 2.4 the market equilibrium condition, in terms of the values of the profit-maximizing outputs of the three companies and the market-clearing price are reported. Moreover, on the basis of the same operating condition, the output of each company is also shown when a perfect competitive market model is supposed.

Table 2.4. Market equilibrium condition

	Oligopolistic market model	Perfect competitive market model
P_1 [MWh]	49.46	49.38
P_2 [MWh]	54.02	82.06
P_3 [MWh]	50.00	78.98
ρ [$/MWh]	8.68	3.975

It can be noted how the value of the market-clearing price is much higher in the oligopolistic market model than in the perfect competitive environment. This is due to the fact that in the oligopolistic environment each firm has a relative market power and its actions, regarding the amount of output to sell, can have a deep effect on the value of the market price. On the contrary, in the perfect competitive environment, each firm knows that it has no power in setting the price and the only choice is to bid at marginal costs.

The proposed model allows, as mentioned earlier, both Reaction Functions and Isoprofit Curves to be determined. In particular, in Figure 2.8, the reaction function of Company 3 has been reported, that is the profit maximising output levels for Company 3 according to the actions of the other two generating companies sharing the market. It should be noted how the actions of the three generating companies to maximise their profit are highly interdependent.

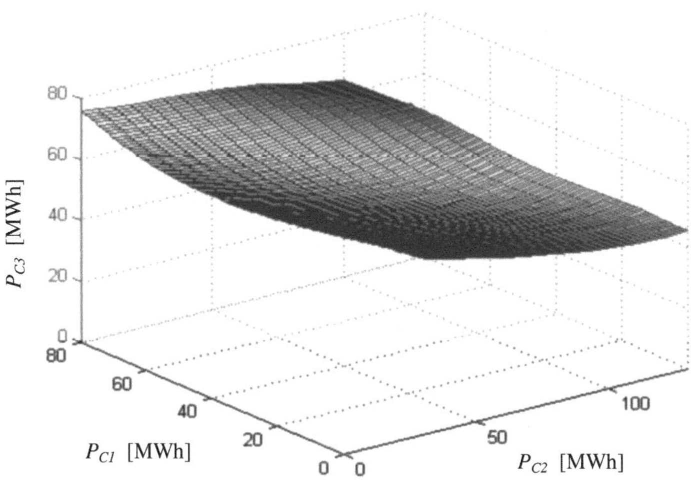

Figure 2.8. Reaction function of company 3

In Figure 2.9, the isoprofit curves for Company 2, with the profit level varying in a range of [1000:6000] ($/MWh), have been illustrated, supposing that the output of Company 3 is zero. In this way it is possible to detect how for the same profit level, different output level are required according to the different amount of real power sold by Company 1.

The influence of private contracts has been examined for the market under study. As an example, it has been supposed that Company 1 is involved in a bilateral agreement and the effects of the amount traded on the spot market equilibrium condition have been investigated. In Figure 2.10 the profit-maximising output of each generating company is reported for different values of the power traded by Company 1 through private contracts.

It should be noted how the optimal output of Company #1 increases for increasing values of the power traded in the private contract, while the rivals' outputs decrease. The reason is that since Company #1 has a bilateral agreement, that are supposed to be carried out as Contracts for Difference, it is willing to sell more power, even if the price is lower, being edged against price variations by the presence of the contracts. In fact, as shown in Figure 2.11, the market-clearing price decreases for increasing values of the power traded in the bilateral contract. Moreover, since the market-clearing price decreases, the total amount of electricity sold in the spot market increases, as illustrated in Figure 2.12. This effect improves the market efficiency, although the composition of market itself has not been changed.

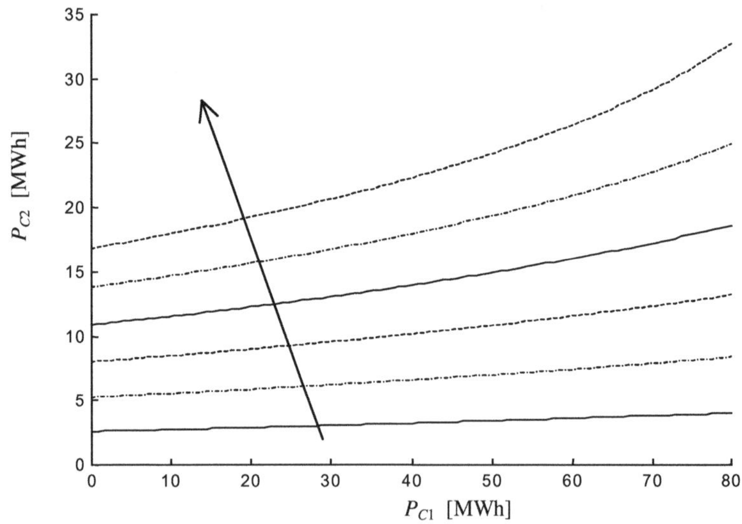

Figure 2.9. Isoprofit curves for company 2

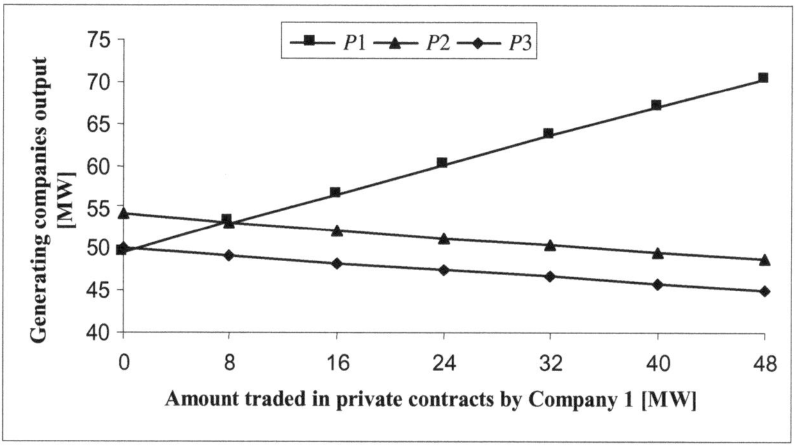

Figure 2.10. Generating companies' output for different values of real power traded in private contracts by company 1

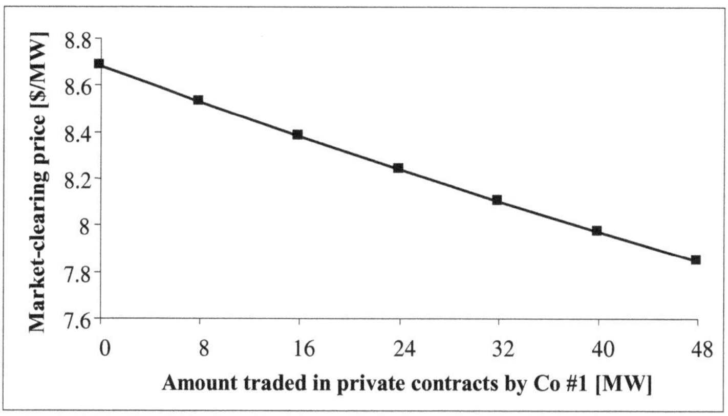

Figure 2.11. Market-clearing Price

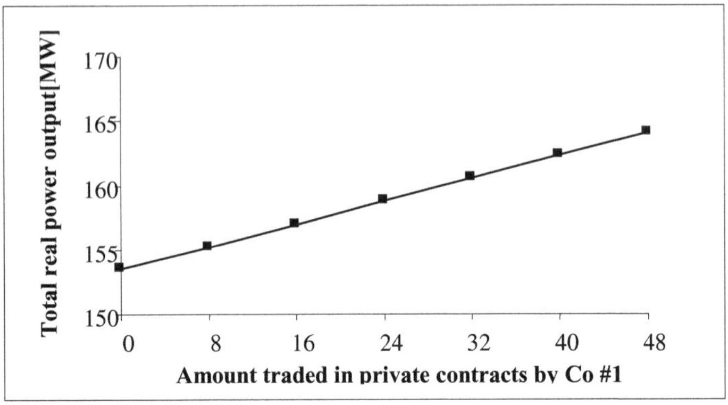

Figure 2.12. Total power supplied in the spot market

2.6 Conclusions

Around the globe, the electricity industry is undergoing a restructuring process moving from the corporate environment towards privatisation and reduced regulation. The expectation is that a market-driven structure for the electricity industry will encourage competition in generation and supply, with open transmission and distribution access to enable it, and that this process will finally lead to efficiency, savings and reduced prices.

As a consequence, while in the case of an integrated or nationalised utility the generation resources are centrally controlled and the objective function is to minimise the total cost made up of capital charges and operating costs, in a deregulated environment emphasis has to be necessarily given to the maximisation of the profit of several actors, such as generators, distribution companies and customers.

However, for each player the possibility of improving its own benefit depends on the market structure and on the type of competition really established. The market structure affects the freedom of action of each player and the nature of the competition determines how much the behaviour of a single player can condition the action of other players.

The main issue addressed in this chapter is the investigation of the oligopolistic aspects of an electricity market, which can influence the competition level.

Strictly interpreted, oligopoly is a market situation where, due to the presence of few firms operating within the market, the actions of a firm have, quite invariably, a significant influence on the behaviour of the remaining firms and, consequently, on their profits [33].

In this chapter, the effects of a non-perfect market structure, caused by the small number of companies sharing the market, are analysed.

For this purpose, the traditional assumptions of a perfect market have to be replaced with more realistic oligopolistic models. A mathematical model is proposed for a wholesale electricity spot market, which takes into account the oligopolistic aspect of the market. The model is developed on the basis of the Cournot paradigm for the analysis of oligopolistic markets and allows the presence of private contracts (bilateral agreements) to be easily taken into account to assess their influence on the profits and on the market-clearing price.

The proposed methodology is supposed to give a useful means to analyse the strategical behaviour carried out by producers and to investigate the market equilibrium conditions revealed.

In order to apply the Cournot model to an electricity market, some of its main features have to be investigated and appropriately taken into account. In particular, three main assumptions have been made:

- the analysis of the hourly electricity market is restricted only to a given time interval;
- the consumers' behaviour is considered from a global point of view, that is the presence of a single equivalent consumer, consuming the total load demand of the system, is supposed;
- the generators belonging to each company are dispatched following the *same marginal criterion*.

These three aspects are examined in detail, showing their impact on the development of the model itself.

The proposed methodology allows several features, typical of an oligopolistic kind of competition in the hourly market of electricity, to be analysed. In particular, after the number of participating generating companies and their operating characteristics are chosen, the following issues will be investigated:

- the evaluation of market equilibrium conditions, in terms of market-clearing price, real power amount sold by each participant and its profit for a given load demand level required for the time interval under consideration;
- the comparison of such a market equilibrium condition with one relative to a perfect competitive market, assuming the same operating conditions;
- the evaluation of the behaviour of the market equilibrium condition, in terms of the same variables, when *two-way contracts for differences* are assumed to be set up;
- the investigation of the opportunity for a generating company to set up bilateral contracts.

The effectiveness of the proposed methodology is tested on the IEEE 30-bus system, assuming that the relative marketplace is composed of three different generating companies.

2.7 References

[1] "Final Order 888: Promoting Wholesale Competition Through Open Access Non-discriminatory Transmission Services by Public Utilities", Federal Energy Regulatory Commission, April 24, 1996.
[2] "Directive 96/92/EC of the European Parliament and the Council concerning Common Rules for the Internal Market in Electricity", Official Journal of the European Commission, L27 (30/1/97), pp. 20-29.
[3] C. Bier, "Network Access in the Deregulated European Electricity Market: Negotiated Third-Party Access vs. Single Buyer", CSLE Discussion paper 9906, University of Saarland, June 1999.
[4] A. Vojdani, F.A. Rahimi, "Electricity Market Structures", Proc. of EPSCOM '98, Zurich, September 23-25,1998.
[5] B. Murray, "Electricity Markets", John Wiley & Sons, New York, 1998.
[6] J. S. Gans, D. Price, K. Woods, "Contracts and Electricity Pool Prices", Australian Journal of Management, Vol. 23, No. 1, pp. 83-96, June 1998.
[7] A. Holmes, L. Plaskett, "The New British Electricity System", Financial Times B.I.S., London 1991.
[8] R. D. Tabors, "Lessons from the UK and Norway", IEEE Spectrum, August 1996.
[9] R. D. Christie, I. Wangensteen, "The Energy Market in Norway and Sweden: the Spot and Future Markets", IEEE Power Eng. Review, March 1998.
[10] S. A. Fismen, "Experience: Norway", IEE Colloquium, 28 June 1996.
[11] A. Kuusela, H. Viheriävaara, "The Role of the Finnish System Operator and the Interactions of Commercial Mechanism", CIGREE Symposium, Tours 1997.
[12] B. R. Arkovich, D. V. Hawk, "Charting a New Course in California", IEEE Spectrum, July 1996.
[13] D. P. Mendes, D. S. Kirschen, "Assessing Pool-based Mechanism in Competitive Electricity Markets, Proc. of "IEEE-PES 2000 Summer Meeting, Seattle", USA, July 2000, pp.2195-2200.
[14] J. Kumar, G. Sheble, "Auction Market Simulator for Price Based Operation", IEEE Transactions on Power Systems, Vol. 13, No. 1, Feb. 1998, pp. 250-255;
[15] R. J. Thomas, T. R. Scheider, "Underlying Technical Issue in Electricity Deregulation", PSERC Publication, 97-18.
[16] D. Laider, S. Estrin, "Introduction to Microeconomics", Philip Allan, 1989.
[17] R. S. Pindyck, D. L. Rubinfeld, "Microeconomics", Prentice-Hall Inc., 1998.
[18] M. C. Sawyer, "Theories of the Firm", Weidenfeld & Nicolson, 1979.
[19] G. J. Stigler, "The Theory of Price", The Macmillan Company, 1967.
[20] R. H. Havenman, K. A. Knopf, "The Market System", John Wiley & Sons, 1981.
[21] A. Dunnett, "Understanding the Market", Longman, 1998.
[22] D. N. McCloskey, "The Applied Theory of Price", Macmillan Publishing Company, 1982.
[23] E. K. Browing, M. A. Zupan, "Microeconomic Theory & Applications", Addison-Wesley, 1999.
[24] Z. Younes, M. Ilic, "Transmission System Constraints in Non-Perfect Electricity Market", Proc. of 18th Annual North American Conference, USAEE/IAEE, 1997, pp. 256-265.
[25] C. D. Wolfram, "Measuring Duopoly Power in the British Electricity Market", MIT Department of Economics, WP, November 1995.
[26] M. Trovato, A. Maiorano: "Modelling and Analysis of Oligopolistic Electricity Markets", Proc. of "ELECO 99- International Conference on Electrical and Electronics Engineering", Bursa, Turkey, 1-5 Dec, 1999, pp. 1-9.

[27] A. Maiorano, Y.H. Song, M. Trovato: "Dynamics of Non-collusive Oligopolistic Electricity Markets", Proc. of "IEEE-PES Winter Meeting 2000", Singapore, 23-27 Jan, 2000.
[28] James D. Weber, Thomas J. Overbye, Peter W. Sauer, Christopher L. de Marco: "A Simulation based Approach to Pricing Reactive Power", Proc. of the Hawaii International Conference on System Sciences, January 6-9, 1998, Kona Hawaii.
[29] James D. Weber, Thomas J. Overbye, Christopher L. de Marco: "Inclusion of Price Dependent Load Models in the Optimal Power Flow", Proc. of the Hawaii International Conference on System Sciences, January 6-9, 1998, Kona Hawaii.
[30] R. Rajaraman, J. V. Sarlashkar, F. L. Alvarado, "The Effect of Demand Elasticity on Security Prices for the Poolco and Multilateral Contract Models", IEEE Trans. On Power Systems, vol. 12, no 3, Aug 1997, pp. 1177-1184.
[31] B. A. Morse, Z. Yu, F. T. Sparrow, "A Market Power Model with Demand Flexibility over Time for Deregulated Electricity Markets", Proc. of IEEE-PES 2000 Summer Meeting, Seattle, USA, July 2000, pp. 2248-2252.
[32] E. Bompard, G. Carpaneto "The Role of Demand Elasticity in Congestion Management and Pricing", Proc. of IEEE-PES 2000 Summer Meeting, Seattle, USA, July 2000, pp. 2229-2234.
[33] A. J. Wood, B. F. Wollenberg, "Power Generation Operation and Control", John Wiley and Sons Inc., 1996.

3. Location-based Marginal Pricing of Electricity and its Decomposition

K. Xie and Y.H. Song

3.1 Introduction

In the emerging electricity market, which relies on price-based competition, an unambiguous, transparent and predictable pricing framework of electricity for both active and reactive power is one of the major issues. Also, with the growing interest in determining the costs of supplying the ancillary services needed to maintain quality and reliable electricity service, the spot price should be decomposed and distributed to different ancillary services.

From the economic point of view, spot pricing based on SRMC (Short Run Marginal Cost) has the potential to provide economic signals for system operation. Various models and approaches [1–17] for determining spot pricing have been proposed, which will be reviewed in Section 3.2 to provide background and motivation for the development of an Integrated Optimal Spot Pricing (IOSP) model in this chapter. Section 3.3 will describe the details of the derivation of the proposed model.

The proposed model includes the detailed derivation of optimal nodal specific real-time prices for active and reactive powers, and the method to decompose them into different components corresponding to generation, loss and many selected ancillary services, such as spinning reserve, voltage control and security control. The features of the proposed model are discussed in relation to existing pricing models and classical economic dispatch. The model is then implemented by extending the IPOPF method developed in previous chapters. Insight is given by case studies on 5-bus, IEEE 30-bus and IEEE 118-bus systems.

3.2 Spot Pricing Models

3.2.1 Review of Some Existing Spot Pricing Models

The concept of spot price was introduced into power systems in the late 1970s. The early years research results were summarised in [2] by Schweppe *et al.* Although the authors did not take into account the pricing of reactive power and most of the ancillary services (congestion alleviation is the sole consideration), it provided the foundation and starting point for most later research. In [2] the concepts of classical economic dispatch and DC load flow were employed to obtain the essential parts of the Spot Price. A tool named WRATES, for evaluating the marginal cost of wheeling was described in [3]. It handled network flows and losses using either a modified DC load flow approximation or an exogenously provided "sensitivity matrix" obtained from the AC power flow. However, due to some assumptions and simplifications, the coupling of different price components was ignored. Reference [4] further developed the spot price model established in [2] by demonstrating the physical meanings and numerical properties of the generation and transmission components of spot price, focused on the interpretation of "slack bus" and "system lambda".

Ray and Alvarado [5] used a modification of the Optimal Power Flow (OPF) model, which allowed for the price responsiveness of real power demand to analyse the effects of spot pricing policies. It was the first trial to use OPF as a spot price calculation tool. Baughman and Siddiqi [6] developed this model by introducing reactive power pricing and revealed that Lagrangian multipliers, corresponding to node power balance equations in Optimal Power Flow, represent the marginal costs of node power injections. This model shows that OPF is a promising tool for spot pricing. However, the reactive prices of the generator buses would be kept to zero until the reactive power generating capacity limits are reached. This fact implies that any expansion of Var capacities is discouraged because no compensation is given for doing so.

Reference [7] took account of the reactive power producing cost by introducing MVAR cost curves, which are a counterpart of the standard MW incremental cost curve. The successive LP method was applied to solve this non-linear reactive power optimisation problem. However, the influence of reactive power on real power loss was ignored. EL-Keib and Ma [8] made progress in reactive power pricing through employing P-Q decoupled OPF to obtain the SRMC of active and reactive power respectively and emphasised that the production cost of reactive power should be accounted for when pricing independent power producers (IPPs) and wheeling transactions. First, the work directly introduced real power loss component into reactive power spot pricing derived from the Q-subproblem whose objective function is real loss minimization. Second, the influence of reactive power on the voltage level appears in the pricing formulas. Reference [9] tried to calculate the wheeling marginal cost of reactive power.

Today, because of the growing interest in determining the costs of supplying ancillary services, not only must the reactive power/voltage support pricing be decoupled from the energy cost, but also the spot price should be decomposed to coupled from the energy cost, but also the spot price should be decomposed to other ancillary services. The spinning reserve pricing problem was addressed in [10], and implemented by the reliability differentiated pricing model (RDP) through combining customers' outage cost into the overall objective function to reflect the idea that outages created by insufficient generation or transmission actually cause a loss of welfare. Reference [11] developed a method to calculate the contribution of each generation unit participating in balancing the system under known time-varying transactions and unknown transmission losses. Essentially, it focuses on the ancillary service of energy imbalance compensation. Congestion alleviation cost, as the basic idea of security pricing, was advanced in [12], and Kaye and Wu [13] extended this approach to consider pricing in the presence of contingencies and to determine, via a feedback mechanism, the optimal societal consumption and optimal prices. Alvarado *et al.* [14] considered the formal quantification of system security by computing the outage cost associated with specific operating points, as well as the influence of actions on this cost.

Considering most ancillary services and incorporating constraints on power quality and environment, an advanced pricing prototype was recently introduced in [15], which combined the dynamic equations for load-frequency control with the static equations of constrained OPF. However, it is still uncertain how to solve this large-scale stochastic optimal control problem. Reference [16] retained the spirit of Reference [6], in which a coupled OPF was used to perform the most fundamental operation of the poolco model. An important result was derived under the satisfaction of certain regularity assumptions and by LP theory: namely the decomposition of Lagrangian multipliers corresponding to power balance equations into two components that represented the sum of generation and losses and system congestion (defined by line flow and voltage). The authors pointed out that the latter can be interpreted as the sensitivity of the congestion constraints to additional load at the bus times the shadow prices for the constraints. When no congestion constraints are active, then the price for power at a bus is strictly due to generation and losses. A prototype program structure based on OPF was introduced to compute the decomposition of spot prices [17]. A sequential quadratic programming algorithm was employed to solve the OPF problem. A special program called constraints decomposition was used to obtain the different parts of spot prices after OPF calculation.

For state-of-the art spot pricing methods, as mentioned in the comments of [16], congestion would appear to be a "go" "no-go" gauge, which means if any constraints are below the established limits, the corresponding congestion part of the spot price is zero. Once constraint violations occur, the spot price will jump to an unacceptable high point. This phenomenon makes spot price changes too volatile to be predicted and responded to at congestion points. A more relaxed approach with smooth increasing congestion cost as the operating constraints approach the limits will be more practical.

3.2.2 The Decomposition Model

The first objective of this chapter, based on existing work, is to develop an integrated optimal spot pricing model for active and reactive power and various ancillary services. In this chapter, the authors prove that λ_p and λ_q, the Lagrangian multipliers corresponding to power balance equations in OPF, play an important role in spot pricing of electricity. Economically, they are the shadow prices of node power injections, and therefore can be adopted as spot prices of active power and reactive power directly for both generators and loads. Furthermore they can be decomposed into different components to reflect the effects of system marginal cost, loss compensation, voltage support and congestion management. They are all important price terms in the deregulated electricity market and can be forwarded to the generators and consumers as control signals to regulate the level of their generation or consumption. Attributed to the introduction of reactive power generation cost, the coupling between the price of active power and the price of reactive power is taken into account.

The second objective of the chapter is to extend the IPOPF method for effectively solving the proposed integrated optimal spot pricing model. The modified Primal-Dual Interior Point method (PDIP) based OPF developed in the previous chapters has several advantages when employed as a pricing tool. First, dual variables provide meaningful economic information because of their relations with shadow prices. Second, the price signals are smoother and more predictable due to the application of logarithmic penalty functions of slack variables as soft constraints. It can avoid the "go" or "no-go" gauge perfectly (a detailed discussion will be given in Chapter 9). Third, because most of the data needed by the modification from an OPF program to pricing software are by-products of IPOPF, the calculation burden of the modification is very limited.

Table 3.1. OPF-based spot pricing terminology

Literature	0 Active power spot price	1 Reactive power spot price	Methods
Reference 2	$\rho_{pi} = \vartheta(1+\frac{\partial P_L}{\partial P_{di}}) + \sum_k \frac{\partial N_i}{\partial P_k}\frac{\partial P_k}{\partial P_{di}} + \sum_k \frac{\partial P_k}{\partial P_{di}}(\pi_{luk}+\pi_{llk})$		ED DC Flow DC OPF
Reference 6	$\rho_{pi} = \lambda_{pi}$ *for generators* $\lambda_{pi} = \frac{\partial C_i(P_{gi})}{\partial P_{gi}} + \pi_{upi} + \pi_{lupi}$	$\rho_{qi} = \lambda_{qi}$ *for generators* $\lambda_{qi} = \pi_{uqi} + \pi_{lqi}$	GRG2
Reference 8	$\rho_{pi} = \lambda_{ps} - \lambda_{ps}\frac{\partial P_L}{\partial P_i} - \sum_{i=1}^{N} \pi_{uli}\frac{\partial P_{li}}{\partial P_i}$	$\rho_{qi} = -\lambda_{ps}\frac{\partial P_L}{\partial Q_i} + \sum_{k=m+1}^{N}\left[\pi_{lvk} + \pi_{vuk}\right]\frac{\partial V_k}{\partial Q_i} + \pi_{lqi} + \pi_{uqi}$	Decoupled OPF Lindo 5.0
Reference 17	$\begin{bmatrix}\rho_{pi}\\ \rho_{qi}\end{bmatrix} = \begin{bmatrix}(\nabla_V F)^T\\ (\nabla_\theta F)^T\end{bmatrix}^{-1}\begin{bmatrix}(\nabla_V f_s)^T\\ (\nabla_\theta f_s)^T\end{bmatrix}\lambda_{ps} + \begin{bmatrix}(\nabla_V F)^T\\ (\nabla_\theta F)^T\end{bmatrix}^{-1}\begin{bmatrix}(\nabla_V G)^T\\ (\nabla_\theta G)^T\end{bmatrix}\pi$		Successive Quadratic Programming
Decomposition Method (18)	$\rho_i^p = \pi_{upi} + \pi_{lpi} + (1 - \frac{\partial P_L}{\partial P_i})\lambda_{ps} - \frac{\partial Q_L}{\partial P_i}\lambda_{qs} - \sum_{j=1}^{l}(\pi_{llj} + \pi_{ulj})\frac{\partial P_{lj}}{\partial P_i}$	$\rho_i^q = \pi_{uqi} + \pi_{lqi} - \frac{\partial P_L}{\partial Q_i}\lambda_{ps} + (1 - \frac{\partial Q_L}{\partial Q_i})\lambda_{qs} - \sum_{j=1}^{l}(\pi_{llj} + \pi_{ulj})\frac{\partial P_{lj}}{\partial Q_i} - \sum_{j=1}^{n}(\pi_{lvj} + \pi_{uvj})\frac{\partial V_j}{\partial Q_i} - \sum_{j=1}^{f}(\pi_{ltj} + \pi_{utj})\frac{\partial t_j}{\partial Q_i}$	Newton OPF Interior Point

3.3 Lagrangian Multipliers, Marginal Cost and Spot Price

The introduction of Lagrangian multipliers allows us to reduce the constrained problem to an unconstrained problem and often provides additional information about the problem that is extremely valuable. In many economic problems, the optimal value of the Lagrangian multiplier λ has an important economic interpretation. For a general optimisation problem:

$$\begin{aligned} &\min f(x) \\ &s.t.\ h(x) = b \end{aligned} \tag{3.1}$$

The Lagrangian for problem (3.1) is:

$$L(x) = f(x) + \lambda(b - h(x)) \tag{3.2}$$

At the optimal (x^*, λ^*), the first-order conditions

$$\frac{\partial f(x_i^*)}{\partial x_i^*} - \lambda^* \frac{h(x_i^*)}{\partial x_i^*} = 0 \text{ for all i=1,.....,}n \text{ and } h(x) - b = 0$$

must hold. These n+1 equations can be solved for n+1 optimal values in terms of parameter b. By writing these parameterised optimal values as $x^*(b) = (x^*{}_1(b), x^*{}_2(b), \cdots\cdots, x^*{}_n(b))$ and $\lambda^*(b)$, consider the change in the indirect objective function caused by a change in parameter b, writing out this change we have:

$$\frac{df(x^*)}{db} = \sum_i \frac{\partial f(x^*)}{\partial x_i^*} \frac{dx_i^*}{db} \tag{3.3}$$

Recalling the n first-order conditions, we can rewrite (3.3) as

$$\frac{df(x^*)}{db} = \lambda^* \sum_i \frac{\partial h(x^*)}{\partial x_i^*} \frac{dx_i^*}{db} \tag{3.4}$$

The remaining first-order condition is $h(x^*) - b = 0$. Differentiation of this condition gives

$$\sum_i \frac{\partial h(x^*)}{\partial x_i^*} \frac{dx_i^*}{db} = 1\,.$$

Substituting this into (3.4) gives the following expression for the Lagrangian multipliers:

$$\frac{df(x^*)}{db} = \lambda^* \tag{3.5}$$

The result in (3.5) says that the optimal value of the Lagrangian multipliers measure the "sensitivities" of the indirect objective function to a change in the value of parameter b.

In Optimal Power Flow, parameter b is the net power injection vector, therefore, the meaning of Lagrangian multipliers is: the change in optimal operating cost if we change the power produced or consumed at a bus in the network. It could be expressed as the following derivative:

$$\frac{dL}{dP_i}$$

We see that the interpretation of the vector of Lagrangian multipliers is that they indicate the increment in optimal cost with respect to small changes in the parameters of the network. In the case of small change in power, either consumed or produced at a bus, the Lagrangian multipliers for that bus then indicate the incremental cost that will be incurred as a result of this change. This cost is also the definition of so called Short Run Marginal Cost (SRMC) which is one of the key concepts of economics and denotes the extra or additional cost of producing one extra unit of output. SRMC, as introduced in Chapter 1, provides the basis of spot pricing of electricity, as a result of the application of spot pricing in which generators are paid at marginal cost of generation buses and consumers are charged at marginal cost of load buses. Strictly, the spot prices for active and reactive power are defined as:

$$\frac{\partial(\text{total generation cost})}{\partial(\text{active power consumption})} \text{ and } \frac{\partial(\text{total generation cost})}{\partial(\text{reactive power consumption})}$$

Therefore, from the above discussion, spot prices or marginal costs are equivalent to the Lagrangian multipliers corresponding to power flow equations, and can be obtained through OPF calculation:

$$\lambda_p^* = \frac{dF(x^*)}{dP_i} \Longleftrightarrow \frac{\partial(\text{total generation cost})}{\partial(\text{active power consumption})}$$

$$\lambda_q^* = \frac{dF(x^*)}{dQ_i} \Longleftrightarrow \frac{\partial(\text{total generation cost})}{\partial(\text{reactive power consumption})}$$

3.4 Integrated Sensitivity Calculation

3.4.1 Loss Sensitivity

Nodal power balance function can be written in the form of total system power balance equation:

$$\sum_{i \neq s} P_i + P_S = P_L \tag{3.6}$$

where s represents the slack bus. P_L is system active power loss. Then define

$$X = (\theta_1, V_1 \cdots \theta_i, V_i \cdots),\ \lambda = (\lambda_{p1}, \lambda_{q1} \cdots \lambda_{pi}, \lambda_{qi} \cdots)$$
$$g = (g_{p1}, g_{q1} \cdots g_{pi}, g_{qi} \cdots)\ \ (i=1, \cdots n\ \ i \neq s)$$

And notice that the system operation is determined by all bus injections except the slack bus. From Equation (3.6), we have

$$\frac{\partial P_L}{\partial X} = \frac{\partial \sum_{i \neq s} P_i}{\partial X} + \frac{\partial P_S}{\partial X} = \sum_{i \neq s} \frac{\partial P_L}{\partial P_i}\frac{\partial P_i}{\partial X} + \sum_{i \neq s} \frac{\partial P_L}{\partial Q_i}\frac{\partial Q_i}{\partial X} \tag{3.7}$$

$$\frac{\partial P_S}{\partial X} = \sum_{i \neq s} \left(-1 + \frac{\partial P_L}{\partial p_i} \right) \frac{\partial P_i}{\partial X} + \sum_{i \neq s} \frac{\partial P_L}{\partial Q_i}\frac{\partial Q_i}{\partial X} \tag{3.8}$$

$$\begin{bmatrix} \vdots \\ \vdots \\ \dfrac{\partial P_s}{\partial \theta_i} \\ \dfrac{\partial P_s}{\partial V_i} V_i \\ \vdots \\ \vdots \end{bmatrix} = [J]^T \begin{bmatrix} \vdots \\ \vdots \\ 1 - \dfrac{\partial P_L}{\partial P_i} \\ -\dfrac{\partial P_L}{\partial Q_i} \\ \vdots \end{bmatrix} \qquad (i = 1 \cdots \cdots n, i \neq s) \tag{3.9}$$

Similarly, for reactive power loss:

$$\begin{bmatrix} \vdots \\ \frac{\partial Q_s}{\partial \theta_i} \\ \frac{\partial Q_s}{\partial V_i} V_i \\ \vdots \end{bmatrix} = [J]^T \begin{bmatrix} \vdots \\ -\frac{\partial Q_L}{\partial P_i} \\ 1 - \frac{\partial Q_L}{\partial Q_i} \\ \vdots \end{bmatrix} \qquad (i = 1 \cdots \cdots n, i \neq s) \tag{3.10}$$

3.4.2 Sensitivity Coefficients of an AC Network Model

The general equations for active and reactive power injections at a bus can be defined as the following:

$$P_i(V,\theta) = \sum_{j=1}^{n} V_i V_j \left|Y_{ij}\right| \cos(\theta_i - \theta_j - \delta_{ij}) \tag{3.11}$$

$$Q_i(V,\theta) = \sum_{j=1}^{n} V_i V_j \left|Y_{ij}\right| \sin(\theta_i - \theta_j - \delta_{ij}) \tag{3.12}$$

Then at each bus:

$$\begin{aligned} P_i(V,\theta) &= P_{gi} - P_{di} \\ Q_i(V,\theta) &= Q_{gi} - Q_{di} \end{aligned} \tag{3.13}$$

The set of equations that represents the first-order approximation of the AC network around the initial point is:

$$\sum_j \frac{\partial P_i}{\partial V_j} \Delta V_j + \sum_j \frac{\partial P_i}{\partial \theta_j} \Delta \theta_j = \Delta P_{gi} \tag{3.14}$$

$$\sum_j \frac{\partial Q_i}{\partial V_j} \Delta V_j + \sum_j \frac{\partial Q_i}{\partial \theta_j} \Delta \theta_j = \Delta Q_{gi} \tag{3.15}$$

This can be placed in matrix form for easier manipulation:

$$J \Delta x = \Delta u \tag{3.16}$$

where x is the state vector of voltages and phase angles, and u is the vector of control variables. The control variables are the generator MW, transformer taps, and generator voltage MVAR (or generator voltage magnitudes). Note that other controls can be easily added to this formation.

Assume that there are several transmission-dependent variables, h, that represent, for example, MVA flows, load bus voltages, line amperes, etc., and we wish to find their sensitivities with respect to changes in the control variables. Each of these equations can be expressed as a function of the state and control variables, for example branch flow between bus i and j:

$$h(V,\theta) = P_{ij} = -V_i^2 G_{ij} + V_i V_j (G_{ij}\cos\theta_{ij} + B_{ij}\sin\theta_{ij})$$

As before, we can write a linear version of these variables around an operating point:

$$\Delta h = \begin{bmatrix} \frac{\partial h_1}{\partial V_1} & \frac{\partial h_1}{\partial \theta_1} & \cdots \\ \frac{\partial h_2}{\partial V_1} & \frac{\partial h_2}{\partial \theta_1} & \cdots \\ \vdots & \vdots & \cdots \end{bmatrix} \begin{bmatrix} \Delta V_1 \\ \Delta \theta_1 \\ \vdots \end{bmatrix} + \begin{bmatrix} \frac{\partial h_1}{\partial u_1} & \frac{\partial h_1}{\partial u_2} & \cdots \\ \frac{\partial h_2}{\partial u_1} & \frac{\partial h_2}{\partial u_2} & \cdots \\ \vdots & \vdots & \cdots \end{bmatrix} \begin{bmatrix} \Delta u_1 \\ \Delta u_2 \\ \vdots \end{bmatrix} \tag{3.17}$$

We can put this into a compact format using the vectors x and u:

$$\Delta h = J_{hx}\Delta x + J_{hu}\Delta u \tag{3.18}$$

Δx can be estimated from Equation (3.16). Then, substituting:

$$\Delta h = J_{hx}J^{-1}\Delta u + J_{hu}\Delta u \tag{3.19}$$

This last equation gives the linear sensitivity coefficients between the transmission system quantities, h and the control variables, u. For most of the cases, $J_{hu} = 0$, then $J_{hx}J^{-1}$ is the sensitivity matrix of h:

$$\frac{\Delta h}{\Delta u} = J_{hx}J^{-1} \tag{3.20}$$

3.5 Decomposition Model of Spot Pricing

3.5.1 Framework

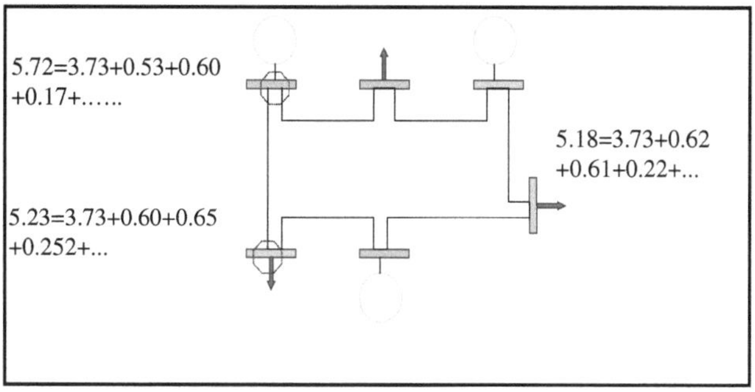

Figure 3.1. Unbundled nodal price

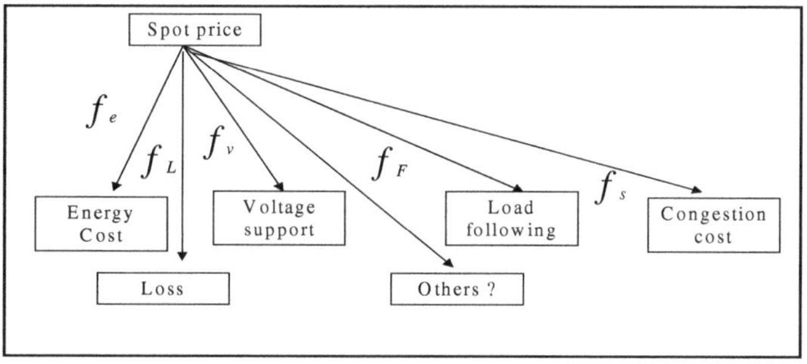

Figure 3.2. Unbundling of spot price

In an ideal deregulation environment, the generators should be paid for every contribution to the power system operation, and consumers should be charged for every service from the system. Thus there are two problems needing to be solved: (1) how to calculate spot prices, and (2) how to decompose spot price into various components. Figures 3.1 and 3.2 illustrate the spot prices and unbundled components. The optimisation problem discussed in Chapters 3–5 can be interpreted from the power markets perspective:

(1) Objectives:
The IOSP problem inherits the objective function from classic economic dispatching with the extension to include the reactive power generation cost:

$$Min\ F = \sum_{i=1}^{m} (f_{pi}(P_{gi}) + f_{qi}(Q_{gi})) \tag{3.21}$$

In general, it is more difficult to obtain an expression of $f_{qi}(Q_{gi})$ than $f_{pi}(P_{gi})$. However, it is very important to compensate generators for their MVAR contribution to the electricity market [7]. The production cost of reactive power should be accounted for in spot pricing model and the quadratic curve expression [7], which is similar to the active power generation cost, is adopted in this chapter. An extra advantage brought by the introduction of $f_{qi}(Q_{gi})$ is that reactive power generations can be treated as basic variables like active power, therefore it enhances the numerical stability of IPOPF.

(2) Constraints—ancillary services

(i) Energy components

Generation real power limits

For short-term operation, the unit commitment is given to the operating condition. Consequently, any no load and out-of-merit cost considerations that might influence the unit commitment are beyond the scope of this chapter.

Loss compensation

System real power loss is dominated by system configuration and operating conditions. Variations in any control variables can cause changes in the loss. The impacts of real and reactive power generation and consumption on real power loss are important in determining spot prices.

(ii) Capacity component

Spinning reserve plays an important role in reliable and secure power system short-term operation. The spinning reserve is defined as the difference between the maximum generation capacity and the present generation output, and the units are not able to provide spinning reserve for more than a certain percentage of their capacity.

The curtailed load is the difference between the maximum load and the actual load, which can be viewed as another type of spinning reserve. In this manner, the curtailed portion of the dispatchable load can be modelled as a generator. The system minimum spinning reserve constraint should also be satisfied.

(iii) Voltage control and reactive support

Voltage limits refer to the requirement for the system bus voltages to remain within a narrow range of levels. Since voltages are affected primarily by reactive power flows, any voltage limits will show up as a premium on the price assigned to reactive power on the bus.

(iv) Security control

Transmission congestion could force the system operator to use more expensive generation instead of cheaper alternatives, therefore it will cause congestion costs that must be assigned to the grid users. The power exchange and bilateral contracts do not take this into account to schedule their transactions. It is the operator's responsibility to detect and solve these problems. The system operator will only modify the schedules produced by the power exchange and the bilateral contracts for this purpose, using the bid prices of the power exchange and incremental prices for the bilateral contracts. Kirchoff's laws require that real and reactive power flows balance at each bus *i* through the system.

In addition, voltage stability and angle stability are important factors, which are not included in the model but can be incorporated by modifying the line flow limits and maximum loadability through running separate stability evaluation software.

3.5.2 Optimal Spot Pricing

The optimal spot pricing problem defined in 3.5.1 can then be generalised as:

$$\begin{aligned} \min \quad & F(z) \\ s.t. \quad & g(z) = 0 \\ & h_l \le h(z) \le h_u \end{aligned} \tag{3.22}$$

As we have done before, slack variables are introduced into the objective function to transform inequality constraints into equality constraints by incorporating them in a logarithmic barrier function:

$$\begin{aligned} \min L = {} & F(z) - \lambda^T g(z) + \pi_l^T (h(z) - s_l - h_l) \\ & + \pi_u^T (h(z) + s_u - h_u) - \mu\left(\sum_{i=1}^{k} \ln s_{li} + \sum_{i=1}^{k} \ln s_{ui}\right) \end{aligned} \tag{3.23}$$

where $s_l \ge 0, s_u \ge 0, \pi_l < 0, \pi_u > 0, \mu > 0$.

Using Lagrangian multipliers of constraints to estimate the cost change for a unit change in each binding constraint, the optimisation problem can be formulated explicitly as:

$$
\begin{aligned}
\min L = & \sum_{i=1}^{m} (f_{pi}(P_{gi}) + f_{qi}(Q_{gi})) \\
& - \sum_{i=1}^{n} \lambda_{pi} g_{pi} - \sum_{i=1}^{n} \lambda_{qi} g_{qi} + \sum_{i=1}^{m} \pi_{lpi}(P_{gi} - s_{lpi} - P_{gi\min}) + \sum_{i=1}^{m} \pi_{upi}(P_{gi} + s_{upi} - P_{gi\max}) \\
& + \sum_{i=1}^{m} \pi_{lqi}(Q_{gi} - s_{lqi} - Q_{gi\min}) + \sum_{i=1}^{m} \pi_{uqi}(Q_{gi} + s_{uqi} - Q_{gi\max}) \\
& + \sum_{i=1}^{l} \pi_{lli}(P_{li} - s_{lli} - P_{li\min}) + \sum_{i=1}^{l} \pi_{uli}(P_{li} + s_{uli} - P_{li\max}) + \pi_{lr}(\sum_{i=1}^{m} R_i - s_{lr} - R_{\min}) \\
& + \sum_{i=1}^{n} \pi_{lvi}(V_i - s_{lvi} - V_{i\min}) + \sum_{i=1}^{n} \pi_{uvi}(V_i + s_{uvi} - V_{i\max}) \\
& - \mu(\sum_{i=1}^{k} \ln s_{ui} + \sum_{i=1}^{k} \ln s_{li})
\end{aligned}
\tag{3.24}
$$

where

$$
s_u = [s_{up} \quad s_{uq} \quad s_{ul} \quad s_{uv}]^T
$$

$$
s_l = [s_{lp} \quad s_{lq} \quad s_{ll} \quad s_{lv} \quad s_{lr}]^T
$$

$$
\pi_u = [\pi_{up} \quad \pi_{uq} \quad \pi_{ul} \quad \pi_{uv}]^T
$$

$$
\pi_l = [\pi_{lp} \quad \pi_{lq} \quad \pi_{ll} \quad \pi_{lv} \quad \pi_{lr}]^T \tag{3.25}
$$

From economic theory, λ_{pi} and λ_{qi} are the shadow prices of real power and reactive power injections at node i at the optimal solution point. π_{li} and π_{ui} represent the marginal costs of associated constraints. Then we can obtain the spot prices for all generators and customers:

$$
\rho_i^P = \left.\frac{\partial L}{\partial P_i}\right|_* = \lambda_{pi} \tag{3.26}
$$

$$
\rho_i^q = \left.\frac{\partial L}{\partial Q_i}\right|_* = \lambda_{qi} \tag{3.27}
$$

For generators, two extra equations are satisfied:

$$\lambda_{pi} = \left.\frac{\partial F(P_{gi})}{\partial P_{gi}}\right|_{*} + \pi_{lpi} + \pi_{upi} + \pi_{lr}\frac{\partial \sum_i R_i}{\partial P_{gi}} \tag{3.28}$$

$$\lambda_{qi} = \left.\frac{\partial F(Q_{gi})}{\partial Q_{gi}}\right|_{*} + \pi_{lqi} + \pi_{uqi} \tag{3.29}$$

where * denotes the optimal solution point.

3.5.3 Spot Price Decomposition

Rewrite Lagrangian function (3.22) as:

$$L = F(P_g, Q_g) - \lambda^T g(X) + (\pi_l + \pi_u)^T h(z) + \sigma \tag{3.30}$$

where:

$$\sigma = \pi_l^T(-s_l - h_l) + \pi_u^T(s_u - h_u) - \mu(\sum_{i=1}^{k} \ln s_{li} + \sum_{i=1}^{k} \ln s_{ui}) \tag{3.31}$$

$$\frac{\partial L}{\partial X} = \frac{\partial F}{\partial X} - \frac{\partial g^T}{\partial X}\lambda + \frac{\partial h^T}{\partial X}(\pi_l + \pi_u) = 0 \tag{3.32}$$

$$\begin{aligned}\lambda &= \left(\frac{\partial g^T}{\partial X}\right)^{-1}\left(\frac{\partial F}{\partial X} + \frac{\partial h^T}{\partial X}(\pi_l + \pi_u)\right) \\ &= \left(\frac{\partial g^T}{\partial X}\right)^{-1}\left(\frac{\partial F}{\partial P_S}\frac{\partial P_S}{\partial X}\right) + \left(\frac{\partial g^T}{\partial X}\right)^{-1}\left(\frac{\partial F}{\partial Q_S}\frac{\partial Q_S}{\partial X}\right) + \left(\frac{\partial g^T}{\partial X}\right)^{-1}\frac{\partial h^T}{\partial X}(\pi_l + \pi_u)\end{aligned} \tag{3.33}$$

According to the sensitivity calculation theory introduced in Section 3.4 and because $\frac{\partial F}{\partial P_S}$ and $\frac{\partial F}{\partial Q_S}$ are scalars, then substitute (3.9), (3.10) and (3.21) into (3.33), λ_{pi} and λ_{qi} ($i = 1 \cdots\cdots n, i \neq s$) can then be decomposed as:

$$\lambda_{pi} = \left(1 - \frac{\partial P_L}{\partial P_i}\right)\frac{\partial F}{\partial P_S} - \frac{\partial Q_L}{\partial P_i}\frac{\partial F}{\partial Q_S} - \sum_{j \notin h_0}\frac{\partial h_j}{\partial P_i}(\pi_{lj} + \pi_{uj}) \tag{3.34}$$

$$\lambda_{qi} = (1 - \frac{\partial Q_L}{\partial Q_i})\frac{\partial F}{\partial Q_s} - \frac{\partial P_L}{\partial Q_i}\frac{\partial F}{\partial P_s} - \sum_{j \notin h_0} \frac{\partial h_j}{\partial Q_i}(\pi_{lj} + \pi_{uj}) \quad (3.35)$$

According to Equation (3.34)-(3.35), the decomposition formulae of active power and reactive power spot prices are obtained as:

$$\rho_i^p = (1 - \frac{\partial P_L}{\partial P_i})\lambda_{ps} - \frac{\partial Q_L}{\partial P_i}\lambda_{qs} - \sum_{j=1}^{l}(\pi_{llj} + \pi_{ulj})\frac{\partial P_{lj}}{\partial P_i} \quad (3.36)$$

$$\rho_i^q = -\frac{\partial P_L}{\partial Q_i}\lambda_{ps} + \lambda_{qs}(1 - \frac{\partial Q_L}{\partial Q_i}) - \sum_{j=1}^{l}(\pi_{llj} + \pi_{ulj})\frac{\partial P_{lj}}{\partial Q_i} - \sum_{j=1}^{n}(\pi_{lvj} + \pi_{uvj})\frac{\partial V_j}{\partial Q_i} \quad (3.37)$$

3.5.4 Implementation

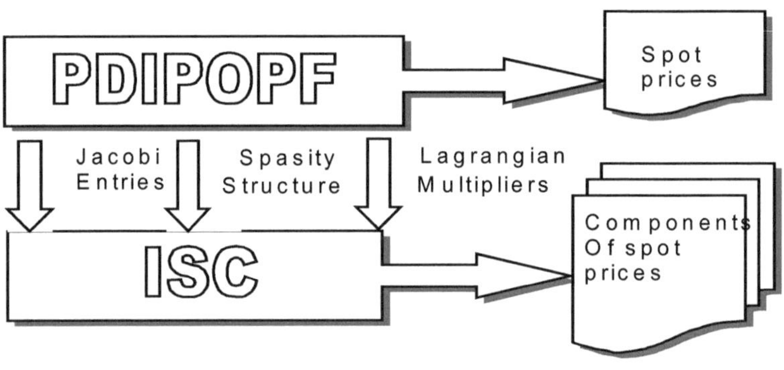

Figure 3.3. The integrated spot pricing scheme

In the decomposition of spot price, sensitivities of power loss and constraints with respect to node injections are essential. The sensitivity calculation method proposed in [26] can be added into IOSP naturally. The main idea of this method is to utilise the extended Jacobian matrix J whose elements have already been obtained during the IPOPF iteration. Once it has been factorised of J, the sensitivities of control variables with respect to state variables are available through forward and backward substitution with the changing of the right hand side vector. Combining different basic sensitivities which are the by-products of PDIPOPF, we can obtain every sensitivity needed. In summary, the framework of the proposed method is given in Figure 3.3, in which ISC stands for Integrated Sensitivity Calculation.

From Equation (3.36) and (3.37), ISC includes four types of sensitivities:

Jacobian sensitivities $(\frac{\partial P}{\partial V}, \frac{\partial P}{\partial \theta}, \frac{\partial Q}{\partial V}, \frac{\partial Q}{\partial \theta})$, control sensitivities

$(\frac{\partial V}{\partial P}, \frac{\partial V}{\partial Q}, \frac{\partial \theta}{\partial P}, \frac{\partial \theta}{\partial Q})$, loss sensitivities $(\frac{\partial P_L}{\partial P}, \frac{\partial P_L}{\partial Q}, \frac{\partial Q_L}{\partial P}, \frac{\partial Q_L}{\partial Q})$ and combination sensitivities $(\frac{\partial P_l}{\partial P}, \frac{\partial P_l}{\partial q})$. In ISC, the greatest computation effort is required to factorise the Jacobian matrix whose elements are obtained from the Hessian matrix. Once factorisation of J is complete, the four types of sensitivities are available through forward and backward substitution with the changing of the right hand side vectors.

Type 1: Jacobian sensitivities $(\frac{\partial P}{\partial V}, \frac{\partial P}{\partial \theta}, \frac{\partial Q}{\partial V}, \frac{\partial Q}{\partial \theta})$

These are elements of the Jacobian matrix obtained directly from the Hessian matrix, i.e.

$$S_1 = J \tag{3.38}$$

Type 2: Control sensitivities $(\frac{\partial V}{\partial P}, \frac{\partial V}{\partial Q}, \frac{\partial \theta}{\partial P}, \frac{\partial \theta}{\partial Q})$

$$S_2 = -J^{-1}E \tag{3.39}$$

E is the unit matrix.

Type 3: Loss Sensitivities $(\frac{\partial P_L}{\partial P}, \frac{\partial P_L}{\partial Q}, \frac{\partial Q_L}{\partial P}, \frac{\partial Q_L}{\partial Q})$

$$S_{pl} = \left(J^{-1}\right)^T S_p \tag{3.40}$$

$$S_{ql} = \left(J^{-1}\right)^T S_q \tag{3.41}$$

S_{pl}, S_p and S_{ql}, S_q stand for the right-hand side term and left-hand side terms respectively in Equation (3.9) and (3.10).

Type 4: Combination sensitivities $(\frac{\partial P_l}{\partial P}, \frac{\partial P_l}{\partial q})$

If line l is between bus m and n, then we have:

$$\frac{\partial P_l}{\partial P_i} = \frac{\partial P_l}{\partial V_m}\frac{\partial V_m}{\partial P_i} + \frac{\partial P_l}{\partial \theta_m}\frac{\partial \vartheta_m}{\partial P_i} + \frac{\partial P_l}{\partial V_n}\frac{\partial V_n}{\partial P_i} + \frac{\partial P_l}{\partial \theta_n}\frac{\partial \theta_n}{\partial P_i} \tag{3.42}$$

which is a combination of first and second type sensitivities.

In (3.38)–(3.42), the most computation effort is to factorise the Jacobian matrix whose elements are picked up from the Hessian matrix of Newton OPF [25]. Once factorisation of J, the four types of sensitivities are available through the forward and backward substitution with the changing of right hand side vectors.

3.6 Features of the Proposed IOSP Model

3.6.1 Comparison with Existing Models

Equations (3.36) and (3.37) indicate that spot prices for both active and reactive power are composed of four parts. The first is marginal generation cost which is equivalent to system lambda. The second part is the loss compensation cost, which is proportional to the system lambda multiplied by the loss sensitivity. The third part is concerned with the coupling between active and reactive power. The last part is associated with security cost, for active power it only represents congestion alleviation cost, but for reactive power, it also includes voltage support cost. Compared with some of the existing models based on OPF (Table 3.1), the following features of IOSP are worth noting:

(1) λ_p and λ_q can be adopted as spot prices for both generators and consumers.

(2)Due to the introduction of reactive power generation cost, the coupling between active power and reactive power can be accounted for. Reactive power spot price has a similar decomposition formula to that of active power.

(3)The generation capacity constraint terms will not appear in either active power price or in reactive power price. However, they do effect the spot prices by influencing the values of λ_{ps} and λ_{qs}. Equations (3.28) and (3.29) established the connection between spot prices, generation costs and generation capacity constraints. π_{upi} ,π_{uqi}, π_{lpi} and π_{lqi} reflect the benefits that the system will gain if the generator relaxes its generation limits.

(4)The proposed model provides a more general decomposition formulation for considering various security costs and has the potential to include other security costs which are not considered in this chapter through simply adding them into the OPF constraints. The security costs could vary depending on system conditions, but typically they are only a small fraction of the marginal generating costs though in some cases these costs could be significantly higher.

(5)The values of π_u and π_l change exponentially due to the application of a logarithmic barrier function as a soft constraint. Therefore high volatility can be avoided. This point is extremely important for a practical SP methodology and will be emphasised in Chapter 9.

From Table 3.1, we can see that IOSP can be viewed as the extended formulation of the model provided in [16], and the derivation and the price signals obtained are more meaningful from an engineering point of view. The first term in Equation (7) of [16] is further decomposed into two parts explicitly corresponding to the system lambda and active power loss respectively. Because of the introduction of reactive power cost, reactive power system lambda and the term associated with reactive power loss appear in both ρ^p and ρ^q. For active power price, (3.36) is similar to Equation (6) in [8] but with a new reactive power loss term. For reactive power pricing, the pricing term, which corresponds to the incremental increase of real power loss attributed to the change of reactive power, is a natural part of ρ^q instead of the introduction of the loss optimisation sub-problem used in [8]. Also, the second term in Equation (7) of [16] is unbundled to reflect the influence of security constraints or control variables explicitly. The contribution of reactive power to system security, including the real power flow on transmission lines, is considered explicitly.

With proper simplification and assumptions, the price models in [6,8,16] can be obtained from the proposed model.

3.6.2 The Inner Connection with Economic Dispatching (ED)

Because $\frac{\partial Q_L}{\partial P_i}\frac{\partial F}{\partial Q_S}$ is negligible, rewrite Equation (3.36) as:

$$\frac{\lambda_{pi} + \sum_j \frac{\partial h_j}{\partial P_i}(\pi_{lj} + \pi_{uj})}{1 - \frac{\partial P_L}{\partial P_i}} = \frac{\partial F}{\partial P_S} = \lambda_{ps} \tag{3.43}$$

It can easily be seen that Equation (3.43) will be the optimal condition of active power dispatch constrained by security limits. If all security constraints are neglected, it will then be in the same form as the coordination equation in classical economic dispatching.

$$\frac{\lambda_{p1}}{1-\frac{\partial P_L}{\partial P_1}} = \frac{\lambda_{p2}}{1-\frac{\partial P_L}{\partial P_2}} = \cdots\cdots = \frac{\lambda_{pi}}{1-\frac{\partial P_L}{\partial P_i}} = \lambda_{ps} \tag{3.44}$$

Similarly, reactive power dispatch satisfies:

$$\frac{\lambda_{qi} + \frac{\partial P_L}{\partial Q_i}\frac{\partial F}{\partial P_S} + \sum_j \frac{\partial h_j}{\partial Q_i}^T (\pi_{lj} + \pi_{uj})}{1 - \frac{\partial Q_L}{\partial Q_i}} = \frac{\partial F}{\partial Q_S} = \lambda_{qs} \tag{3.45}$$

Equation (3.44) shows that, under normal operating conditions, the nodal price of reactive power is determined by reactive power generation incremental cost and real power loss sensitivity. λ_{qs} provides the marginal unit cost information for reactive power. From (3.45), if the cost of reactive power generation is not accounted for, it is apparent that the price of reactive power will be null when the generation capacity constraints are not violated. Thus we can reach the conclusion drawn in [6].

3.7 Numerical Studies

3.7.1 Case Study 1: Insight View of a 5-bus System

A simple 5-bus test system as shown in Figure 3.4 is used to first gain an insight into the spot price decomposition method proposed in this chapter. G_1 and G_2 are utility controlled units. Their parameters are showed in Table 3.2. Quadratic cost curves are adopted for both active and reactive power production. All prices are in ¢.

$$f_{pi}(P_{gi}) = C_{pi}P_{gi}^2 + B_{pi}P_{gi} + A_{pi} \;,\; f_{qi}(Q_{gi}) = C_{qi}Q_{gi}^2 + B_{qi}Q_{gi} + A_{qi}$$

Node 5 is selected as the reference bus. Voltage magnitude is in per unit. Voltage angle, active power and reactive power are in radians, MW and MVAR respectively. All bus voltage upper limits are 1.05, lower limits 0.95. System minimum spinning reserve capacity $R_{\min}$ is 50MW. Maximum active power reserve factor for every unit is 0.2. Loads are kept fixed: D1 has a demand of 50MW and 40MVAR. D2 has a demand of 100MW and 20MVAR. Therefore customer outage cost and load following price are zero. Line parameters and power flow results are summarised in Tables 3.6 and 3.3. All Lagrangian multipliers obtained are given in Table 3.4. Table 3.5 provides all sensitivities needed in the decomposition of spot price.

The results of spot price and components are presented in Table 3.7, where:

$$\lambda_{Pi}^{*} = \lambda_{ps}(1 - \frac{\partial P_L}{\partial P_i})$$

$$\lambda_{Pi}^{**} = \lambda_{Pi}^{*} - \sum_j (\pi_{llj} + \pi_{ulj}) \frac{\partial P_{lj}}{\partial P_i}$$

$$\lambda_{qi}^{*} = \lambda_{qs}(1 - \frac{\partial Q_L}{\partial Q_i})$$

$$\lambda_{qi}^{**} = \lambda_{qi}^{*} - \lambda_{ps}(\frac{\partial P_L}{\partial Q_i})$$

$$\lambda_{qi}^{***} = \lambda_{qi}^{**} - \sum_j (\pi_{llj} + \pi_{ulj}) \frac{\partial P_{lj}}{\partial Q_i} - \sum_j (\pi_{lvj} + \pi_{uvj}) \frac{\partial V_j}{\partial Q_i}$$

The dominant elements in determining components of spot price are shown bold in the following tables. The optimal active power output of G2 is close to the lower limit, reactive power is close to the upper limit, therefore π_{lp2} and π_{uq2} diverge from zero.

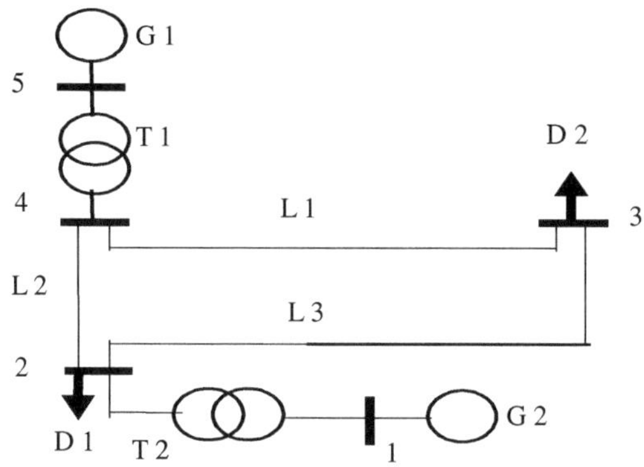

Figure 3.4. 5-bus test system

We have: λ_{P1}=40.891

$$(2C_{p2} * P_{g2} + B_{p2}) + \pi_{lp1} = 2 * 0.2 * 110 + 1.0 + (-4.109) = 40.891 ,$$

$$\lambda_{q1} = 2.1421$$

$$2C_{q2}Q_2 + B_{q2} + \pi_{uq1} = 2 * 0.02 * 50.5 + 0.1 + 0.0226 = 2.1426$$

For G1, its outputs are far from limits, therefore π_{lp5}, π_{up5}, π_{lq5} and π_{uq5} all nearly zero. The spot prices are controlled by the incremental generation cost:

$$(2C_{p1} * P_{g1} + B_{p1}) + \pi_{lp5} = 2 * 0.5 * 41.418 + 1.2 = 42.618$$

$$2C_{q1}Q_{g1} + B_{q1} + \pi_{uq5} = 2 * 0.05 * 17.435 + 0.12 = 1.8635$$

$$\lambda_{p2} = 42.618$$

$$\lambda_{q2} = 1.864$$

Therefore, (3.28) and (3.29) are followed exactly.

According to the results in Table 3.7, the sum of the components of spot price equals λ_p and λ_q. This supports the decomposition method presented in this chapter. It can be seen that active power results are more accurate than those for reactive power because of the stronger non-linearity characteristics of reactive power. No voltage violations occur in this example. Both generators are type 2 units. The total active reserve is 60MW. λ_{ps} and the loss compensation dominates the active power spot price. The first two terms in (3.36) are key elements in reactive power pricing. In this example, line flow of L3 is close to its upper limit, and this has an influence on both active and reactive spot prices at all nodes. For nodes 2 and 3, the spot price at the sending node drops, and it increases at the receiving node. The active and reactive production cost of G1 are 1009.1 and 10.173. For G2 the active and reactive production costs are 1025.3 and 10.515 respectively. Table 3.8 gives the percentage errors of the decomposition models, which is calculated by:

$$Err\%_p = \frac{\lambda_p - \lambda_p^{**}}{\lambda_p} \times 100\% \text{ and } Err\%_q = \frac{\lambda_q - \lambda_q^{***}}{\lambda_q} \times 100\%$$

From Table 3.8, we can see the error in the decomposition model is rather small. The error for active power decomposition is smaller than that for reactive power decomposition.

The final spot prices and its components for every node are illustrated in Figure 3.5.

Table 3.2. Unit parameters of 5-bus system

Unit	P	Q	P_{max}	P_{min}	Q_{max}	Q_{min}	Cp	Bp	Ap	Cq	Bq	Aq
G1	41.42	13.435	100	0	100	-20	0.5	1.2	10^3	0.05	0.12	10
G2	110	50.499	200	110	50.5	-20	0.2	1.0	10^3	0.02	0.1	10

Table 3.3. Power flow results of 5-bus system

Nodes	V	θ	Line	P_{ij}	Q_{ij}	P_{ji}	Q_{ji}
1	1.0276	6.306E-02	L1 (Bus4~Bus3)	62.108	12.685	-61.610	-14.441
2	0.9993	-3.252E-03	L2 (Bus4~Bus2)	-20.696	3.534	20.992	-4.442
3	0.9770	-4.725E-02	L3 (Bus2~Bus3)	38.998	6.357	-38.385	-5.559
4	0.9895	-2.589E-02	T1 (Bus5~Bus4)	41.418	13.431	-41.4167	-16.181
5	1	0	T2 (Bus1~Bus2)	109.999	50.	-109.991	-41.913

Table 3.4. Lagrangian multipliers of 5-bus system

π_{up}	π_{lp}	π_{uv}	π_{lv}	π_{ul}	π_{ll}	π_{uq}	π_{lq}
1.53E-05	-4.109	3.06E-05	-2.49E-05	0.0027	-4.38E-04	0.0226	-4.12E-08
0	0	3.24E-05	-4.42E-05	5.96E-04	-9.32E-04	0	0
0	0	2.2E-05	-1.17E-04	1.23	-5.16E-04	0	0
0	0	2.7E-05	-5.78E-05	0	0	0	0
2.33E-05	-3.9E-05	2.156	-1.69E-05	0	0	2.31E-08	-2.52E-08

Table 3.5. Sensitivities of 5-bus system

Nodes	$\frac{\partial P_L}{\partial P_i}$	$\frac{\partial P_{l1}}{\partial P_i}$	$\frac{\partial P_{l2}}{\partial P_i}$	$\frac{\partial P_{l3}}{\partial P_i}$	$\frac{\partial P_L}{\partial Q_i}$	$\frac{\partial P_{l3}}{\partial Q_i}$	$\frac{\partial Q_L}{\partial Q_i}$
1	0.0287	0.419	0.558	0.43	-0.0049	-0.019	-0.018
2	0.0222	0.411	0.567	0.417	-0.0051	-0.023	-0.063
3	-0.019	-0.876	-0.138	-0.142	-0.0073	-0.021	-0.056
4	-0.0023	0.00009	-0.00002	-0.00334	-0.002	-0.002	-0.044
5	0	0	0	0	0	0	0

Table 3.6. Line parameters price decomposition

Line	P_{ijmax}	R(P.U.)	X(P.U.)	
L1	100	4.298E-3	2.198E-2	
L2	100	2.521E-2	6.917E-2	
L3	39	3.390E-3	1.777E-2	
XF1	100	0	0.06189	
XF2	200	0	0.06189	

Table 3.7. Spot price and decomposition of 5-bus system

Nodes	λ_{ps}	λ_{pi}^{*}	λ_{pi}^{**}	λ_{pi}	λ_{qs}	λ_{qi}^{*}	λ_{qi}^{**}	λ_{qi}^{***}	λ_{qi}
1	42.618	41.396	40.882	40.891	1.864	1.898	2.108	2.131	2.142
2	42.618	41.680	41.186	41.189	1.864	1.982	2.199	2.230	2.245
3	42.618	43.428	43.597	43.593	1.864	1.968	2.280	2.305	2.291
4	42.618	42.716	42.719	42.719	1.864	1.946	2.032	2.035	2.037
5	42.618	42.618	42.618	42.618	1.864	1.864	1.864	1.864	1.864

Table 3.8. Errors of spot price

Node	Err $(\%)_p$	Err $(\%)_q$
1	0.02	0.5
2	0.007	0.67
3	-0.009	-0.61
4	0	0.098
5	0	0

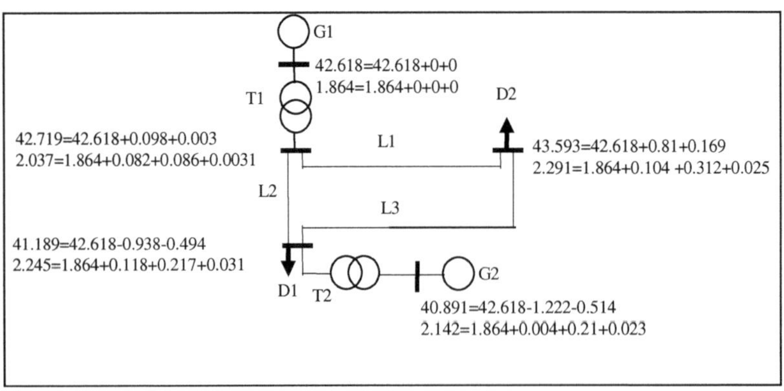

Figure 3.5. Spot price of active power and its components

3.7.2 Case Study 2: IEEE 30-bus System

The IEEE 30-bus system has six generators and four ULTC transformers. A loading condition of 324.29MW and 114.19MVAR is used to perform the IOSP and calculate spot prices. The spot prices of active and reactive power and their components under normal operation conditions are shown in Figures 3.6–3.9. Figure 3.7 clearly reveals that active power spot price is dominated by system lambda and the network loss compensation term under normal operating conditions. The congestion and security terms are relatively small. This conclusion is the same as [8]. Figure 3.9 shows the components of reactive power price and reveals that reactive power spot prices at all buses are dominated by reactive power generation cost and the real power loss compensation item. Compared with active power, reactive power spot price is much smaller due to the adopted reactive power generation cost. Under normal operation condition, the security prices for both active and reactive power are negligible. Compared with active power, reactive power spot price is much more volatile.

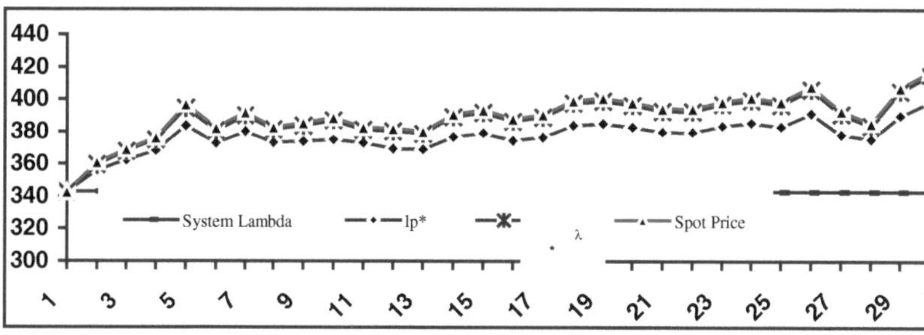

Figure 3.6. Active power spot price under normal operating conditions

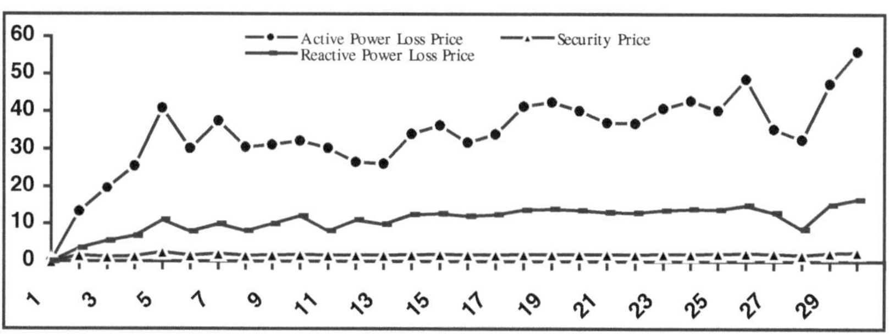

Figure 3.7. The components of active power spot price under normal operating conditions

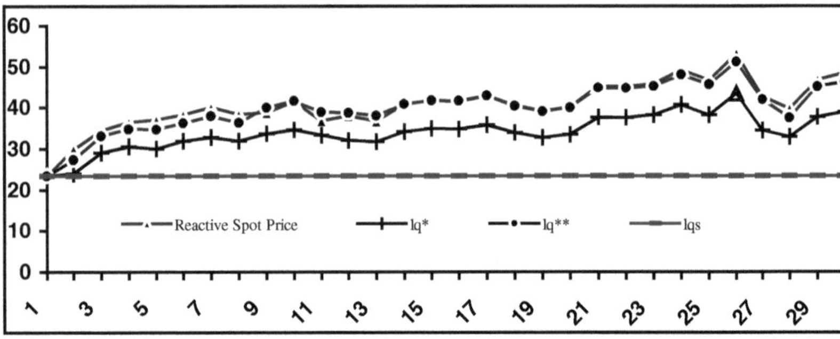

Figure 3.8. Reactive power spot price under normal operating conditions

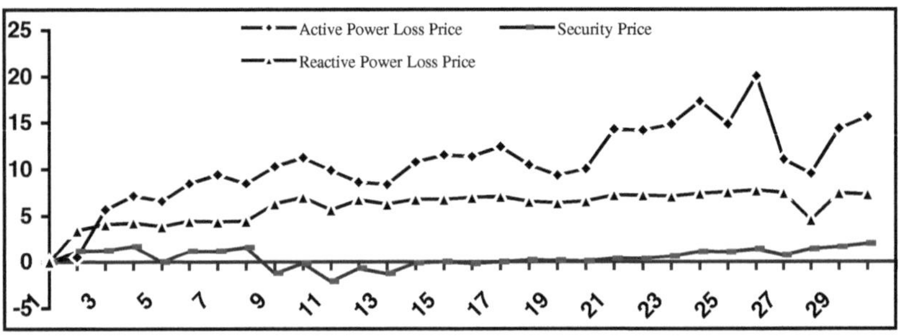

Figure 3.9. Components of reactive power spot price under normal operating conditions

3.7.3 Case Study 3: 118-bus System

Spot prices are computed for the 118-bus system and are plotted in Figures 3.10 and 3.11. System λ_{Ps} and λ_{qs} for active power and reactive power which are not shown in Figs 3.10 and 3.11, are 19223 and 327 respectively. Similar features can be observed to those associated with the 30-bus system. The active loss component dominates the actual active power spot price, and network loss related components dominate the SRMC of reactive power. It is noted that actual spot prices for reactive power can be positive or negative, which is in agreement with previous studies. Depending on the operating conditions, the magnitudes of reactive power spot prices can be relatively high as shown in Figure 3.11, in which the curve for λ_q^{***} is omitted, because the security component is too small.

The influence of initial point

Due to the importance of pricing electricity, the spot prices obtained by the proposed method have to be immune to the perturbation of initial points. Three experiments, flat power flow, converged power flow and warm start, have been done to test the stability of spot prices. Err%$_p$ and Err%$_q$ are the relative percentage errors of spot price perturbations for active power and reactive power. BS stands for bus number. The results show that $Err\%_p$ and Err%q are very small, their averages are 0.33% and 3.25% respectively. The perturbation of reactive power spot price is greater than that of active power, which is also due to the strong non-linearity of reactive power.

The influence of reference bus selection

Another test is designed to test the effect of selection of the reference bus. Unlike in the power flow calculations, the reference bus in OPF is just a reference for voltage angle and is involved in the optimisation process. Three cases with Bus 1, 10 and 12 as reference respectively are tested; the spot prices obtained are depicted in Figures 3.12 and 3.13. In Figure 3.12, three curves almost overlap; the effects of changing reference on active power spot price are negligible. Again, the percentage changes of reactive power are relatively larger than those of active power, but also not significant.

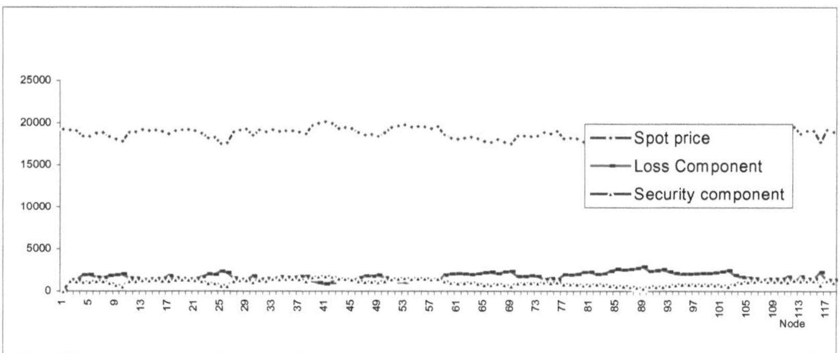

Figure 3.10. Active spot price for 118-bus system

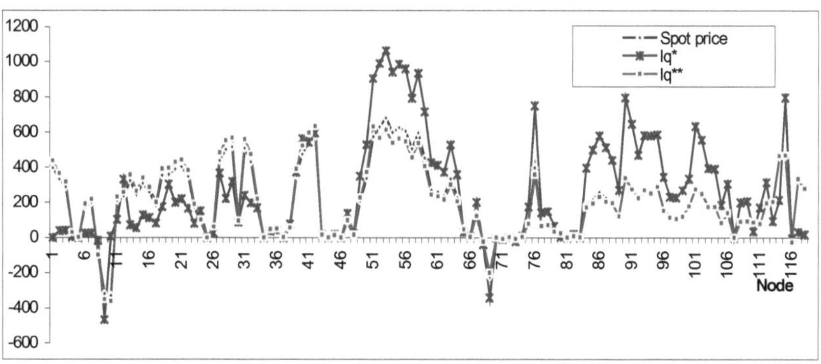

Figure 3.11. Reactive power spot price for 118-bus system

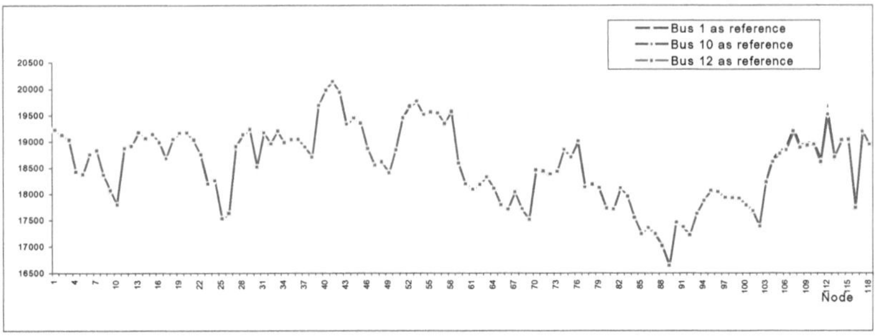

Figure 3.12. The effect of reference bus on active power spot price.

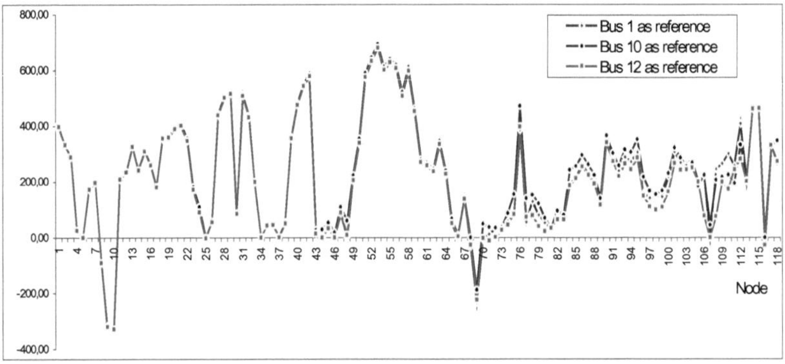

Figure 3.13. The effect of reference bus on reactive power spot price

3.8 Summary

After reviewing the relevant existing work on spot pricing models, the chapter proposes and develops an integrated method to calculate both real and reactive power spot price and to decompose them into the prices of selected ancillary services. The derivation of this model is based on economic theory and optimal power flow concepts and is given in detail. It has a clear relationship with conventional economic dispatching. The model is then implemented by extending IPOPF. Case studies on 5-bus, 30-bus and 118-bus systems are reported to illustrate the effectiveness and robustness of the proposed method.

3.9 List of Principal Symbols

a_p, b_p, c_p	coefficients of active power generation cost curve
a_q, b_q, c_q	coefficients of reactive power generation cost curve
B	customer benefit function
F	objective function
f_i	maximum active power reserve factor of unit i
f_{p_i}	active power generation cost function of unit i
f_{q_i}	reactive power generation cost function of unit i
$g(z)$	equality constraints vector
g_p, g_q	active power and reactive power mismatch vector
H	Hessian matrix
$h(z)$	inequality constraints vector
h_l	lower limits vector for inequality constraints
h_u	upper limits vector for inequality constraints
J	Jacobian matrix
k	the number of total inequality constraints
L	Lagrangian function
n_s	the number of curtailable loads
n	the number of nodes
nb	the number of branches
nc	the number of generation contracts
nd	the number of curtailable loads
nf	the number of transformers
ng	the number of generators
nt	the number of time periods for the research horizon
P_{cui}	upper limit of generation contract i
P_{Cli}	lower limit of generation contract i
P_d, Q_d	active load and reactive load vector
P_{ds}, Q_{ds}	curtailable active load and reactive load vector
P_g, Q_g	active generation and reactive generation vector
P_{Rli}	ramping rate lower limit of unit i
$P_{gi\max}, P_{gi\min}$	real power output lower and upper limit of unit i
P_{ij}	active power flow on line ij
P_{Rui}	ramping rate upper limit of unit i
$Q_{gi\max}$	reactive power output upper limit of unit i
$Q_{gi\min}$	reactive power output lower limit of unit i

R_i	active power spinning reserve of unit i
$R_{\min}$	system minimum spinning reserve capacity.
s_l , s_u	lower and upper limits slack variables vector
$[s_l]$, $[s_u]$, $[\pi_l]$, $[\pi_u]$	diagonal matrix of s_{li} s_{ui} π_{li} π_{ui}
t	time index
V_i , θ_i	bus voltage and angle of node i
Y_{ij}	the ij^{th} term of the bus admittance matrix
z	all variables in Newton method OPF
δ_{ij}	phase angle of the admittance Y_{ij}
λ_p	Lagrangian multipliers vector for active power flow
λ_q	Lagrangian multipliers vector for reactive power flow
μ	barrier parameter
μ_T	damping factor for inter-temporal constraints
π_l , π_u	lower and upper limits dual variables vector
s_{Tl} , s_{Tu}	vectors of slack variables for inter-temporal constraints
π_{Tl} , π_{Tu}	vectors of Lagrangian multipliers for inter-temporal constraints
ζ_i	demand price elasticity of load i
∇	differentiation operation
Δ	change in variables
Ω	set of category one unit

3.10 References

[1] M. Einborn, R. Siddiqi, "Electricity Transmission Pricing and Technology", Kluwer Academic Publishers, 1996

[2] F.C. Schweppe, M.C. Caramanis, R.D. Tabors, and R.E. Bohn, "Location based Pricing of Electricity", Kluwer Academic Publishers, Boston, MA, 1988

[3] M.C. Caramanis, R.E. Bohn and F.C. Schweppe, "WRATES: A Tool for Evaluating the Marginal Cost of Wheeling,", IEEE Transactions on Power Systems, Vol. 4, No. 2, May 1989

[4] M. Rivier, I. Perez-Ariaga, "Computation and decomposition of location based price for transmission pricing", 11 PSCC Conference, 1991

[5] D. Ray, and F. Alvarado, "Use of an Engineering Model for Economic Analysis in the Electric Utility Industry", Presented at the Advanced Workshop on Regulation and Public Utility Economics, Rutgers University, May 25-27, 1988

[6] M.L. Baughman, S.N. Siddiqi, "Real Time Pricing of Reactive Power: Theory and Case Study Results", IEEE Transactions on Power Systems, Vol. 6, No. 1, February 1993

[7] N. Dandachi, M. Rawlins, O. Alsac, M. Prais, B. Stott, "OPF For Reactive Pricing Studies On The NGC System", IEEE Transactions on Power Systems, Vol. 11, No. 1, February 1996
[8] A.A. EL-Keib, X. Ma, "Calculating of Short-Run Marginal Costs of Active and Reactive Power Production", IEEE Transactions on Power Systems, Vol.12, No. 1, May 1997
[9] Y.Z .Li, A.K. David, "Wheeling Rates of Reactive Flow under Marginal Cost Theory", IEEE Transactions on Power Systems, Vol.10, No. 3, August 1993
[10] S.N. Siddiqi, M.L. Baughman, "Reliability Differentiated Pricing of Spinning Reserve", IEEE Transactions on Power Systems, Vol.10, No. 3, August 1995
[11] A. Zobian, M.D. Ilic, "Unbundling of Transmission and Ancillary Services", IEEE Transactions on Power Systems, Vol.12, No. 1, February 1997
[12] M.C. Caramanis, R.E. Bohn, F.C. Schweppe, "System Security Control and Optimal Pricing of Electricity", Electrical Power & Energy Systems", Vol.9, No. 4, October 1987
[13] R. Kaye, F.F. Wu, "Security Pricing of Electricity", IEEE Transactions on Power Systems, Vol.4, No. 3, August 1995
[14] F. Alvarado, Y. Hu, D. Ray, R. Stevenson, E. Cashman, " Engineering Foundations for Determination of Security Costs", IEEE Transactions on Power Systems, Vol.6, No. 3, August 1991
[15] M.L. Baughman, S.N. Siddiqi, J.W. Zarnikau, "Advanced Pricing in Electricity Systems", IEEE Transactions on Power Systems, Vol.12, No. 1, February 1997
[16] J.D. Finney, H.A. Othman, W.L. Rutz, "Evaluating Transmission Congestion Constraints in System Planning", IEEE Transactions on Power Systems, Vol. 12, No. 3, August 1997
[17] L Willis, J Finney , G. Ramon, " Computing the Cost of Unbundled Services", October 1996, Computer Applications in Power
[18] K. Xie, Y. H. Song, J. Stonham, E. Yu, G. Liu, "Decomposition Model and Interior Point Methods for Optimal Spot Pricing of Electricity in Deregulated Environments", IEEE Transactions on Power Systems, Vol. 15, No. 1, February 2000
[19] D.R. Bobo, D.M. Mauzy, "Economic Generation Dispatch With Responsive Spinning Reserve Constraints", IEEE Transactions on Power Systems, Vol. 9, No.1, February 1994
[20] J.A. Momoh, R.J. Koessler, M.S. Bond, B. Stott, D. Sun, A. Papalexopoulos, P.Ristanovic, "Challenges to Optimal Power Flow", IEEE Transactions on Power Systems, Vol. 12, No. 1, February 1997
[21] S. Granville , "Optimal Reactive Dispatch Through Interior Point Methods", IEEE Transactions on Power Systems, Vol. 9, No. 1, February 1994
[22] G.D. Irisarri, X. Wang, J. Tong, S. Mokhtari, "Maximum Loadability of Power Systems Using Interior Point Non-Linear Optimization Method", IEEE Transactions on Power Systems, Vol. 12, No. 1, February 1997
[23] C.N. Lu, M.R. Unum, "Network Constrained Security Control Using An Interior Point Algorithm", IEEE Transactions on Power Systems, Vol. 8, No. 3, August 1993
[24] J.A. Monmoh, S.X. Guo, E.C. Ogbuobiri, R. Adapa, "The Quadratic Interior Point Method Solving Power System Optimization Problems", IEEE Transactions on Power Systems, Vol. 9, No. 3, August 1994
[25] D.I. Sun, B. Ashley, B. Brewer, A. Hughes, W.F. Tinney, "Optimal Power Flow by Newton Approach", IEEE Transactions on Power Systems, Vol. PAS-103, No.10, October 1984
[26] Y. Hao, "Study of Optimal Power Flow and Short-term Reactive Power Scheduling", PhD Thesis, Electric Power Research Institute (China), 1997

[27] G.A. Maria, J.A. Findlay, "A Newton Optimal Power Flow Program for Ontario Hydro EMS" , IEEE Transactions on Power Systems, Vol.2, No.3, August 1987

4. Coordinated Real-time Dispatch of Unbundled Electricity Markets

X. Wang, Y.H. Song and M. Tan

As with other commodities markets, electricity markets should also be fully open and encourage competition among participants. However, because electrical energy cannot be stored in large amounts for a long time, deregulated unbundled electricity markets need a centralised control, which is the Independent System Operator (ISO), to keep the power system operating in line with security, economy, and reliability criteria. The functions of the ISO have been presented in Chapter 1.

It is well known that the generation and consumption of electricity must occur essentially at the same time. Therefore, real-time operations and the associated markets and pricing approaches are crucial to the design and implementation a successful competitive wholesale electricity market. The basic tasks of the ISO during real-time operation should at least include:

- meeting the imbalance between the real-time and scheduled load and generation;
- and relieving real-time network congestion due to unexpected contingencies.

One of the solutions to the real-time dispatch of electricity markets is to establish a real-time balancing market and to encourage all market participants to take part in the competition in this balancing market. The Federal Energy Regulatory Commission (FERC) of the United State recognised the importance of these markets in its Order 2000 on regional transmission organisations (RTOs) [1], where FERC wrote:

... an RTO must ensure that its transmission customers have access to a real-time balancing market that is developed and operated by either the RTO itself or another entity that is not affiliated with any market participant. We have determined that real-time balancing markets are necessary to ensure non-discriminatory access to the grid and to support emerging competitive energy markets. Furthermore, we believe that such markets will become extremely important as states move to broad-based retail access, and as generation markets move toward non-traditional resources, such as wind and solar energy, that may operate only intermittently.

In this chapter, the major differences between power system operation in vertically integrated utilities and operation in deregulated electricity markets are analysed. An efficient dispatch framework for unbundled electricity markets through a real-time balancing mechanism is then presented. This framework is able to meet the real-time energy imbalance caused by unexpected contingencies and unpredictable load fluctuation with all the available adjustment bids in the real-time balancing market.

4.1 Power System Operation

4.1.1 Operation in Vertically Integrated Utilities

Traditionally, the three main objectives of power system operation are: power balance, system security/reliability, and economy. Generally speaking, the power system operation can be categorised into two basic stages [2,3]:

- Scheduling, which takes place a day, week, or year in advance. Scheduling includes several functions.
  (1) Hydro scheduling. The System Operator (SO) determines the schedule for next year's or next week's reservoir level.
  (2) Maintenance scheduling. The SO determines when to shut down power plants for maintenance.
  (3) Interchange scheduling between companies.
  (4) Unit Commitment (UC). The SO determines which generators to start up or to shut down for the operating day.
- Dispatch, which is done in real time. Dispatch includes two main functions:
  (1) Economic Dispatch (ED). The SO increases or decreases generation output levels according to fuel costs, typically performed in minute intervals.
  (2) Automatic Generation Control (AGC). The power system is balanced with AGC on a second-to-second basis.

Although this chapter focuses on real-time dispatch, UC, the most important scheduling task, should be discussed. UC is a process to decide which units to operate on the operating day, and when to turn these units on and off, in order to minimise the overall costs of electricity production. The characteristics of the individual generating units are needed as inputs of this large, complex computer program. These characteristics include current unit status, minimum and maximum output levels, ramp rate limits, startup and shutdown costs and times, minimum runtimes, and unit fuel costs at various output levels. In addition, the day-ahead forecast of system loads, hour by hour, as well as any scheduled wholesale sales or purchases for the following day, should be provided to the model. If the network model is considered, the inputs may also include expected network topology information of the transmission system for the operating day. This optimisation program is then run with all these inputs to obtain the least-cost solution to meet the following day's electricity demands while being able to withstand the loss of any single generation or transmission element.

Solving this optimisation problem is mathematically difficult because of the inter-temporal constraints of generators. With the existing UC algorithms, the final solution may not be exactly least-cost. For a vertically integrated utility, this near optimal solution may not be a problem because its profitability depends on the portfolio level of its generators, not on the performance of only one or two of them.

Once generators are committed, they are available to participate in ED to meet the fluctuating system load and reliability requirements. ED can be described as a security-constrained least-cost dispatch model and is typically run every five minutes. ED requires short-term load forecasting to decide how much generation is needed to increase or decrease in the next 5-minute interval. ED may also have the ability to decide the commitment of some fast-start/shutdown units, such as combustion turbines and hydro generators. The output levels of individual units obtained from ED are passed to AGC as the base points of these units during operation.

4.1.2 Operation in Competitive Electricity Markets

Power system operation becomes much more complex when generation, transmission and system control are owned and operated by different entities. The primary difference between the operation of a deregulated electricity market and a vertically integrated power system is that the ISO owns no generation resources. Then who should make unit commitment decisions in an electricity energy market, generators or the ISO? The answer to this question decides if generators should bid into energy markets with simple energy-only bids or multi-part bids having separate prices for startup, no-load, and energy costs.

The design of power system operation, particularly if a centralised UC is still required, is mainly determined by the market architecture. Basically, the possible architectures of electricity energy markets can be categorised into three elementary models:

(1) Pure centralised spot market.
(2) Spot market with pre-signed bilateral energy transactions.
(3) Bilateral energy market with centralised real-time balancing market.

In a pure centralised spot market, UC performed by the ISO is indispensable for scheduling. Here, all generations in the market must trade with the ISO and no physical bilateral contracts are allowed. The old England & Wales Pool is a typical example of this type of market. In a spot market built on top of bilateral contracts, suppliers can select if they want to self-commit or they want the ISO to commit their units. PJM, NYISO, and ISONE can be put into this category. In bilateral energy markets with centralised real-time balancing mechanism, such as ERCOT in the USA and the new UK market, UC is no longer one of the scheduling tasks of the ISO. All generators in the energy market must be responsible for committing their own units and reporting the commitment results to the ISO. Although the scheduling process is very different between these three different types of energy market, the common part of the market designs is a centralised real-time balancing mechanism.

As Angelidis and Papalexopoulos summarised in [20], there are two different structures that can be used to design a mechanism to manage the real-time energy imbalances and real-time transmission congestion. Under the first structure, the ISO dispatches resources according to their bids and rolls the incurred costs into an uplift that is allocated pro rata to all market participants. This structure seeks to minimise the uplift costs by imposing penalties on resources that have uninstructed deviations from their schedules established in the forward market, rather than make use of any real-time price signals. This structure can be found in the electricity market of England and Wales.

Under the second structure, the ISO operates a real-time market with transparent market clearing prices. These prices are intended to provide incentives for market participants to operate consistently with the goal of reducing energy imbalances to zero. The regulating markets in Norway and Sweden and the real-time market in California are examples of this structure. PJM adopts a similar dispatch structure. In addition, PJM provides Locational Marginal Price (LMP) for every node in its system.

4.2 Coordinated Real-time Dispatch through Balancing Mechanism

With the deregulation of the power industry, the main services in power systems have been unbundled into several separate markets, such as the pool auction energy market where the schedule of generation can be arranged to meet the system load; the bilateral contract market where the generators and consumers can sell or buy electricity by themselves; and the ancillary services market where the ISO can procure the necessary services like system reserves and voltage support to maintain system security.

Some research has been undertaken on the dispatch problem in electricity markets. In [4], Wu and Varaiya proposed a decentralised optimal dispatch method with the objective of maximising social welfare under the coordinated multilateral trade model. In [5], Singh, Hao, and Papalexopoulos made some comparisons between the approaches of transmission congestion management in the pool model and the bilateral model respectively. In [6,7], David and Fang provided some useful curtailment strategies based on minimising deviations from transaction requests made by market participants in a structure dominated by bilateral and multilateral contracts. Singh and Papalexopoulos introduced the basic idea of auction market for ancillary services in California in [8]. The dispatch of ancillary services and the interaction between the various markets were also discussed briefly. In [9,10], optimal scheduling methods were proposed, in which the procurement of necessary operating reserves was combined with the procurement of the energy (Joint Dispatch) while network constraints were taken into account.

However, a problem still remaining for an ISO to resolve is how to use all the possible resources efficiently and in a coordinated way to ensure system security during the real-time execution of various electricity commodity contracts. The main difficulties occurring during real-time coordinated dispatch may include:

- how to utilise the signed operating reserve contracts and the supplementary energy bids in the balancing market to obtain the optimal dispatch solution;
- with the trend to more and more bilateral contracts being used to trade electricity, how to eliminate network congestion if the resources in the balancing market are not enough;
- to maintain a certain system security level, how to obtain replacement operating reserves in time if any of the pre-arranged operating reserves are called upon to provide energy for real-time system balancing or congestion management.

To resolve these difficulties, a new framework for real-time dispatch of unbundled electricity markets is proposed in this chapter. Under this framework, almost all the contracts in the various electricity markets can be dispatched in coordination by submitting their adjustment bids to the balancing market. In particular, some bilateral contracts can be adjusted if transmission congestion is very serious [11,12]. Demand-side participants are encouraged to take an active role in the competition of the real-time balancing market. A modified P-Q decoupled Optimal Power Flow (OPF) is applied to solve this problem. The objective of the P-subproblem is to coordinate dispatch among the pool auction contracts, the bilateral contracts and the operating reserve contracts, in light of the bids submitted by these contracts to the balancing market. The spot pricing and the two possible settlement methodologies for real-time dispatch are analysed as well. The IEEE 30-bus test system is studied to illustrate the proposed framework and its mathematical solution.

In the proposed framework, which is shown in Figure 4.1, there are four unbundled markets, the Bilateral Contract Market (BCM), Pool Day-ahead Energy Auction Market (PEAM), Pool Ancillary Services Auction Market (PAAM) and Real-time Balancing Market (RBM), respectively.

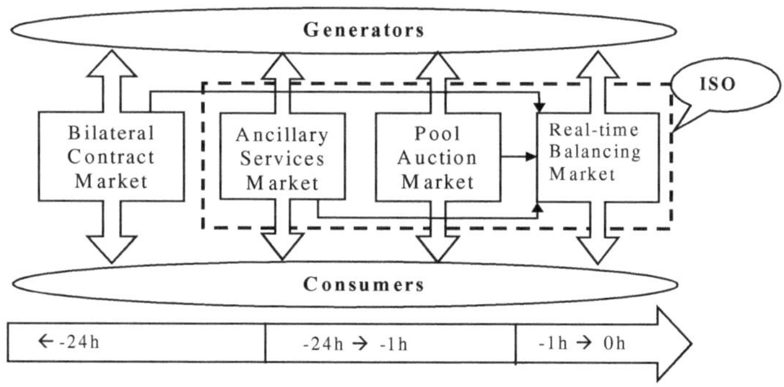

Figure 4.1. Proposed framework of real-time coordinated dispatch

4.2.1 Bilateral Contract Market

In BCM, generators and consumers arrange physical electrical energy trades with each other based on their own financial interests. Instead of letting the ISO know the prices of their contracts, participants must report the details of their bilateral contracts to the ISO before the opening of the day-ahead energy auction market.

4.2.2 Pool Day-ahead Energy Auction Market

In PEAM all the contracts must be signed with the ISO, who receives energy bids from generators and consumers and then selects the cheapest generations to supply the demands of the whole system while all the network constraints must be satisfied. PEAM opens after BCM closes. It gives market participants another opportunity to buy or sell sufficient electricity to meet their requirement if they cannot conclude all the necessary transactions in BCM.

4.2.3 Pool Ancillary Services Auction Market

Instead of electrical energy, the ancillary services are the commodities traded in PAAM. The ISO can determine the amount of required ancillary services by the load forecast and the preferred schedule. The ancillary services traded here include regulation reserves (AGC), spinning reserves, non-spinning reserves and replacement reserves, which are shown in Figure 4.2. The other two services, voltage support and black start, can be procured in some special long-term markets. Each participant in PAAM must submit both a capacity bid and an energy bid for each service [8]. The energy bid will be used for real-time dispatch in RBM. The coordinated bid clearing strategies between PEAM and PAAM were given for New Zealand in [9] and for New England in [10]. Since the subject of this paper is real-time dispatch, only the operating reserves, which include AGC, spinning reserves and non-spinning reserves, are taken into account.

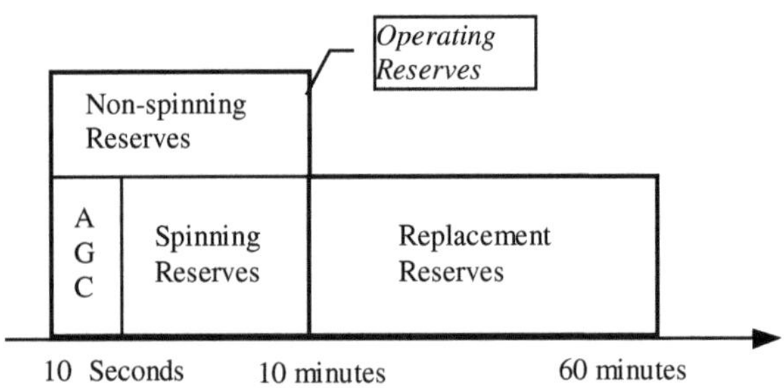

Figure 4.2. Four types of reserves traded in PAAM

4.2.4 Real-time Balancing Market and Coordinated Dispatch

RBM plays a key role in real-time operation of electricity markets. In RBM the ISO is responsible for dispatching all the available resources to meet imbalances between actual and scheduled load and generation and alleviate network congestion. The ISO will select the least-cost resources to meet these imbalances. In addition, the ISO may also need to purchase replacement ancillary services if services arranged in advance are used to provide balancing energy. The purpose of RBM is to establish a fully open market-based mechanism for all market participants to take part in real-time competition. In RBM, all generators and consumers can submit to the ISO their incremental and decremental bids for providing balancing energy and their capacity bids for the replacement of operating reserves which are used as balancing energy. In the NETA of the UK market, these incremental/decremental bids are called "pairs of offer and bid" [13].

Because the ISO does not know the price information of bilateral contracts and the modification of bilateral contracts could involve both sides of the contract, it is very difficult to find a suitable way to change them because of transmission congestion. One method is for both parties to submit their own supplemental bids in RBM separately, just like the other participants in the pool. The ISO does not take into account the content of the bilateral contract during settlement. However, in the event of a high percentage of bilateral contracts but not enough voluntary supplemental bids from them, or if both parties of a bilateral contract are willing to be curtailed at the same amount to simplify settlement, other methods are needed. In the proposed framework, every bilateral contract should submit a sort of compensative price that both parties to the contract are willing to accept if curtailment is imposed by the ISO during periods of congestion. In accordance with such information, the ISO can reduce the scheduled bilateral contract if the other available resources in RBM are not enough to eliminate the congestion.

Some typical active power contracts associated with a generator at bus i and their bids submitted to the ISO in RBM are shown in Figure 4.3, in which the adjustment of generation at bus i is divided into four separate parts based on three unbundled electricity contracts.

In Figure 4.3:

i,j	Index of network buses.
P_i^0	Scheduled MW generation of generator or MW load of consumer at bus i.
P_i^{ij}	Total MW amount of bilateral contracts between buses i and j.
$P_i^{\min}$, $P_i^{\max}$	Low and high limit of MW at bus i either for generator or for consumer.
P_i^{Res}	Operating reserves procured by ISO from participant at bus i in ancillary services market.

ΔP_i^+ , ΔP_i^- Incremental and decremental MW changes of participant at bus i in real-time balancing market.

ΔP_i^{ij} Curtailment of bilateral contract between buses i and j.

$\Delta P_i^{\mathrm{Res}}$ Operating reserves called upon by ISO in the real-time balancing market out of P_i^{Res} .

$\Delta P_i^{\mathrm{Rep}}$ Reserves at bus i procured by ISO in real-time balancing market to replace the used operating reserves.

b_i^+ , b_i^- Incremental and decremental bidding prices of participant at bus i in RBM ($/MW).

b_i^{ij} Compensative price for the curtailment of bilateral contract between buses i and j ($/MW).

b_i^{En} Bidding price of the energy output from P_i^{Res} ($/MW).

b_i^{Rep} Bidding price to provide $\Delta P_i^{\mathrm{Rep}}$ at bus i ($/MW).

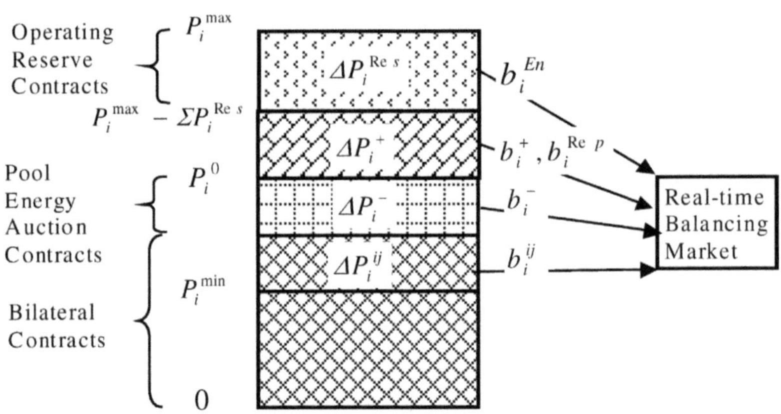

Figure 4.3. Typical contracts associated with a generator at bus i and their adjustment bids to RBM

One advantage of this framework is that the ISO can have more available real-time resources to dispatch in accordance with various adjustment bids in RBM to maintain system security. The cost of real-time dispatch can be allocated to all other market participants according to real-time Locational Marginal Price (LMP) or through a system uplift.

4.3 Mathematical Model of the Proposed Framework

A modified P-Q decoupled OPF is applied to implement the proposed framework. Both P- and Q sub-problems are presented, and the P sub-problem is analysed in more detail. Before the trade period, scheduling of unbundled markets should have been done. The principle of the balancing mechanism is to follow the schedules as closely as possible and to minimise the cost of real-time dispatch. After linearising at the scheduled operating point, the model of real-time coordinated optimal dispatch can be written as follows.

4.3.1 P Sub-problem

4.3.1.1 Objective

To make the settlement of unbundled markets clear and simple, the objective of real-time active power dispatch can be decomposed into four parts.

$$\text{Min: } C_p = C_{p1} + C_{p2} + C_{p3} + C_{p4} \tag{4.1}$$

where

$$C_{p1} = \sum_{i \in PEAM} (b_i^+ \Delta P_i^+ + b_i^- \Delta P_i^-) \tag{4.2}$$

is the total cost for the adjustment of PEAM contracts whose supplemental bids are accepted in RBM. But for accepted decremental bids, participants cannot get payment for this part of the reduced energy from the ISO in PEAM settlements.

$$C_{p2} = \sum_{i \in PAAM} b_i^{En} \Delta P_i^{\text{Res}} \tag{4.3}$$

is the cost for calling upon energy from the signed operating reserves contracts in RBM.

$$C_{p3} = \sum_{i \in BCM} \sum_{\substack{j \in BCM \\ j<i}} (b_i^{ij} \Delta P_i^{ij}) \tag{4.4}$$

is the cost for curtailment of bilateral contracts in RBM. This curtailment should be done on both sides of the bilateral contract.

$$C_{p4} = \sum_{i \in RBM} (b_i^{\text{Re}\,p} \Delta P_i^{\text{Re}\,p}) \tag{4.5}$$

is the cost of reserves re-procured by ISO in RBM to replace the used operating reserves. In the event of an unscheduled increase in system load, the ISO will be required to purchase additional reserves. In addition, if any of the pre-arranged operating reserves are used to meet a real-time imbalance, the ISO will be required to purchase replacement operating reserves. The reserve market and the balancing energy market will still be settled separately. Embedding the cost of replacement of operating reserves in the objective function of real-time dispatch provides a global optimal solution. It can be regarded as the real-time joint dispatch of energy and reserves.

Obviously, the objective is to minimise modifications to all scheduled contracts in accordance with the associated dispatch cost. ΔP_i^+, ΔP_i^-, ΔP_i^{ij}, $\Delta P_i^{\mathrm{Res}}$ and $\Delta P_i^{\mathrm{Rep}}$ are treated as independent control variables during the optimisation process but their upper/lower limits are coupled with each other. All the bidding curves could be multi-step. The incremental bidding price is higher than the decremental bidding price, while the curtailment price of bilateral contracts is much higher than the other two. The reason is that increasing output requires greater fuel costs and curtailment of a bilateral contract will affect the financial interests of both parties.

4.3.1.2 Equality Constraints

$$(-1)^{\beta}[\Delta P_i^+ - \Delta P_i^- - \sum_{\substack{j\in BCM \\ j\neq i}} (\Delta P_i^{ij})] + \Delta P_i^{\mathrm{Res}} + \sum_{\substack{j=1 \\ j\neq s}}^{n} B'_{ij}\Delta\theta_j - \Delta P_i^{Loss} = 0 \tag{4.6}$$

$$\beta = 0, \text{if } i \in G;\ \beta = 1, \text{if } i \in C.$$

is the nodal active power flow balance equation of bus i, where G is the set of buses connected with generators, C is the set of buses connected with consumers, n is the total number of network buses, s is the index of the slack bus, B'_{ij} is an element of matrix $\boldsymbol{B}'$, which is the inverse of reactance of branch ij. ΔP_i^{Loss} is the summed change of losses on branches that are connected to bus i and flows on the branches are flowing to bus i. Using the piece-wise linear loss model, ΔP_i^{Loss} can be written as $\Delta P_i^{Loss} = \sum_{j\in\Theta_i} LF_{ij}(\Delta P_{ij})$. Θ_i is the set of buses which have branches connected with bus i and flows on the branches flow to bus i.

$$\sum_{i\in RBM} \Delta P_i^{\mathrm{Rep}} = \alpha \sum_{i\in PAAM} \Delta P_i^{\mathrm{Res}} \tag{4.7}$$

is the requirement of re-procurement to replace used operating reserves. α is decided by ISO according to how many new operating reserves are procured to compensate the used reserves in real-time dispatch; $0 \le \alpha \le 1$.

4.3.1.3 Inequality Constraints

$$0 \le \Delta P_i^{+} + \Delta P_i^{\mathrm{Re}\,p} \le \Delta P_i^{+,\max} = P_i^{\max} - P_i^{0} - P_i^{\mathrm{Re}\,s} \tag{4.8}$$

$$0 \le \Delta P_i^{-} \le \Delta P_i^{-,\max} = P_i^{0} - \max(\sum_{\substack{j=1 \\ j\ne n}}^{n} P_i^{ij}, P_i^{\min}) \tag{4.9}$$

$$0 \le \Delta P_i^{\mathrm{Re}\,s} \le P_i^{\mathrm{Re}\,s} \tag{4.10}$$

$$0 \le \Delta P_i^{ij} \le P_i^{ij}, 0 \le \sum_{\substack{j=1 \\ j\ne i}}^{n} \Delta P_i^{ij} \le \min(\sum_{\substack{j=1 \\ j\ne i}}^{n} P_i^{ij}, P_i^{\min}) \tag{4.11}$$

are changing ranges of the various control variables.

$$-P_l^{\max} - P_l^{0} = \Delta P_l^{\min} \le \Delta P_l \le \Delta P_l^{\max} = P_l^{\max} - P_l^{0} \tag{4.12}$$

is the constraint for real power flow change on branch *l*.

4.3.1.4 Pricing for Real-time Active Power Dispatch

From (4.6), the system active power balance equation can be written as:

$$\sum_{i=1}^{n} \Delta P_i - \Delta P^{Loss} = 0 \tag{4.13}$$

where ΔP_i is the total change of active power at bus *i* and ΔP^{Loss} is the total change of active power losses. $\Delta P^{Loss} = \sum_{i=1}^{n} \Delta P_i^{Loss}$, so the Lagrangian function of the primary optimisation problem is:

$$\begin{aligned} L = C_p - \lambda(\sum_{i=1}^{n} \Delta P_i - \Delta P^{Loss}) \\ - \mu^{\mathrm{Re}\,p} (\sum_{i \in RBM} \Delta P_i^{\mathrm{Re}\,p} - \alpha \sum_{i \in PAAM} \Delta P_i^{\mathrm{Re}\,s}) \\ + \sum_{i \in PEAM} \mu_i^+ (\Delta P_i^+ + \Delta P_i^{\mathrm{Re}\,p} - \Delta P_i^{+,\max}) \\ + \sum_{i \in PEAM} \mu_i^- (\Delta P_i^- - \Delta P_i^{-,\max}) \\ + \sum_{i \in PAAM} \mu_i^{\mathrm{Re}\,s} (\Delta P_i^{\mathrm{Re}\,s} - P_i^{\mathrm{Re}\,s}) + \sum_{i \in BCM} \mu_i^{ij} (\Delta P_i^{ij} - P_i^{ij}) \\ + \sum_{l \in B} \mu_l^{\min} (\Delta P_l^{\min} - \Delta P_l) + \sum_{l \in B} \mu_l^{\max} (\Delta P_l - \Delta P_l^{\max}) \end{aligned} \tag{4.14}$$

According to the KKT first-order conditions, we have:

$$\frac{\partial C_p}{\partial \Delta P_i^+} - (-1)^{\beta} \lambda (1 - \frac{\partial \Delta P^{Loss}}{\partial \Delta P_i^+}) + \mu_i^+ + \sum_{l \in B} (\mu_l^{\max} - \mu_l^{\min}) \frac{\Delta P_l}{\Delta P_i^+} = 0 \tag{4.15}$$

$$\frac{\partial C_p}{\partial \Delta P_i^-} + (-1)^{\beta} \lambda (1 + \frac{\partial \Delta P^{Loss}}{\partial \Delta P_i^-}) + \mu_i^- + \sum_{l \in B} (\mu_l^{\max} - \mu_l^{\min}) \frac{\Delta P_l}{\Delta P_i^-} = 0 \tag{4.16}$$

$$\frac{\partial C_p}{\partial \Delta P_i^{ij}} + (-1)^{\beta} \lambda \frac{\partial \Delta P^{Loss}}{\partial \Delta P_i^{ij}} + \mu_i^{ij} + \sum_{l \in B} (\mu_l^{\max} - \mu_l^{\min}) \frac{\Delta P_l}{\Delta P_i^{ij}} = 0 \tag{4.17}$$

$$\begin{aligned} \frac{\partial C_p}{\partial \Delta P_i^{\mathrm{Re}\,s}} - \lambda (1 - \frac{\partial \Delta P^{Loss}}{\partial \Delta P_i^{\mathrm{Re}\,s}}) + \alpha \mu^{\mathrm{Re}\,p} + \mu_i^{\mathrm{Re}\,s} \\ + \sum_{i \in B} (\mu_l^{\max} - \mu_l^{\min}) \frac{\Delta P_l}{\Delta P_i^{\mathrm{Re}\,s}} = 0 \end{aligned} \tag{4.18}$$

$$\frac{\partial C_p}{\partial \Delta P_i^{\mathrm{Re}\,p}} - \mu^{\mathrm{Re}\,p} + \mu_i^+ = 0 \tag{4.19}$$

It is obvious that

$$\frac{\partial \Delta P^{Loss}}{\partial \Delta P_i} = (-1)^{\beta} \frac{\partial \Delta P^{Loss}}{\partial \Delta P_i^+} = \frac{\partial \Delta P^{Loss}}{\partial \Delta P_i^{\mathrm{Re}\,s}} = (-1)^{\beta+1} \frac{\partial \Delta P^{Loss}}{\partial \Delta P_i^-} = (-1)^{\beta+1} \frac{\partial \Delta P^{Loss}}{\partial \Delta P_i^{ij}} \tag{4.20}$$

and

$$\frac{\partial \Delta P_l}{\partial \Delta P_i} = (-1)^{\beta} \frac{\partial \Delta P_l}{\partial \Delta P_i^{+}} = \frac{\partial \Delta P_l}{\partial \Delta P_i^{\mathrm{Re}\,s}} = (-1)^{\beta+1} \frac{\partial \Delta P_l}{\partial \Delta P_i^{-}} = (-1)^{\beta+1} \frac{\partial \Delta P_l}{\partial \Delta P_i^{ij}} \tag{4.21}$$

Using the theory of spot pricing [14], the LMP of participant (either a generator or a consumer) at bus i is:

$$\rho_i^P = \lambda - \frac{\partial \Delta P^{Loss}}{\partial \Delta P_i}\lambda - \sum_{l \in B}(\mu_l^{\max} - \mu_l^{\min})\frac{\partial \Delta P_l}{\partial \Delta P_i} \tag{4.22}$$

Shown in equation (4.22) the real-time spot price at bus i can be decomposed into three parts: the system lambda, the active power losses and the congestion management cost.

From Equations (4.15) to (4.18), we can have different forms of ρ_i^P at bus i at which the participant has made some contribution to the real-time dispatch.

$$\rho_i^P = \begin{cases} (-1)^{\beta}(\dfrac{\partial C_p}{\partial \Delta P_i^{+}} + \mu_i^{+}); \text{ or} \\ (-1)^{\beta+1}(\dfrac{\partial C_p}{\partial \Delta P_i^{-}} + \mu_i^{-}); \text{ or} \\ \dfrac{\partial C_p}{\partial \Delta P_i^{\mathrm{Re}\,s}} + \alpha\mu^{\mathrm{Re}\,p} + \mu_i^{\mathrm{Re}\,s}. \end{cases} \tag{4.23}$$

Although adjustment of the contracts of a participant is divided into four independent control variables (two increasing and two decreasing), at the optimal point only one of them can be active (on adjustment). From Equation (4.23) we can determine the effects of the costs and changing ranges of these control variables on λ.

From Equations (4.17) and (4.23), it is noticed that the curtailment of bilateral contracts has little effect on the system λ. It is due to the item ΔP_i^{ij}, which exists in two nodal active power balance equations with different sign, but does not appear in the system balance Equation (4.13). The various cases of bilateral contract curtailment and its pricing will be discussed in the following section.

4.3.1.5 Meeting Real-time Imbalance of Market under Normal Operating Conditions

If no serious contingency occurs, the operating reserves and the supplemental energy in RBM should be enough to follow system load fluctuations. The curtailment of bilateral contracts and load shedding are not necessary in this case. To model this problem, the objective in (4.1) should be rewritten as:

$$\text{Min: } C_p = C_{p1} + C_{p2} + C_{p4} \tag{4.24}$$

and all constraints on ΔP_i^{ij} should be removed.

Given the system load fluctuation ΔP^{sys}, the bus load change can be expressed as:

$$BL_i = \eta_i \Delta P^{sys} \tag{4.25}$$

where η_i is bus load allocation factors, $\sum_{i \in C} \eta_i = 1$.

To meet this system imbalance, the right side item of Equation (4.6) changes to BL_i from 0. Considering that rapid response generation units are enough to meet the normal system imbalance, to reduce the number of control variables, the nodal active power balance Equations (4-6) for pure load buses can be rewritten as:

$$\sum_{\substack{j=1 \\ j \neq s}}^{n} B'_{ij} \Delta\theta_j - \Delta P_i^{Loss} = BL_i \tag{4.26}$$

and the system balance Equation (4.13) should be changed to:

$$\sum_{i=1}^{n} \Delta P_i - \Delta P^{Loss} = \Delta P^{sys} \tag{4.27}$$

Without any branch limit violation, the LMP at bus i is:

$$\rho_i^P = \lambda - \frac{\partial \Delta P^{Loss}}{\partial \Delta P_i} \lambda \tag{4.28}$$

4.3.1.6 Replacement of Operating Reserves

From Equation (4.19), the marginal price of the replacement of operating reserves is:

$$\rho^{\mathrm{Re}\,p} = \mu^{\mathrm{Re}\,p} + \mu_i^+ \tag{4.29}$$

The cost of replacement of operating reserves affects the LMPs through the item $\alpha\mu^{\mathrm{Re}\,p}$ instead of appearing in Equation (4.22) or (4.28) directly.

When calling up energy from operating reserves during real-time dispatch, the ISO must pay for not only the energy but also the procurement of replacement. Therefore, if the supplementary energy in RBM is fast and cheap enough, the reserve contracts signed in PAAM could be left unchanged. As a result, the real-time balancing mechanism expands the concept of operating reserves and gives market participants more competitive opportunities.

4.3.1.7 Curtailment of Bilateral Contracts

According to the characteristics of bilateral contracts, curtailment should be on both parties to the contract as has been discussed above. However, the problem in pricing is that the system λ and the LMPs cannot reflect the cost of curtailment of bilateral contracts. During the settlement of RBM, the ISO should pay the owners of a bilateral contract its bidding price while the ISO will allocate this cost to all other participants by uplift or according to the ratios decided by Equation (4.22).

However, sometimes the load in a bilateral contract is too important to be curtailed or the response of the generation unit in a bilateral contract is not fast enough to decrease its output to mitigate network congestion. In these two cases, the energy imbalance caused by single sided curtailment will be taken by the other available resources in RBM. Here, the bilateral contract party without curtailment will not receive the compensation payment from the ISO. On the contrary, this participant must pay for the dispatch cost of this imbalance. Because one ΔP_i^{ij} has been removed from the nodal power balance Equation (4.6), Equation (4.17) changes to:

$$\frac{\partial C_p}{\partial \Delta P_i^{ij}} + (-1)^{\beta} \lambda (1 + \frac{\partial \Delta P^{Loss}}{\partial \Delta P_i^{ij}}) + \mu_i^{ij} + \sum_{l \in B} (\mu_l^{\max} - \mu_l^{\min}) \frac{\Delta P_l}{\Delta P_i^{ij}} = 0 \tag{4.30}$$

Then the equation (4.23) has the form:

$$\rho_i^P = (-1)^{\beta+1} (\frac{\partial C_p}{\partial \Delta P_i^{ij}} + \mu_i^{ij}) \tag{4.31}$$

In these two cases, the LMPs can reflect curtailment of this bilateral contract.

4.3.2 Q Sub-problem

The Q Sub-problem of real-time coordinated optimal dispatch can be formulated as:

$$\text{Minimise: } \sum_{i=1}^{n} (w_i^+ \Delta Q_i^+ + w_i^- \Delta Q_i^-) + \sum_{k=1}^{T} r_k (\Delta t_k^+ + \Delta t_k^-) \tag{4.32}$$

Subject to:

$$\Delta Q_i^+ - \Delta Q_i^- + \sum_{j=1}^{n} B_{ij}'' \Delta V_j + \sum_{k=1}^{T} {\partial Q_i}/{\partial t_k} (\Delta t_k^+ - \Delta t_k^-) = 0 \tag{4.33}$$

$$0 \le \Delta Q_i^+ \le Q_i^{\max} - Q_i^0, 0 \le \Delta Q_i^- \le Q_i^0 - Q_i^{\min} \tag{4.34}$$

$$0 \le \Delta t_k^+ \le t_k^{\max} - t_k^0, 0 \le \Delta t_k^- \le t_k^0 - t_k^{\min} \tag{4.35}$$

$$V_i^{\min} - V_i^0 \le \Delta V_i \le V_i^{\max} - V_i^0 \tag{4.36}$$

where w_i^+ and w_i^- are the reactive power incremental and decremental bidding prices of participant *i* in RBM respectively, ΔQ_i^+ and ΔQ_i^- are its increasing output and decreasing output of reactive power respectively, Q_i^0, $Q_i^{\min}$, $Q_i^{\max}$ are its current reactive power output or load, minimum reactive power and maximum reactive power respectively. *T* is the number of transformers, whose tap positions are adjustable, in the system. Δt_k^+ and Δt_k^- are the increasing turns ratio and the decreasing turns ratio of transformer *k*. r_k is the bidding price for the adjustment of tap positions of transformer *k* if there are any independent transmission companies and the transmission sector is also competitive. $t_k^0, t_k^{\min}, t_k^{\max}$ are the current tap ratio, minimum and maximum tap ratio of transformer *k*, respectively. ΔV_i is the change in voltage magnitude at bus *i*, $V_i^0, V_i^{\min}$, and $V_i^{\max}$ are its current value, lower limit and upper limit. B''_{ij} is an element of the matrix $\boldsymbol{B}''$.

The Q sub-problem looks similar to the P sub-problem. Because the reactive power cannot be sent long distances, there is no bilateral contract for it. The first item in the objective represents all the controllable reactive power injections, such as generators, capacitors, reactors, STATCOM, SVC and so on. These devices may belong to generation companies, transmission companies, distribution companies or large consumers. The second item in the objective is the cost for adjusting the taps of controllable transformers that belong to the transmission or distribution companies. Equation (4.33) is the nodal reactive power balancing equation of bus *i*. The constraints on the control variables of reactive power are given in Formulae (4.34) and (4.35). formula (4.36) is the voltage constraint on bus *i*.

4.4 Imbalance Settlement Methodologies

Basically, imbalance settlement should be post-event, limited volumes and mandatory for most market participants. There are two possible methods to settle the dispatch cost in RBM:

- Pay as bid. The ISO pays the participants whose bids have been accepted in RBM for their measured adjustment according to their bidding prices. Generators whose decremental bids have been accepted by the ISO will receive payment in RBM, but these reduced outputs should not be included in the settlement of PEAM. On the other hand, the ISO will allocate the dispatch cost to those participants who have caused the system imbalance. The advantage of this method is its simplicity, while the disadvantage is that all participants are treated equally even though they have very different effects on network congestion or the change of system loss.

- Pay as LMPs. The ISO will use the LMPs obtained to make the payment as well as to collect the dispatch cost. The principle of this method is to settle the cost in RBM in accordance with the participants' contribution to network congestion and system loss, as analysed in 4.3. Because the settlement of RBM should use *ex post* pricing, the proposed coordinated dispatch method should be run again after the ISO obtains the measurement from SCADA to provide the exact LMPs for the settlement.

4.5 Implementation

To solve the problem rapidly and reliably, AC Power Flow and a Primal-dual Interior Point (IP) Linear Program (LP) are used to implement the above OPF algorithm.

There have been some discussions regarding which method is more suitable for OPF, an LP-based or Newton-based method. Regarding the real-time optimal dispatch problem in the electricity market, LP seems promising for the following reasons:

- Multi-step bidding price curves instead of second-order cost curves are used to form the objective function. Linearising the objective function, which is one of the main problems with LP-based OPF [15,16], does not exist any longer.
- LP-based OPF is more reliable than Newton-based OPF. LP-based OPF can detect infeasible problems quickly and can deal with any sort of constraint easily. These features are very important to real-time dispatching in the electricity market.
- New developments in the IP method have made the LP-based OPF much faster when solving large-scale problems and more suitable for real-time applications.
- LMPs can be obtained from the shadow prices of nodal equation constraints, which are the by-products of Prime-dual LP-based OPF.

The procedure for the decoupled OPF implemented in this chapter is as follows:

Step1: Run the AC Power Flow to get the initial state of the power system.
Step2: Compute the necessary sensitivities and linearise the constraints.

Step3: Select the LP control variables according to the bids in RBM. Decide if bilateral contract curtailment is needed and which curtailment strategy should be used.
Step4: Run LP to solve the P sub-problem.
Step5: Run LP to solve the Q sub-problem.
Step6: Correct the control variables, then run AC Power Flow to get the new state of the power system.
Step7: Check if all the constraints have been satisfied. If yes, continue; if no, go to step2.
Step8: Obtain the optimal coordinated dispatch strategy.

4.6 Test Results

The IEEE 30-bus system shown in Figure 4.4 is used here to illustrate the proposed coordinated dispatch method. The network parameters and injection data can be found in [17].

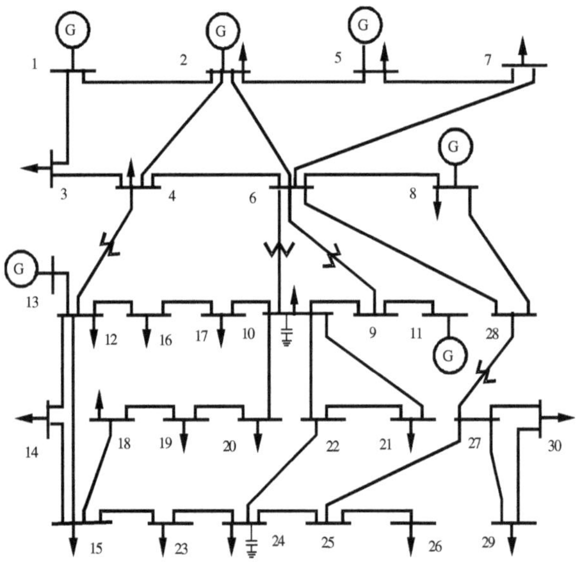

Figure 4.4. IEEE 30-bus test system

The various bids of participants, including generators, consumers and bilateral contracts, are given in Table 4.1, in which G-1 means generator at bus 1 and C-24 means consumer at bus 24.

4.6.1 Coordinated Dispatch without Network Congestion

Assume that there is a 100MW increase in system load, which has been distributed to individual buses according to their current load shares. Set $\alpha = 1$, then the optimal dispatch strategy obtained to meet this load fluctuation is shown in Table 4.2. Figure 4.5 reveals the two components of LMPs under normal operating conditions, system lambda and network losses.

Table 4.1. Bids of participants in RBM

Partici-pants	Supplementary bids		Operating Reserves			Base Point (MW)	MAX MW	MIN MW
	Incr. Bids ($/MWh)	Decr. Bids ($/MWh)	Cap. Bids ($/MW)	En. Bids ($/MWh)	Amount (MW)			
G-1	35	15	3.5	35	/	138.53	200	50
G-2	15	8	2.5	15	/	57.56	100	20
G-5	15	8	1.5	15	/	24.56	100	10
G-8	30	12	1.5	15	30	35.00	65	10
G-11	25	10	2.5	25	/	17.93	50	10
G-13	15	5	1.5	15	/	16.91	50	5
C-24	/	40	/	/	/	8.70	15	3

Bilateral contract	From	To	Contract Amount (MW)	Curtailment Bids ($/MW)
B1	G-13	C-30	10.6	50

Table 4.2. Optimal dispatch strategy to meet the load fluctuation

Partici-pants	PEAM contracts		Calling Upon PAAM contracts (MW)	Replacement of Operating Reserves (MW)
	Increase (MW)	Decrease (MW)		
G-5	75.44	0	0	0
G-8	0	0	24.72	0
G-13	0	0	0	24.72

The cost of replacement of used operating reserves is not a component of LMPs in Equation (4.28). However, it can affect the value of system lambda. Figure 4.6 demonstrates the change in LMPs under different replacement procurements of used operating reserves by setting $\alpha = 0, \alpha = 0.5, \alpha = 1$.

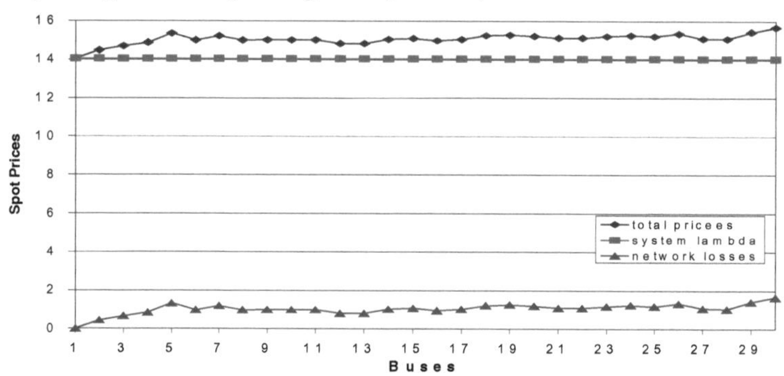

Figure 4.5. Components of spot prices under normal operating conditions

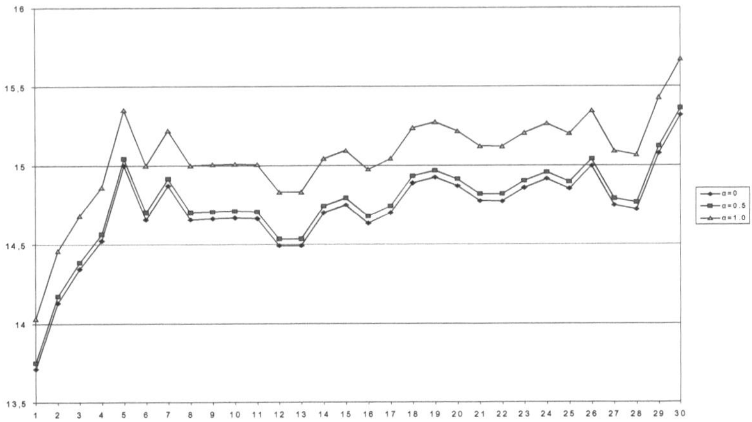

Figure 4.6. Effects of different replacement procurements of operating reserves on LMPs

4.6.2 Coordinated Dispatch with Network Congestion

Based on the same load fluctuation and the same operating reserves replacement level ($\alpha = 1$) as those in the previous section, the active power flow on line 36 (from bus-28 to bus-27) is 24.79MW. Now reduce the limit of active power flow on this line to 24MW; the optimal dispatch strategy to eliminate this light congestion while meeting the given load change is shown in Table 4.3. In this case, three components of the LMP given in Equation (4.22), system lambda, network losses and congestion management, are illustrated in Figure 4.7.

Table 4.3. Optimal dispatch strategy with active power flow violation on line 36

Partici-pants	PEAM contracts		Calling Upon PAAM contracts (MW)	Replacement of Operating Reserves (MW)
	Increase (MW)	Decrease (MW)		
G-5	75.44	0	0	0
G-8	0	0	13.22	0
G-13	11.63	0	0	13.22

Compared with Figure 4.5, the spot prices in Figure 4.7 have not changed much at most buses except buses 25-27, 29-30. This means that these several have higher sensitivity to the congestion line 36 than the other buses. In other word, the consumers at these buses should pay most of the cost caused by network congestion. Because the congestion is very slight, the associated Lagrangian multiplier $\mu_{36}^{\max} = -2.21$ is not very large.

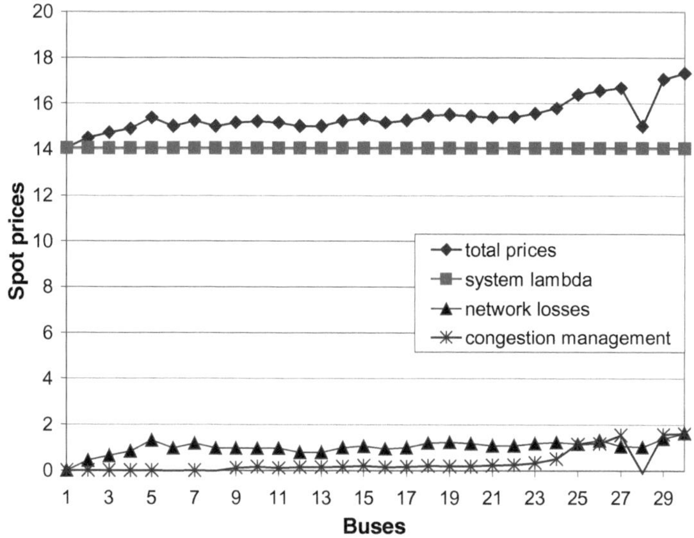

Figure 4.7. Components of spot prices with network congestion

Voltage limit violation. Increase the active load at bus 6 from 3.5 MW to 33.5 MW. As a result, the voltage at bus 6 decreases to 0.82 p.u. and violates the lower limit 0.85 p.u. Reactive congestion management is run to eliminate the voltage violation. Figure 4.8 shows the nodal prices of reactive power, which are the shadow prices of the nodal reactive power balancing equality constraints of the Q subproblem. It is obvious that the reactive power price at bus 6 is prohibitively high. This penalty is reasonable, because the voltage violation is caused by the heavy active load at bus 6. This price signal constitutes planning information that reactive power compensation devices should be installed at or near this spot.

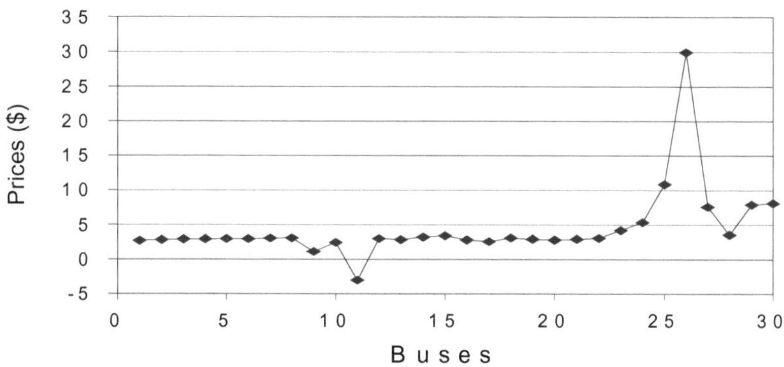

Figure 4.8. Effects of voltage violation on nodal prices

4.6.3 Comparison Between the RSLP and PDIPLP

Figures 4.9 and 4.10 show the number of iteration and the CPU time for linear programming to solve the real-time dispatch problem in 5-bus, IEEE 30-bus, IEEE 57-bus and IEEE-118 bus test systems using Primal-dual Interior Point Linear Programming (PDIPLP) and Revised Simplex Linear Programming (RSLP). Both algorithms have been coded in Visual FORTRAN 6.0 and implemented on a PII/400 PC. The advantages of PDIPLP can be seen clearly from these two figures. With increasing problem size, the number of iterations of RSLP increases very quickly while the number of iterations for PDIPLP remains almost unchanged. The CPU time for PDIPLP is also much less than that for RSLP. The bigger the system, the better the performance of PDIPLP will over RSLP. Therefore, applying PDIPLP to the real-time dispatch of the electricity market can meet its time requirement very well.

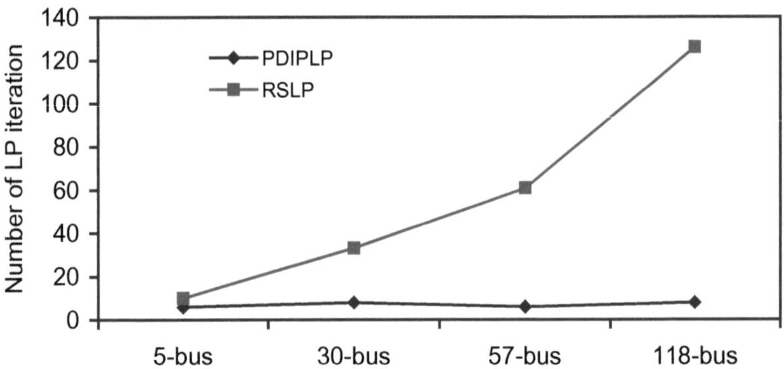

Figure 4.9. Comparison of number of iterations for PDIPLP and RSLP

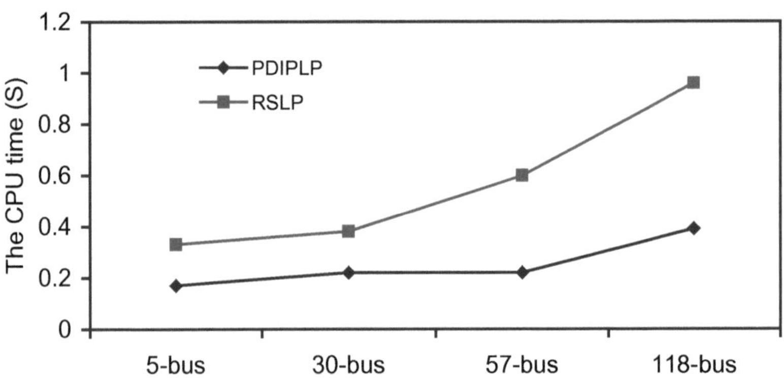

Figure 4.10. Comparison of CPU time for PDIPLP and RSLP.

4.7 Conclusions

In this chapter, a coordinated real-time optimal dispatch method for unbundled electricity markets has been proposed. The main features of this method are the following:

- Three types of electricity contract, bilateral energy contract, pool energy auction contract and ancillary services contract, have been taken into account in the coordinated dispatch.
- Balancing mechanisms play the key role in the proposed framework, where the ISO can meet system imbalance and mitigate network congestion by using various bids.
- The adjustment of bus injection has been divided into several independent control variables according to unbundled contracts to embed all the possible bids in RBM into the objective of active power optimisation.
- The curtailment strategies of bilateral contracts have been integrated into the proposed method.
- The economical meaning of Lagrangian multipliers, which decide the LMPs of real-time dispatch, has been analysed in detail.

Test results demonstrate that the proposed coordinated dispatch method implemented with a modified OPF can deal with system imbalance and network congestion simultaneously and successfully. Comparison between PDIPLP and RSLP shows that PDIPLP is much more efficient when dealing with large systems with a huge number of control variables.

4.8 References

[1] FERC, 'Order No. 2000, Regional Transmission Organization (RTO)-Final Rule', http://www.ferc.fed.us/news1/rules/pages/rulemake.htm, December 20, 1999

[2] A. J. Wood and B. F. Wollenberg, Power Generation, Operation, and Control, second edition, John Wiley & Sons, New York, NY, 1996

[3] E. Hirst: "Real-time balancing operations and markets: Key to competitive wholesale electricity markets", Report prepared for the Edison Electric Institute, Washington, D.C., and the Project for Sustainable FERC Energy Policy, Alexandria, Virginia, 2001

[4] F. F. Wu, and P. Varaiya: "Coordinated Multilateral Trades Of Electric Power Networks: Theory and Implementation", Electrical Power And Energy Systems 21 (1999) 75-102

[5] H. Singh, S. Hao, A. Papalexopoulos: "Transmission Congestion Management in Competitive Electricity Markets", IEEE Trans. Power Systems, Vol. 13, No.2, pp. 672-679, May 1998

[6] K. David: "Dispatch Methodologies for Open Access Transmission Systems", IEEE Trans. Power Systems, Vol.13, No.1, pp.46-53, February 1998

[7] R. S. Fang, and A. K. David: "Optimal Dispatch Under Transmission Contracts", IEEE Trans. Power Systems, Vol.14, No.2, pp.732-737, May 1999

[8] H. Singh and A. Papalexopoulos: "Competitive Procurement of Ancillary Services by an Independent System Operator", IEEE Trans. on Power Systems, Vol.14, No.2, May 1999, pp. 498-504
[9] T. Alvey, D. Goodwin, X. Ma, D. Streiffert, and D. Sun: "A Security-constrained Bid-clearing System for the New Zealand Wholesale Electricity Market", IEEE Trans. Power Systems, Vol. 13, No. 2, pp. 340-346, May 1998
[10] K. W. Cheung, P. Shamsollahi, D. Sun: "Energy and Ancillary Service Dispatch for the Interim ISO New England Electricity Market", in Proc. the 21st International Conference on Power Industry Computer Applications, Santa Clara, California, May 1999
[11] Xing Wang and Y.H. Song: "Advanced Real-Time Congestion Management through Both Pool Balancing Market and Bilateral Market", IEEE Power Engineering Review, Vol.20, No.2, pp.47-49, Feb 2000
[12] Xing Wang and Y.H. Song: "A Coordinated Real-Time Optimal Dispatch Method for Unbundled Electricity Markets", IEEE Transactions on Power Systems, Vol.17, No.2, pp.482-490, May 2002
[13] OFFER, The New Electricity Trading Arrangements, July 1999, UK
[14] F.C. Schweppe, M.C. Caramanis, R.D. Tabors, and R.E. Bohn: Spot Pricing of Electricity, Kluwer Academic Publishers, 1988
[15] Stott, and J. L. Marinho: "Linear Programming for Power-system Network Security Applications", IEEE Trans. Power Apparatus and Systems, Vol. PAS-98, No. 3, pp. 837-848, May/June 1979
[16] O. Alsac, J. Bright, M. Prais, and B. Stott: "Further Developments in LP-Based Optimal Power Flow", IEEE Trans. Power Systems, Vol. 5, No. 3, pp. 697-711, August 1990
[17] The IEEE 30-bus test system. Available: http://www.ee.washington.edu/research/pstca/pf30/pg_tca30bus.htm
[18] R.E. Mehrotra: "On the Implementation of a Primal-dual Interior Point Method", SIAM Journal on Optimization, Vol.2, No.4, 1992, pp.575-601
[19] X.YAN, V. QUINTANA: "An Infeasible Interior-Point Algorithm for Optimal Power-flow Problems", Electric Power Systems Research, Vol.39, 1996, pp.39-46
[20] G. A. Angelidis and A. D. Papalexopoulos: "Challenge in real-time electricity market design" Sixth IASTED International Conference on Power and Energy Systems (Euro-PES 2001) July 3–6, 2001 Rhodes, Greece

Appendix A: Primal-dual Interior Point Linear Programming Method

Karmarkar's algorithm is significantly different from Dantzig's simplex method, which solves a linear programming problem starting with one extreme point along the boundary of the feasible region and skips to a better neighbouring extreme point along the boundary, finally stopping at an optimal extreme point. Karmarkar's interior point rarely visits too many extreme points before an optimal point is found. The IP method stays in the interior of the polytope and tries to position a current solution as the "centre of the universe" in finding a better direction for the next move. By properly choosing the step lengths, an optimal solution is achieved after a number of iterations. Although this IP approach requires more computational time to find a move direction than the traditional simplex method, better move directions are achieved resulting in fewer iterations.

Figure A.1 illustrates how the two methods, the IP method and simplex method, reach an optimal solution. In this small problem, the projective scaling algorithm requires about the same number of iterations as the simplex method. But for a large problem this method would require only a small fraction of the number of iterations that the simplex method requires.

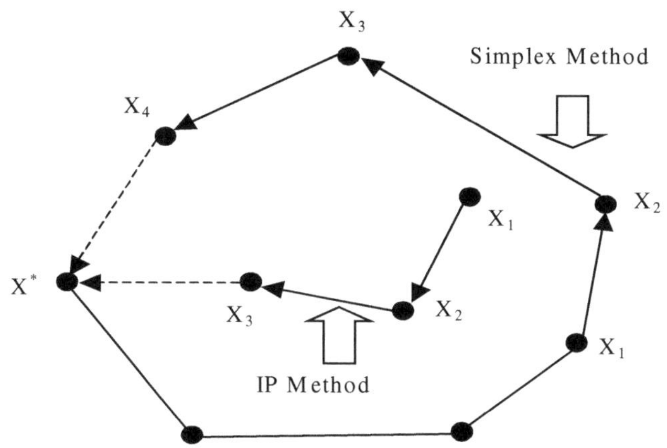

Figure A.1. Comparison between IP and simplex methods

Since Karmarkar's discovery of the interior point method and its reported speed advantage over the simplex method, many variants of the IP method have evolved in an attempt to solve linear programming problems. As one of the most efficient interior point approaches, a second-order primal-dual predictor-corrector method, based on Mehrotra's method [18], is introduced here briefly.

The linearised OPF problems can be written as the following standard primal linear programming problem:

$$\text{Minimise } c^T x \tag{A.1}$$

$$\text{Subject to} \quad Ax = b$$

$$x + s = u$$

$$x, s \geq 0$$

where $c, x, s, u \in R^n, b \in R^m, A \in R^{m \times n}$. Its dual problem is:

$$\text{Maximise } b^T y - u^T w \tag{A.2}$$

$$\text{Subject to} \quad A^T y + z - w = c$$

$$z, w \geq 0$$

where $z, w \in R^n$, $y \in R^m$.

The logarithmic barrier function is given by adding the non-negativity of constraints in the primal formulation into the objective function as logarithmic barrier penalty items:

$$L(x, s, \mu) = c^T x - \mu(\sum_{j=1}^{n} \ln x_j + \sum_{j=1}^{n} \ln s_j) \tag{A.3}$$

The first-order Karush–Kuhn–Tucker (KKT) optimality conditions for (A.3) are:

$$Ax = b \tag{A.4}$$

$$x + s = u$$

$$A^T y + z - w = c$$

$$XZe = \mu e$$

$$SWe = \mu e$$

where X, S, Z, W are diagonal matrices with the elements x_j, s_j, z_j, w_j, respectively, e is the n-vector of all ones, and μ is a barrier parameter.

So the Newton's direction can be obtained by solving:

$$\begin{bmatrix} A & 0 & 0 & 0 & 0 \\ I & 0 & I & 0 & 0 \\ 0 & A^T & 0 & I & -I \\ Z & 0 & 0 & X & 0 \\ 0 & 0 & W & 0 & S \end{bmatrix} \begin{bmatrix} \Delta x \\ \Delta y \\ \Delta s \\ \Delta z \\ \Delta w \end{bmatrix} = \begin{bmatrix} b - Ax \\ u - x - s \\ c - A^T y - z + w \\ \mu e - XZe \\ \mu e - SWe \end{bmatrix} \tag{A.5}$$

Once $(\Delta x, \Delta y, \Delta s, \Delta z, \Delta w)$ (denoted Δ below) has been calculated, the maximum step sizes α_P and α_D, which maintain the non-negativity of variables in the primal and dual spaces, are found. Next, the variables are updated by:

$$\begin{aligned} x^{k+1} &= x^k + \alpha_0 \alpha_P \Delta x \\ s^{k+1} &= s^k + \alpha_0 \alpha_P \Delta s \\ y^{k+1} &= y^k + \alpha_0 \alpha_D \Delta y \\ z^{k+1} &= z^k + \alpha_0 \alpha_D \Delta z \\ w^{k+1} &= w^k + \alpha_0 \alpha_D \Delta w \end{aligned} \tag{A.6}$$

where α_0 is the step-reduction factor. Then, the barrier parameter μ is updated and the iteration process is repeated.

Factorisations of the KKT matrix can take 60% to 90% of the total CPU time to solve a problem. So it is important to look for some way to reduce the number of iterations. Mehrotra's predictor-corrector technique [18] incorporates high-order information when approximating the central trajectory and computing the direction step. The second-order variant of this technique has been used widely and proved to be efficient.

The predictor-corrector technique regards the direction step Δ as two parts:

$$\Delta = \Delta_A + \Delta_C \tag{A.7}$$

where Δ_A is the affine-scaling component, which is the predictor term and is responsible for optimisation to reduce the primal and dual infeasibilities and the duality gap, Δ_C is the centring component, which is the corrector term and keeps the current iteration away from the boundary of the feasible region (keeps it close to the central path ideally) .

Δ_A is obtained by solving (A.5) with $\mu = 0$. Δ_C is the solution of (A.5) with the right-hand side replaced by

$$(0,0,0, \mu e - XZe, \mu e - SWe)^T$$

where $\mu > 0$ is centring parameter ($\mu = \dfrac{x^T z + s^T w}{2n}$, for example). The maximum step sizes in the primal and in the dual spaces preserving the non-negativity of (x, s) and (z, w), respectively, are determined, and the predicted complementarity gap

$$g_a = (x + \alpha_{Pa} \Delta x)^T (z + \alpha_{Da} \Delta z) + (s + \alpha_{Pa} \Delta s)^T (w + \alpha_{Da} \Delta w)$$

is computed. It is then used to determine the barrier parameter

$$\mu = (\frac{g_a}{g})^2 \frac{g_a}{n} \tag{A.8}$$

where $g = x^T z + s^T w$ is the current complementarity gap. Next, the second-order component of the predictor-corrector is computed.

$$(X + \Delta x)(Z + \Delta z) = \mu e \Rightarrow Z\Delta x + X\Delta z = -XZ + \mu e - \Delta x \Delta z$$

When solving (A.5) the second-order term $\Delta x \Delta z$ is neglected. To obtain Δ_C, $\Delta x \Delta z$ can be estimated using the affine-scaling direction $\Delta x_A \Delta z_A$. The same process is carried out for the other second-order term $\Delta s \Delta w$. So for such a μ in (A.8) the Δ_C can be calculated by solving:

$$\begin{bmatrix} A & 0 & 0 & 0 & 0 \\ I & 0 & I & 0 & 0 \\ 0 & A^T & 0 & I & -I \\ Z & 0 & 0 & X & 0 \\ 0 & 0 & W & 0 & S \end{bmatrix} \begin{bmatrix} \Delta x_C \\ \Delta y_C \\ \Delta s_C \\ \Delta z_C \\ \Delta w_C \end{bmatrix} = \begin{bmatrix} 0 \\ 0 \\ 0 \\ \mu e - \Delta X_A \Delta Z_A e \\ \mu e - \Delta S_A \Delta W_A e \end{bmatrix} \tag{A.9}$$

Finally, the direction Δ in (A.7) is determined.

In the above method, a single iteration of the second-order predictor-corrector primal-dual method needs two solutions of the same linear system with two different right-hand sides. The advantage of this method is that the barrier parameter μ can be estimated very well and a high-order approximation is applied to the central path.

One advantage of the primal-dual method is that it allows for separate step lengths in the primal and dual spaces as shown in (A.6). This has been proven highly efficient in practice, significantly reducing the number of iterations required for ocnvergence. The step lengths (α_P, α_D) are determined in the following way so that the non-negativity conditions $x, s \geq 0$ and $z, w \geq 0$ are preserved, respectively

$$\alpha_P = \min\left\{1, -\frac{x_j}{\Delta x_j}, -\frac{s_j}{\Delta s_j}, \Delta x_j < 0, \Delta s_j < 0\right\} \tag{A.10}$$

$$\alpha_D = \min\left\{1, -\frac{z_j}{\Delta z_j}, -\frac{w_j}{\Delta w_j}, \Delta z_j < 0, \Delta w_j < 0\right\} \tag{A.11}$$

According to in [19], the step-reduction factor α_0 in (A.6) can be initially set as $\alpha = 0.95$, then may be increased to $\alpha = 0.9995$ when the primal and dual infeasibility is less than a certain value (say 10^{-2}). This will be far more efficient than using a constant value.

The iteration terminates when the following feasibility and optimality conditions are all satisfied.

- Primal feasibility: $$\frac{\|Ax-b\|}{1+\|x\|} \le \varepsilon_f$$
- Bound feasibility: $$\frac{\|x+s-u\|}{1+\|x\|+\|s\|} \le \varepsilon_f$$
- Dual feasibility: $$\frac{\|A^T y - w + z - c\|}{1+\|y\|+\|w\|+\|z\|} \le \varepsilon_f$$
- Optimality: $$\frac{\|c^T x - (b^T y - u^T w)\|}{1+\|b^T y - u^T w\|} \le \varepsilon_0$$

where ε_f and ε_0 are the convergence tolerances for feasibility and optimality conditions, respectively.

5. Available Transfer Capability Evaluation

Y. Xiao, Y.H. Song and Y.Z. Sun

Deregulation of the electricity industry throughout the world aims at creating a competitive market to trade electricity, which generates a host of new technical challenges to market participants and power system researchers. For transmission systems, it requires non-discriminatory open access to transmission resources [1]. Therefore, for better transmission services support and full utilisation of transmission assets, one of the major challenges is to accurately gauge the transfer capability remaining in the system for further transactions, which is termed Available Transfer Capability (ATC) [2]. It is crucial to develop an appropriate ATC determination methodology that enables one to evaluate the realistic transmission transfer capability by accounting for all related important requirements. To address this issue, an optimal power flow (OPF) based stochastic model is presented in this chapter for ATC evaluation, with respect to the major uncertainties in power systems associated with ATC level.

5.1 Definition and Application of ATC

5.1.1 Definition of ATC

Open access to transmission systems places a new emphasis on more intensive shared use of interconnected networks reliably by utilities and independent power producers (IPPs) to improve of economy and security. In these circumstances, it is very important to obtain a clear understanding of how much unused capacity is available on a transmission interface.

Historically, to define meaningful measures of the transmission transfer capability of interconnected transmission networks, the electricity industry has used standard terms, including First Contingency Incremental Transfer Capability (FCITC) and First Contingency Total Transfer Capability (FCTTC). As defined by the North American Electric Reliability Council (NERC) [3], FCITC is the amount of power, incremental above normal base power transfers, that can be transferred over the transmission network in a reliable manner, based on all of the following conditions:

1. For the existing or planned system configuration, and with normal (pre-contingency) operating procedures in effect, all facility loadings are within normal ratings and all voltages are within normal limits.
2. Electric systems are capable of absorbing the dynamic power swings, and remaining stable, following a disturbance resulting in the loss of any single electric system element, such as a transmission line, transformer, or generating unit.
3. After the dynamic power swings subside following a disturbance that results in the loss of any single electric system element as described in condition 2 above, and after the operation of any automatic operating systems, but before any post-contingency, operator-initiated system adjustments are implemented, all transmission facility loadings are within emergency ratings and all voltages are within emergency limits.

FCTTC is defined as the total amount of electric power (net of normal base power transfers plus first contingency incremental transfers) that can be transferred between two areas of the interconnected transmission systems in a reliable manner based on conditions 1, 2, and 3 in the FCITC definition above. As analysed in [4], major weaknesses of the definitions are: subjective interpretation is required; laborious studies by skilled specialists are needed to measure them and so on.

In 1995, in response to a NERC strategic initiative to satisfy electricity industry needs on uniform and practical definitions of usable transmission system transfer capability, a new term—Available Transfer Capability was proposed and defined by the Federal Energy Regulatory Commission (FERC) [2]. *Available Transfer Capability Definitions and Determinations* [5] - a 1996 NERC report further establishes a framework for determining ATCs of interconnected transmission networks for a commercially viable wholesale market, in which the principles for ATC calculation are also defined. According to the report – *Notice of Proposed Rulemaking* (NOPR) [2], ATC is defined as below:

Available Transfer Capability is a measure of the transfer capability remaining in the physical transmission network for further commercial activity over and above already committed uses.

Another important term related to the ATC is Total Transfer Capability (TTC), which is defined as the amount of electric power that can be transferred over the interconnected transmission network in a reliable manner while meeting all of a specific set of defined pre- and post-contingency system conditions. Mathematically, ATC is defined as the TTC less the sum of existing transmission commitments, less the Transmission Reliability Margin (TRM), and the Capacity Benefit Margin (CBM). That is:

ATC = TTC – TRM – Existing Transmission Commitments (including CBM)

where TRM is to measure the amount of transmission transfer capability necessary to ensure that the interconnected transmission network is secure under a reasonable range of uncertainties in system conditions. TRMs calculated and used by Independent System Operators (ISOs) should consider the following factors, including load forecast and load distribution variabilities, variations in generation dispatch, unaccounted for parallel path flows, generation reserve sharing and transmission facility maintenance and so forth [5]. In accordance with the terms and conditions of the transmission provider's tariff, TRMs may be sold as recallable transmission services provided that services can be recalled within the time frame necessary for system operators to readjust the system after a contingency or short-term rating of the limiting facilities. Here, recallable service is defined as a transmission service that can be interrupted by a transmission provider for any reason, including economic, that is consistent with FERC policy and the transmission provider's transmission service tariffs or contract provisions[5].

CBM is defined as that amount of transmission transfer capability reserved by load serving entities to ensure access to generation from interconnected systems to meet generation reliability requirements. CBM allows a load serving entity to reduce the generating reserves that are necessary to maintain an acceptable reliability level of service. The corresponding reservations may be sold on a recallable basis. Existing transmission commitments are composed of native load uses, prudent reserves, existing commitments for purchase/exchange/deliveries/sales, existing commitments for transmission service and other pending potential uses of transfer capability [6].

These terms, particularly the ATC, form the basis of a transmission service reservation system that will be used for reserving transmission services, scheduling recallable and non-recallable energy transactions and arranging emergency transfers between areas of an interconnected power network in the competitive electricity market. Concerning both operating horizon and planning horizon, the mathematical definitions and relationships of TTC, ATC and related terms in the transmission service reservation system are depicted in Figure 5.1.

5.1.2 Industrial Applications of ATC

In order to foster generation competition and customer choice, and facilitate wide area coordination throughout the whole transmission network, the FERC mandates that ATC information of some specific interfaces must be accessed by electricity market participants and system operators hourly through an Open Access Same-Time Information System (OASIS) [5]. Calculated by ISOs, ATC information usually includes the quantities of ATC, TTC, non-recallable and recallable ATC and so on. The information would be organised by path and would generally be posted daily for each hour of the next seven days, with hourly updates as required, and posted less frequently for longer future periods. The OASIS is used as a forum for system operators and transmission customers to exchange values of ATC and other reservation margins, outage, load, generation, ancillary services, curtailments and other information on a scheduled basis.

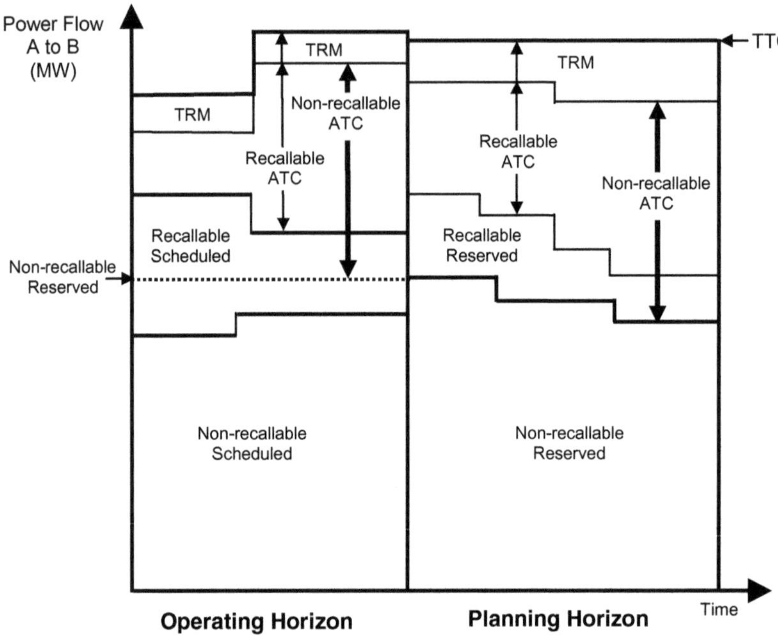

Figure 5.1. Relationships of ATC and related terms in transmission service reservation system [5]

As a primary quantitative indicator of the availability of transmission service, ATC is used to help determine whether requests for transmission service can be accepted. If the current ATC value is sufficient, and there are no known reliability problems, the transmission service request is approved. Once the ISO has accepted the request, the ATC posting is updated to reflect the new transmission service reservation. On the other hand, for transmission customers, the ATC values are useful in their own analysis to determine the feasibility prior to requesting transmission service. Particularly, for transmission services with multiple control areas involved along the desired contract path, the transmission customers must review each transmission provider's OASIS to determine the feasibility [7].

5.2. Criteria for ATC Evaluation

As recognised, ATC is a measure bridging the technical characteristics of how interconnected transmission networks perform to the commercial requirements associated with transmission service requests. Therefore, according to a 1996 report of NERC [5], the calculation and application must satisfy certain principles balancing both technical and commercial issues, among them the major ones are:

1. ATC calculations must produce commercially viable results. ATCs produced by the calculations must give a reasonable and dependable indication of transfer capabilities available to the electric power market.
2. ATC calculations must recognise time-variant power flow conditions on the entire interconnected transmission network. In addition, the effects of simultaneous transfers and parallel path flows throughout the network must be addressed from a reliability viewpoint.
3. ATC calculations must recognise the dependency of ATC on the points of electric power injection, the directions of transfers across the interconnected transmission network, and the points of power extraction. All entities must provide sufficient information necessary for the calculation of ATC.
4. Regional or wide-area coordination is necessary to develop and post information that reasonably reflects the ATCs of the interconnected transmission network.
5. ATC calculations must conform to NERC, regional, subregional, power pool, and individual system reliability planning and operating policies, criteria, or guides.
6. The determination of ATC must accommodate reasonable uncertainties in system conditions and provide operating flexibility to ensure the secure operation of the interconnected network.

The intention of full use of the transmission resources for better support of transmission services relies heavily on the ability to determine transfer capability in the network accurately. In order to build an appropriate model for ATC assessment, and to construct a valid algorithm to solve the model efficiently, there are several essential criteria that should be taken into account [8]:

5.2.1 Accuracy

ATC must accurately reflect the physical realities of the transmission network. Obviously, if the quantitative value of ATC is directly translated into dollars in a market where millions of dollars are at stake for the participants, the demand for accurate models will be imperative. Any conservative estimate of ATC will result in lost trading opportunity, i.e. loss of economic efficiency, and unnecessary capital costs incurred in terms of transmission planning, whereas an over-estimate of ATC may lead to system operation risk, such as shutdowns and blackouts.

5.2.2 Dependability

As far as short-term operational planning is concerned, the determination of ATC must accommodate reasonable uncertainties in system conditions, to ensure the secure operation of the interconnected network [5]. It is fully recognised that there are numerous uncertainties associated with ATC determination, including projected customer demands, generation dispatch, system configuration, scheduled transfers as well as system contingencies. Particularly in a privatised and deregulated industry structure where the system situation is becaming less predictable and it is more difficult to gain the cooperation of all possible resources immediately during an emergency. Consequently, transmission operators face more risks, with greater penalties for making mistakes.

5.2.3 High Efficiency

The two demands aforementioned are posed to the ATC evaluation model. On the other hand, to meet the demand of updating OASIS postings hourly, rapid execution time is an important requirement for the method applied to solve the model.

5.3 Review of Existing Methodologies for ATC Evaluation

5.3.1 Existing Methodologies

The commercial and technological significance of ATC in an open transmission environment led to an explosion of interest in ATC evaluation among power systems researchers and engineers. In recent years, various approaches have been proposed to model and calculate ATC. So far, there are three major approaches suggested for the calculation of ATC:

5.3.1.1 Sensitivity Analysis

Sensitivity analysis is the earliest solution proposed for the ATC approximation value calculation [5, 9–11]. Based on linear incremental power flow, the sensitivity factors are simple to define and easy to calculate. It can vastly improve the computation time required. Therefore, with the merit of providing a reasonably accurate approximation of ATC with fast speed, the method is very suitable for real-time and immediate future ATC calculation.

The effect of uncertainties can be considered by solving the program under postulated different contingency criteria, however, the execution time will be increased very significantly. Only considering thermal-constrained ATC, the linearised approach does not take into account the non-linear effects of reactive power and voltage. Moreover, with a large quantity of ATC value calculated under different conditions, it is still difficult to determine which is the most dependable and reasonable figure.

5.3.1.2 Continuation Power Flow

Another major approach is based on continuation power flow (CPF) algorithms that trace the power flow solution curve, starting at a base load, leading to the steady-state voltage stability limit or the critical maximum loading point of the system. In 1998, Ejebe *et al.* [12] presented a CPF approach to determine ATC for each specified transfer, in which an adaptive localisation is introduced with the intention to enhance its speed in processing a large number of contingencies. However, it is still time consuming when considering all the possible contingencies one by one. The CPF method requires an increase in the loading factor for a specific cluster of generators and loads, which may produce a conservative ATC since the optimal distribution of generation and loading is ignored.

5.3.1.3 Optimal Power Flow

The OPF approach is a modification of the CPF approach, but can directly incorporate steady-state contingencies. In [13], based on the OPF formulation, a neural network solution methodology is proposed for the problem of power transfer capability calculations. A few recent attempts have been made to include dynamic security considerations. Viewing stability as a constraint of the OPF model, Gan *et al.* presents a step-by-step integration (SBSI) technique based approach [14]. Then, by converting the differential equations into numerically equivalent algebraic equations, the model can be solved by standard non-linear programming techniques. In [15], from both operation horizon and planning horizon's viewpoint, Rosales *et al.* described an integrated package for ATC calculation considering dynamic security constraints, together with preventive control schemes to stabilise the system operation state. The approach combines an OPF model and a hybrid direct-time domain method called SIngle Machine Equivalent (SIME) to achieve the transient stability assessment and preventive control. Generally speaking, a daunting challenge to the calculation of stability constrained ATC is how to deal with the stability constraints involved. One possible approach is based on the discretisation of the differential equations. With the stability condition described into limits on trajectories, the resulting algebraic equations can be treated in the same way as static security constraints. Using the idea of extended equal area criteria to transform the stability criterion is another viable approach.

Besides conventional methodologies for ATC calculation mentioned above, various approaches have been proposed with consideration of key uncertain factors affecting the value. Monte Carlo simulation is introduced for ATC assessment to deal with the unavailability of generators in the system [16,17]. A large number of OPF problems have to be calculated under different postulated system conditions. Unarguably, it is too time and labour consuming to promptly answer transmission user requests and fulfil OASIS posting requirements. In [18], Gravener and colleagues apply a first-order sensitivity to quantify the effect of network uncertainties, such as load forecast error and simultaneous transfers. However, it can only deal with continuous uncertainties. And considering one uncertainty at a time, this analysis can only measure the impact in a limited way. Additionally, the probabilistic characteristics of uncertainties have been ignored, which obviously affect the accuracy of the calculated value. How to evaluate ATC more rationally and efficiently is still a big issue.

5.3.2 ATC Evaluation in Industry

As a matter of fact, evaluating the power transfer limits of a transmission grid has long been a goal of transmission planners and system operators in electric industries. Because a large number of contingencies have to be considered in ATC assessment, together with frequently changing power levels, practically, linear DC power programs have been the most widely used solution in the past. Nevertheless, using such linear models gives only an approximate solution and cannot include voltage constraints.

Owing to the commercial and technical significance of ATC in the new electricity market, more and more institutes and utilities have shown increased interest in, and are undertaking studies of evaluation and enhancement of the ATC and related terms.

5.3.2.1 EPRI [19]

In 1996, the Electric Power Research Institute (EPRI) introduced the TRAnsfer Capability Evaluation program (TRACE) to assist power system planners and operators calculating the ATC and TTC. Two major software modules, Security Analysis (SA) and Security Constrained Optimal Power Flow (SCOPF), are used. The former enables one to quickly identify critical contingencies, the latter calculates the maximum power that can be safely transferred, subject to the critical contingencies identified by the SA module. Taking this off-line version as a foundation, the EPRI further developed an on-line system TRACE, to meet the requirements of the FERC for real-time transfer capability information posting on OASIS. This system allows operators to calculate the ATC and TTC on posted paths using real-time Energy Management System (EMS) state estimation data. These results are used in system operations to optimise power transactions in light of security considerations. In October 1999, a workshop - "TTC/ATC-Transmission Transfer Capability", was organised by the EPRI, Southern Co., ABB and PCA at Southern Co., Alabama, with the major objective of providing an understanding of the value of on-line calculation of ATC and TTC.

5.3.2.2 ECAR [20]

To ensure the validity and consistency of ATC values, the East Central Area Reliability Coordination Agreement (ECAR) Transmission Providers (TPs) employ a two-step distributed calculation/coordination procedure. First, each TP is required to calculate ATC quantities for relevant potential transactions as limited by facilities under its own operational control. In the second step, with the submitted individual calculated ATC values, a coordination system selects the lowest values to ensure that the most limiting facility is recognised to determine the wide-area ATC. An ATC Posting Conflict Advisory Procedure (APCAP) is established to deal with situations when the TPs do not agree on the TTC or ATC for a given transmission path. Providing an informed "second opinion," the APCAP offers assistance to resolve posting conflicts and disputes in the quantification and coordination of a wide-area ATC.

5.3.2.3 PJM [7]

In PJM, the calculation of ATCs is generally based on simulations of the operation of the interconnected transmission network under three specific time frames: near-term (hour 0 to hour 168), mid-term (8 to 30 days in the future) and long-term (1 to 13 months in the future). Typically, for each posted path, the ATC is determined by raising the source area generation and the sink area load until a system constraint is reached. In addition, the impacts of generation and transmission contingencies on the ATC are also taken into account. TTC and ATC uncertainty, resulting from variations in generation patterns, are accounted for by determining transfer capability with critical generator(s) out of service. To capture both thermal and reactive constraints, an AC power flow program is utilised.

The models for calculating ATC vary depending on the time period. The near-term ATC is calculated using both an on-line ATC program by Siemens and an off-line power flow application program developed by PTI - PSS/E (Power System Simulator for Engineering). Generally speaking, TTC, recallable ATC for hours 0 to 168, and daily non-recallable and recallable ATC for days 1 through 7 are calculated periodically. On the basis of the real-time state estimation at hour 0, base cases for the next 168 hours are derived by incorporating the PJM operating plan, which includes generation commitment and outage schedules, load forecasts, transmission outage schedules and all scheduled interchange transactions. Mid-term ATC is calculated using an off-line power flow. The results include daily TTC, non-recallable and recallable ATC for days 8 through 30. With significant changes to forecasted system conditions, new base cases may need to be developed. To develop each base case, besides the off-line load flow base case, the following information is used: load forecast, generator maintenance, generator outages, transmission maintenance, transmission reservation schedules, and so forth. The off-line power flow is also used to calculate long-term ATC. The calculations include monthly TTC, non-recallable and recallable ATC. For the purpose of system reliability, an operations reliability review is conducted to check the calculated non-recallable ATC values.

5.3.2.4 NYISO [21]

The New York (NY) ISO assesses ATC during day-ahead and hour-ahead scheduling and real-time system dispatching. The ATC evaluation is conducted based on base system loading and an assessment of critical contingencies on the transmission system. Base system conditions, including projected customer demand, anticipated transmission system facility availability, accepted energy transactions for the New York control area (NYCA), and information about neighbouring regions that affect the area's ATC, are identified and modelled in the ATC determination. Consistent with NERC principles, the ATC is calculated through the performances of Security Constrained Unit Commitment (SCUC), Security Constrained Dispatch (SCD), and the Balancing Market Evaluation (BME).

5.4 Proposed Stochastic Model for ATC Evaluation

5.4.1 Overview of Proposed Approach

It is highlighted in [5] that one of the most important principles for ATC determination is consideration of the reasonable uncertainties in systems, to ensure the security and reliability of the interconnected network operation under a broad range of changing system conditions. Therefore, before presenting the approach proposed for ATC evaluation in detail, it is important to identify the major uncertain factors that should be paid attention to in measuring ATC. In practice, the uncertainties that affect ATC evaluation accuracy may include load, generation and transmission maintenance schedules and dispatch plans, dynamic line ratings, transactions of third parties, long-term transmission resources planning, and so on. In addition, time horizon is also an important factor affecting the accuracy of the estimated ATC. Undoubtedly, the further into the future that projections of ATC are made, the greater the degree of uncertainty in the assumed conditions and hence, the less the validity of the calculated transfer capability.

Generally, in the scope of short-term operational planning, uncertainties can mainly be attributed to the following factors that cannot be predicted precisely:

1. Generator availability

As stated in [7], generator availability is perhaps the most influential variable on path transfer capability. Specific generators or combinations of generators may have a substantial influence on flows to limiting facilities.

2. Outage of transmission line

From the definition of ATC, it is certain that the outage of transmission lines, particularly of relatively heavily loaded circuits electrically close to the transfer interface of the ATC being studied, may result in considerable decrease in the transfer capability.

3. Customer load demand

It is recognised that, affected mainly by weather conditions and human behaviour patterns, customer load demand is fluctuating all the time. Consequently, load forecast errors are unavoidable. Since ATC is the power transfer capability remaining apart from already committed capacity, which is represented by the forecasted load quantity, load forecast inaccuracy could also have significant impacts on ATC value.

4. Transactions of neighbouring control area

It has been widely observed that uncertainty in neighbouring control area transactions can also have effects on ATC estimation of the studied area [4–18]. In an interconnected meshed transmission network, power distribution of one path could be affected by the loading on other paths, therefore, transfer capabilities in one area may also be increased or decreased by power transfers in other areas.

In this chapter, with respect to the major uncertainties of power systems associated with ATC value, a stochastic model for ATC evaluation is formulated [22]. In the model, availability of generators and circuits are considered as random discrete variables with binomial distribution. In addition, fluctuations of loads are also taken into consideration by the use of normal distributed variables. However, with the introduction of probability distributions for each of the uncertain parameters, particularly continuous random variables, the equivalent deterministic model of a large-scale practical problem usually has huge and intractable dimensions. To deal with the complex problem formulated with both discrete and continuous variables involved efficiently, a hybrid stochastic approach, which takes advantage of both two-stage stochastic programming with recourse (SPR) and chance constrained programming (CCP), is proposed. The necessity to consider the impact of uncertainties on ATC assessment is demonstrated clearly by using the IEEE 118-bus standard system. The numerical results also illustrate the salient performance of the proposed methodology.

5.4.2 Modelling Uncertainties

In the formulated ATC evaluation model, three major uncertain factors are taken into account: generator forced outage, transmission line outage and load variations. According to relevant statistics, the availability of circuit and generator can be considered as random variables following binomial distributions, which can be described by Equations (5.1) and (5.2).

$$p(\tilde{\eta}_l = 0) = p_{\eta_l}$$

$$p(\tilde{\eta}_l = 1) = 1 - p_{\eta_l} \tag{5.1}$$

$$p(\tilde{\xi}_i = 0) = p_{\xi_i}$$

$$p(\tilde{\xi}_i = 1) = 1 - p_{\xi_i} \tag{5.2}$$

where

$\tilde{\eta}_l = 0$: circuit l is out of service;

$\tilde{\eta}_l = 1$: circuit l is in operation;

$\tilde{\xi}_i = 0$: generator i is unavailable;

$\tilde{\xi}_i = 1$: generator i is available;

$p(\bullet)$: probability of the corresponding state occurrence;

variables with superscript ~ represent random parameters.

Generally speaking, load is a random variable that fluctuates around the forecast value. Therefore, under reasonable assumption, the load of bus i, $\tilde{\omega}_i$, can be modelled by a normal distribution variable $N(\mu_{\omega i}, \sigma_{\omega i}^2)$, where the mean of load i, $\mu_{\omega i}$, is the forecast value, and $\sigma_{\omega i}$ is the corresponding standard deviation of actual load value.

5.5 Formulated ATC Evaluation Model

As stated in [5], the ability of an interconnected transmission network to reliably transfer power through prescribed interfaces may be limited by thermal limits, voltage limits and stability limits. As the studies carried out here concentrate on the steady-state operation of power systems, the value is calculated in respect of transmission thermal limits and voltage limits.

According to the definition, ATC is determined as a function of increases in power transfer between different systems through prescribed interfaces. Thus, the ATC evaluation model is formulated to obtain the maximum extra power transfer from current system operating situations, by increasing all complex loads with current power factor using loading factor λ_* , i.e. for load i, $i = 1,\ldots,N_L$,

$$
\begin{aligned}
\tilde{\omega}_{Pi*} &= \lambda_* \tilde{\omega}_{Pi} \\
\tilde{\omega}_{Qi*} &= \lambda_* \tilde{\omega}_{Qi}
\end{aligned}
\tag{5.3}
$$

until line thermal limits or nodal voltage limits are attained, where N_L is the total number of loads. This is usually referred to as a critical situation.

5.5.1 Objective Function

The objective to maximise the uncommitted active transfer capacity of the prescribed interface, which is represented by:

$$
F = \sum_{l \in \Omega} (P_{l*} - P_l) \tag{5.4}
$$

where

Ω : set of all the lines across the studied interface, in which the active power shares the same specific direction;

P_l : active power flow of line l, $l \in \Omega$.

Variables with subscript * represents those of the critical equilibrium point, while variables without subscript * denote those of the current operating point.

5.5.2 Operating Constraints

The constraints are categorised as follows:

- Equality constraints

At the current operating points:

For PQ bus i : $i = 1,\ldots,N_1$

$$\tilde{\xi}_i P_{iG} - \tilde{\omega}_{Pi} - V_i \sum_{j=1}^{N-1} \tilde{\eta}_{Lij} V_j (G_{ij} \cos\theta_{ij} + B_{ij} \sin\theta_{ij}) = 0$$

$$\tilde{\xi}_i Q_{iG} - \tilde{\omega}_{Qi} - V_i \sum_{j=1}^{N-1} \tilde{\eta}_{Lij} V_j (G_{ij} \sin\theta_{ij} - B_{ij} \cos\theta_{ij}) = 0$$

For PV bus i: $i = N_1 + 1, \ldots, N-1$

$$\tilde{\xi}_i P_{iG} - \tilde{\omega}_{Pi} - V_i \sum_{j=1}^{N-1} \tilde{\eta}_{Lij} V_j (G_{ij} \cos\theta_{ij} + B_{ij} \sin\theta_{ij}) = 0 \tag{5.5}$$

On the other hand, the critical operating point corresponding to the loading factor λ_* is expressed as follows:

For PQ bus $i: i = 1, \ldots, N_1$

$$\tilde{\xi}_i P_{iG*} - \tilde{\omega}_{Pi} \lambda_* - V_{i*} \sum_{j=1}^{N-1} \tilde{\eta}_{Lij} V_{j*} (G_{ij} \cos\theta_{ij*} + B_{ij} \sin\theta_{ij*}) = 0$$

$$\tilde{\xi}_i Q_{iG*} - \tilde{\omega}_{Qi} \lambda_* - V_{i*} \sum_{j=1}^{N-1} \tilde{\eta}_{Lij} V_{j*} (G_{ij} \sin\theta_{ij*} - B_{ij} \cos\theta_{ij*}) = 0$$

For PV bus $i: i = N_1 + 1, \ldots, N-1$

$$\tilde{\xi}_i P_{iG*} - \tilde{\omega}_{Pi} \lambda_* - V_{i*} \sum_{j=1}^{N-1} \tilde{\eta}_{Lij} V_{j*} (G_{ij} \cos\theta_{ij*} + B_{ij} \sin\theta_{ij*}) = 0 \tag{5.6}$$

- Inequality constraints:

Nodal voltage limits: (for PQ bus i, $i = 1, \ldots, N_1$)

$$V_{i,\min} \leq V_i, V_{i*} \leq V_{i,\max} \tag{5.7}$$

Line thermal limits: (for line l, $l = 1, \ldots, M$)

$$TM_l, TM_{l*} \leq TM_{l,\max} \tag{5.8}$$

Generator capacity limits: (for generator at bus i)

$$P_{iG,\min} \leq P_{iG}, P_{iG*} \leq P_{iG,\max}$$
$$Q_{iG,\min} \leq Q_{iG}, Q_{iG*} \leq Q_{iG,\max} \tag{5.9}$$

where
N : number of buses;
M : number of lines;
N_1 : number of PQ buses;
N_G : number of generators;
P_{iG}, Q_{iG} : active and reactive power generations at bus i ;
$\tilde{\omega}_{P_i}, \tilde{\omega}_{Q_i}$: active and reactive loads at bus i ;
λ_* : loading factor of the critical situation; it is a scalar parameter representing the increase in load.

Subscript min and max represents the minimum and maximum limits of the variable respectively.

To make the description of the model conversion more concise, we rewrite this model as:

$$P1:\ Max \quad F(\mathbf{X}, \lambda_*, \mathbf{X}_*) = \sum_{l \in \Omega} (P_{l*}(\lambda_*, \mathbf{X}_*) - P_l(\mathbf{X}))$$

$$\text{s.t.: } \mathbf{G}(\mathbf{X}, \tilde{\xi}, \tilde{\eta}) = \tilde{\boldsymbol{\omega}}$$

$$\mathbf{G}_*(\lambda_*, \mathbf{X}_*, \tilde{\xi}, \tilde{\eta}) = \tilde{\boldsymbol{\omega}}\, \lambda_*$$

$$\mathbf{H}_{\min} \leq \mathbf{H}(\mathbf{X}), \mathbf{H}_*(\lambda_*, \mathbf{X}_*) \leq \mathbf{H}_{\max} \tag{5.10}$$

where
X : state vector;
G : power flow equations;
L : operating constraints.

5.6 Proposed Hybrid Stochastic Approach

The principle of all stochastic optimisation techniques is to convert the stochastic problem into an equivalent deterministic model so that conventional optimization methodologies can be applied to find the optimum solution. There are two main stochastic approaches, two-stage stochastic programming with recourse (SPR) and chance constrained programming (CCP) [23], which are introduced in detail in Appendix A and B, respectively.

Two-stage SPR was proposed by Dantzig in 1955 [24]. Theoretically in two-stage SPR, all realisations of random variables should be included to calculate the expected costs. Therefore, one of the major criticisms of SPR is that it leads to an enormous size expansion of the model, which may substantially increase the computational effort, especially for problems with continuous variables, as the general method for them is carried out by discretising the probability distribution. Roughly speaking, for uncertain load fluctuations in the model, with SPR, the number of related variables and constraints has to be multiplied by the number of scenarios, which obviously results in an enormous dimensions.

The CCP was originally proposed to solve problems with chance constraints, which are constraints having finite probability of being satisfied (or violated) [25]. An important advantage of CCP is that the dimension of the formulated deterministic problem is not larger than the original stochastic model. However, because the CCP utilises the cumulative density function of continuous random variables, it can only deal with continuous variables.

Generally speaking, the challenge posed by all applications of stochastic approaches is how to formulate an equivalent deterministic model with as few as possible dimensions. In this respect, the CCP seems to be an ideal method for dealing with stochastic problems. However, for such a model with both discrete and continuous variables involved, the CCP cannot be applied directly. Considering the fact that SPR has the merit of tackling any kind of distribution, in order to solve the model efficiently, a novel hybrid stochastic methodology, which takes advantages of both two-stage SPR and CCP, is proposed to deal with the model conversion. The process of converting the original stochastic model to an equivalent deterministic model is described step by step in the following section.

5.6.1 Application of SPR to Deal with Discrete Variables

In the first phase, based on the two-stage SPR technique, and aimed at the discrete variables involved, the original model $P1$ is transformed into an interim stochastic model $P2$ with only continuous variables, which is delineated as follows:

The fundamental idea behind two-stage SPR is to divide the whole problem into two stages, thereafter, to optimise the objective of the first-stage decision, as well as the cost of the second-stage recourse decision as a consequence of the first-stage decision [24]. For the problem formulated, in stage 1, assuming the intactness of the system and no load forecast error, the "here and now" decision variable λ_* is calculated before the actual values of the random variables are known. While in stage 2, with consideration of the situation resulting from the determined stage 1 decision and the possible realisations, the adjustment of λ_* is taken as the decision variable. Let the discrepancy between $\mathbf{G}(\mathbf{X}, \tilde{\xi}^s, \tilde{\eta}^s)$ and $\tilde{\omega}$ be $|Y|$ and f be the penalty associated with this discrepancy, where the discrete variables with superscript s indicate linkage to corresponding realisations. Then the stage 2 problem is formulated to minimise the expected value of the penalty, which is referred to as the recourse cost Q. Combining the stage 1 and stage 2 problems, the original problem $P1$ formulated is transformed into a two-stage optimisation model $P2$:

$$P2: \text{stage 1: } Max\ F(\mathbf{X}, \lambda_*, \mathbf{X}_*) \tag{5.11}$$

$$\text{s.t.: } \mathbf{G}(X) = E(\tilde{\boldsymbol{\omega}}) \tag{5.12}$$

$$\mathbf{G}_*(\lambda_*, \mathbf{X}_*) = E(\tilde{\boldsymbol{\omega}})\lambda_* \tag{5.13}$$

$$\mathbf{H}_{\min} \leq \mathbf{H}(\mathbf{X}), \mathbf{H}_*(\lambda_*, \mathbf{X}_*) \leq \mathbf{H}_{\max} \tag{5.14}$$

$$\text{stage 2: } Min\ Q = E_{\xi\eta}\left(f(Y^{+s} + Y^{-s}) + f_*(Y_*^{+s} + Y_*^{-s}) \right) \tag{5.15}$$

$$\text{s.t.: } Y^{+s} - Y^{-s} = \mathbf{G}(\mathbf{X}, \tilde{\xi}^s, \tilde{\eta}^s) - \tilde{\boldsymbol{\omega}} \tag{5.16}$$

$$Y_*^{+s} - Y_*^{-s} = \mathbf{G}_*(\lambda_*, \mathbf{X}_*, \tilde{\xi}^s, \tilde{\eta}^s) - \tilde{\boldsymbol{\omega}}\lambda_* \tag{5.17}$$

$$Y^{+s}, Y^{-s}, Y_*^{+s}, Y_*^{-s} \geq 0\ \forall s, s \in Sc \tag{5.18}$$

where $E_{\xi\eta}$ represents the expectation value with respect to $\tilde{\xi}$ and $\tilde{\boldsymbol{\eta}}$. Sc is the set of indices representing all possible discrete outcomes under contingency situations and $S_d = |Sc|$. Normally, in the model formed, if considering the scenarios of two-element outage simultaneously, which includes the situations of two generator outage, one generator one circuit outage and two circuit outage, $S_d = C_M^2 + MN_G + C_{N_G}^2$; when accounting for the scenarios of one element outage at a time, $S_d = M + N_G$.

5.6.2 Application of CCP to Deal with Continuous Variables

There are only continuous variables left in the interim stochastic model $P2$. Aiming at the continuous variables, the second phase is to transform the model to a completely deterministic form. Since $\tilde{\omega}$ follows a normal distribution, according to equation (5.16), it is apparent that Y^{+^s} and Y^{-^s} are also continuous random variables. Considering the different realisations of $\tilde{\omega}$ in any scenario s, minimising the discrepancy between $G(X,\tilde{\xi}^s,\tilde{\eta}^s)$ and $\tilde{\omega}$ is equal to minimizing the maximum value of Y^{+^s} and Y^{-^s}. Therefore, the objective function of stage 2 can be transferred to:

$$Q = E_{\xi\eta}\left(f((\underset{\omega}{Y_{\max}}^{+^s} + \underset{\omega}{Y_{\max}}^{-^s})) + f_*((\underset{\omega}{Y_{*\max}}^{+^s} + \underset{\omega}{Y_{*\max}}^{-^s})) \right) \tag{5.19}$$

It is obvious that with the determined stage 1 decision, for any scenario s, the recourse cost Q is affected directly by the limit values of $\tilde{\omega}$. In respect of this, it is crucial to get the probable maximum and minimum values of $\tilde{\omega}$ first. Theoretically, the normally distributed variable $\tilde{\omega}$ is unbounded. However, in practice it is with small probability to reach its limits. As such, our consideration is only restricted within their respective β confidence intervals, where generally $0.9 \le \beta < 1$. Therefore, to calculate the limits of $\tilde{\omega}$, a chance constraint is introduced to interpret it as satisfying the constraints with a probability, which is:

$$p(\omega_{\min} \le \tilde{\omega} \le \omega_{\max}) = \beta \tag{5.20}$$

The limits of $\tilde{\omega}$ can be calculated using the CCP. The process is described as follows:

First, to standardise the normally distributed variable $\tilde{\omega}$, transform equation (5.20) to (5.21) as below:

$$p\left(\frac{\omega_{\min} - E(\tilde{\omega})}{\sqrt{\mathrm{var}(\tilde{\omega})}} \le \frac{\tilde{\omega} - E(\tilde{\omega})}{\sqrt{\mathrm{var}(\tilde{\omega})}} \le \frac{\omega_{\max} - E(\tilde{\omega})}{\sqrt{\mathrm{var}(\tilde{\omega})}}\right) = \beta \tag{5.21}$$

where $E(\bullet)$ and $var(\bullet)$ represents the expectation value and the variance of a random variable. Because ω_{min} and ω_{max} are symmetrical about $E(\tilde{\omega})$, the constraint can be stated as

$$p\left(\frac{\tilde{\omega} - E(\tilde{\omega})}{\sqrt{\mathrm{var}(\tilde{\omega})}} \le \frac{\omega_{\max} - E(\tilde{\omega})}{\sqrt{\mathrm{var}(\tilde{\omega})}}\right) = \frac{1+\beta}{2} \tag{5.22}$$

and

$$p\left(\frac{\omega_{\min} - E(\tilde{\omega})}{\sqrt{\mathrm{var}(\tilde{\omega})}} \le \frac{\tilde{\omega} - E(\tilde{\omega})}{\sqrt{\mathrm{var}(\tilde{\omega})}}\right) = \frac{1+\beta}{2} \tag{5.23}$$

From Equation (5.22), the probability of realising $\tilde{\omega}$ less than or equal to ω_{max} can be written as:

$$p(\tilde{\omega} \le \omega_{\max}) = \Phi\left(\frac{\omega_{\max} - E(\tilde{\omega})}{\sqrt{\mathrm{var}(\tilde{\omega})}}\right) \tag{5.24}$$

where $\Phi(z)$ represents the cumulative density function of the standard normal variable evaluated at z. If $K_{\frac{1+\beta}{2}}$ denotes the value of the standard normal variable at which $\Phi(K_{\frac{1+\beta}{2}}) = \frac{1+\beta}{2}$, then the constraint can be stated as:

$$\Phi\left[\frac{\omega_{\max} - E(\tilde{\omega})}{\sqrt{\mathrm{var}(\tilde{\omega})}}\right] = \Phi(K_{\frac{1+\beta}{2}}) \tag{5.25}$$

This equation will be satisfied if

$$\omega_{\max} = \sqrt{\mathrm{var}(\tilde{\omega})}\, K_{\frac{1+\beta}{2}} + E(\tilde{\omega}) \tag{5.26}$$

Similarly, transfer Equation (5.23) to a deterministic formulation:

$$\omega_{\min} = -\sqrt{\mathrm{var}(\tilde{\omega})}\, K_{\frac{1+\beta}{2}} + E(\tilde{\omega}) \tag{5.27}$$

With the achieved ω_{min} and ω_{max}, it is easy to obtain formulae for $Y_{\underset{\omega}{\max}}^{+s}$ and $Y_{\underset{\omega}{\max}}^{-s}$ for any scenario s. In a similar way, $Y_{*\underset{\omega}{\max}}^{+s}$ and $Y_{*\underset{\omega}{\max}}^{-s}$ can be derived as well.

Thus, converted from the interim model, a completely deterministic model $P3$, which is essentially equivalent to the original model $P1$, is formed as follows:

$P3$: Max $F(\mathbf{X}, \lambda_*, \mathbf{X}_*)$

$$- E_{\xi\eta}(f(Y_{\underset{\omega}{\max}}^{+s} + Y_{\underset{\omega}{\max}}^{-s}) + f_*(Y_{*\underset{\omega}{\max}}^{+s} + Y_{*\underset{\omega}{\max}}^{-s}))$$

$$\text{s.t.: } \mathbf{G}(X) = E(\tilde{\boldsymbol{\omega}})$$

$$\mathbf{G}_*(\lambda_*, \mathbf{X}_*) = E(\tilde{\boldsymbol{\omega}})\lambda_*$$

$$\mathbf{H}_{\min} \le \mathbf{H}(\mathbf{X}), \mathbf{H}_*(\mathbf{X}_*, \lambda_*) \le \mathbf{H}_{\max}$$

$$Y_{\underset{\omega}{\max}}^{+s} = \begin{cases} G(X, \tilde{\xi}^s, \tilde{\eta}^s) - \omega_{\min} & \text{If } G(X, \tilde{\xi}^s, \tilde{\eta}^s) > \omega_{\min} \\ 0 & \text{If } G(X, \tilde{\xi}^s, \tilde{\eta}^s) \le \omega_{\min} \end{cases}$$

$$Y_{\underset{\omega}{\max}}^{-s} = \begin{cases} -G(X, \tilde{\xi}^s, \tilde{\eta}^s) + \omega_{\max} & \text{If } G(X, \tilde{\xi}^s, \tilde{\eta}^s) < \omega_{\max} \\ 0 & \text{If } G(X, \tilde{\xi}^s, \tilde{\eta}^s) \ge \omega_{\max} \end{cases}$$

$$Y_{*\underset{\omega}{\max}}^{+s} = \begin{cases} -G_*(X_*, \tilde{\xi}^s, \tilde{\eta}^s) + \omega_{\min}\lambda_* & \text{If } G_*(X_*, \lambda_*, \tilde{\xi}^s, \tilde{\eta}^s) > \omega_{\min}\lambda_* \\ 0 & \text{If } G_*(X_*, \lambda_*, \tilde{\xi}^s, \tilde{\eta}^s) \le \omega_{\min}\lambda_* \end{cases}$$

$$Y_{*\underset{\omega}{\max}}^{-s} = \begin{cases} -G_*(X_*, \lambda_*, \tilde{\xi}^s, \tilde{\eta}^s) + \omega_{\max}\lambda_* & \text{If } G_*(X_*, \lambda_*, \tilde{\xi}^s, \tilde{\eta}^s) < \omega_{\max}\lambda_* \\ 0 & \text{If } G_*(X_*, \lambda_*, \tilde{\xi}^s, \tilde{\eta}^s) \ge \omega_{\max}\lambda_* \end{cases}$$

$$Y_{\underset{\omega}{\max}}^{+s}, Y_{\underset{\omega}{\max}}^{-s}, Y_{*\underset{\omega}{\max}}^{+s}, Y_{*\underset{\omega}{\max}}^{-s} \ge 0 \ \forall s, s \in Sc \tag{5.28}$$

The sequential two-step process of converting the formulated stochastic model to a completly deterministic model is illustrated in Figure 5.2. Comparison between the resultant models is shown in Table 5.1. It is noticed that the size of the problem described above is determined by the number of equality constraints. In the original model *P1*, assuming the number of equality constraints is C_o. With unfolding the discrete random variables in the first step, the corresponding constraint number of *P2* is expanded to $C_0 \times (S_d + 1)$. In the second step, with the application of the CCP to eliminate continuous random variables, the number of equality constraints in *P3* is $C_0 \times (2 \times S_d + 1)$. However, if the SPR is applied to deal with all the uncertainties instead of the hybrid method, and each load forecast error is simulated by using a 7-step approximation according to [26], the number of relevant constraints will reach $7 \times N_L \times C_0 \times (S_d + 1)$ in the equivalent deterministic model. This obviously results in enormous dimensions for a practical problem. From the significant difference between the sizes of the equivalent problems, undoubtedly, the hybrid stochastic programming proposed is more progressive than normal stochastic programming.

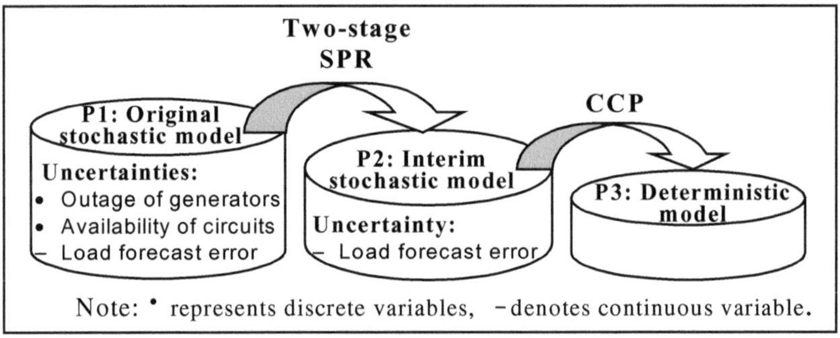

Figure 5.2. Process of converting stochastic model to deterministic model

Table 5.1. Comparison between resultant models

Characteristics		Original stochastic model (P1)	Interim stochastic model (P2)	Deterministic model (P3)
Uncertainty	Discrete variables	Circuit availability Generator outage	/	/
	Continuous variable	Load forecast error	Load forecast error	
Equality constraints number	Hybrid method (SPR+CCP)	C_0	$C_0 \times (S_d + 1)$	$C_0 \times$ $(2 \times S_d + 1)$
	Normal stochastic method (SPR)		/	$7 \times N_L \times C_0$ $\times (S_d + 1)$

5.7 Implementation

Besides the equivalent deterministic model conversion described above, the program involves the development and integration of another two major modules: an AC power flow and an optimisation algorithm for ATC calculation.

With the increase of loading factor, systems ill conditioning will be aggravated, which will result in a long time oscillation before convergence during the power flow calculation process, sometimes leading to an unsolvable system.. As stated in [27], an improved optimal multiplier Newton–Raphson (OMNR) method performs more vigorously in handling ill-conditioned power systems than the normal NR power flow method and the original OMNR algorithm, it is employed to solve the power flow calculation involved.

It is widely recognised that, compared with the conventional Simplex method for large-scale LP problems, the predictor-corrector primal-dual interior point linear programming (PCPDIPLP) is much faster and more reliable for achieving feasibility and convergence [28]. It leads to significant reduction in the number of iterations, especially for practical problems that usually have enormous dimensions. Therefore, the method is used to solve the equivalent deterministic model of the stochastic ATC evaluation problem. The non-linear objective and constraints must be piecewise linearised to utilise the LP algorithm to calculate the incremental values of the loading factor. An iterative procedure is needed.

The overall procedure is sketched in Figure 5.3, where ε is a given threshold value for convergence.

5.8 Case Studies and Interpretation of Results

A series of tests are conducted on various systems to investigate the effectiveness of the proposed method and computation efficiency. In this section, case studies of the IEEE 118-bus system with 200 branches are presented [29]. The studied period of ATC evaluation is focused on one day. The hourly peak load curve for a typical winter day is shown in Figure 5.4. The corresponding computations of ATC are performed hourly. First, power flow is calculated with the daily peak load, which is at hour 18-19.

According to the structure of the network and the power flow results, the whole system is divided into two zones. The studied interface for ATC calculation is shown in Figure 5.5. This also indicates that the West-East transmission corridor is very important, with a huge amount of active power transferred.

The following criteria and assumptions are applied in the case study:

- All nodal voltages to be within the range 0.90-1.10 p.u. under normal and contingency situations.
- For simplicity, assume independence between random variables. Since practically There is a very small probability of two-element contingency simultaneously, ATC calculations are based upon a single contingency

situation only, that is, no more than one generator or circuit outage happens in one hour.

- Under reasonable assumption, forced outage rates (FORs) of units in the system are all 0.01. And the FOR of all circuits is 0.005. Deviation of load forecast error is 2% of the mean [26].
- Line thermal limits and active generation limits are not available for the IEEE 118-bus system. Under reasonable assumptions, thermal limits are given in Appendix C. Based on realistic situations, the generation limit of each unit is assumed as 140% of the corresponding current output.

Taking into consideration the major uncertain factors, ATC of the studied interface is calculated as case 1. For comparison, ATC is also computed without considering the uncertainties as case 2. Both results are given in Table 5.2. From the considerable difference of the ATC values for the two cases, which is 9.37% of the actual ATC value, it is evident that neglecting the effects of uncertainties on ATC evaluation inevitably will lead to system operating risk.

It is to be noted that the uncertainty parameters themselves also vary depending on weather conditions and maintenance situations etc. To study the effects of varying uncertain operating conditions, in case 3, the FORs of circuits and generators are doubled, as well as the deviations of load forecast error. The ATC value calculated is also given in Table 5.2. The results demonstrate that the ATC value is highly dependent on system conditions. Therefore, knowledge of the impacts of the variations of uncertainties on ATC value can provide important information, hence giving more confidence in making decision on power transaction scheduling and delivery.

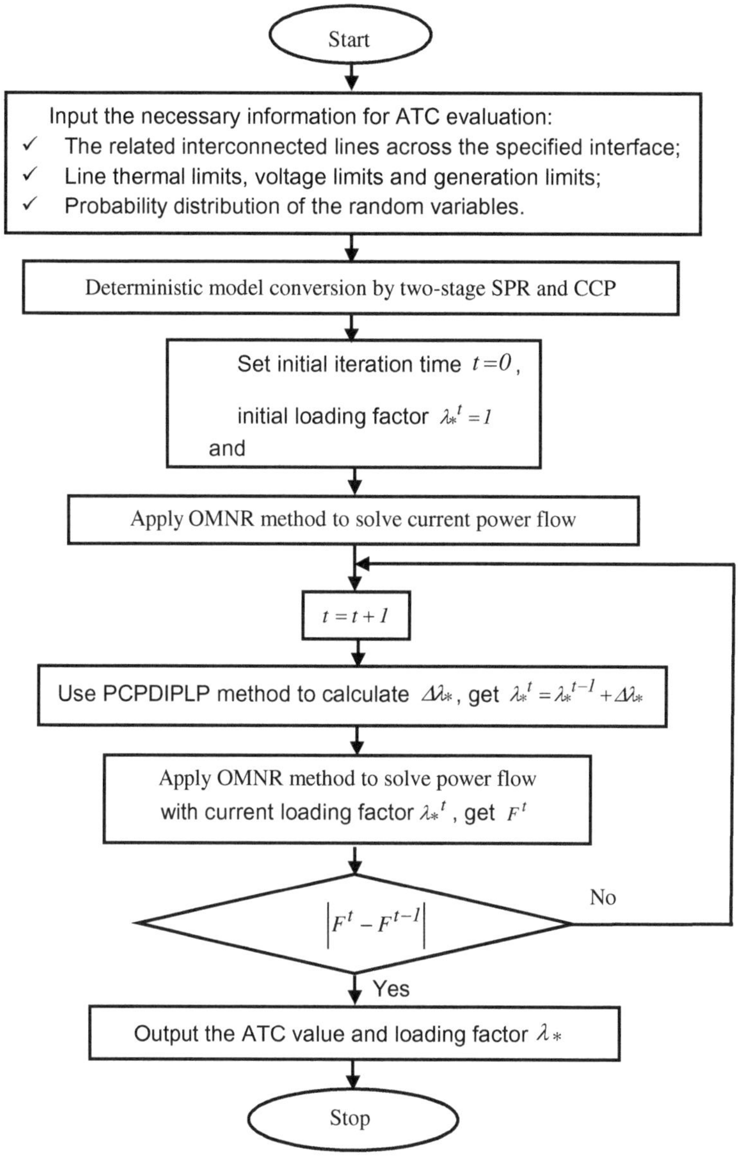

Figure 5.3. Flowchart of proposed approach

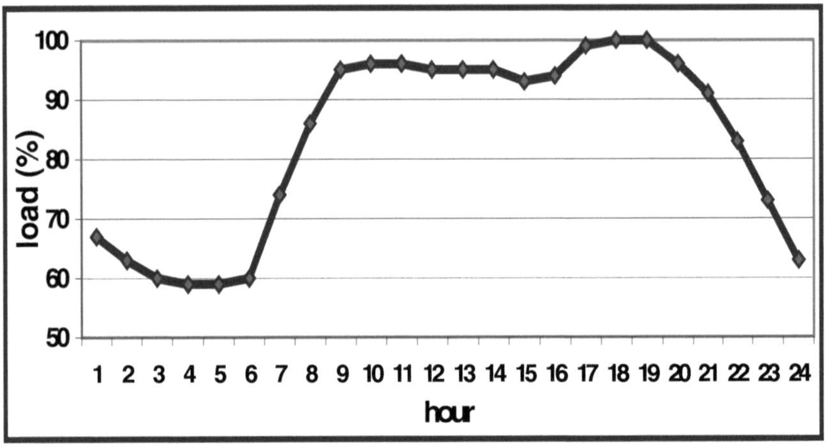

Figure 5.4. Hourly peak load curve as percentage of daily peak for typical winter day

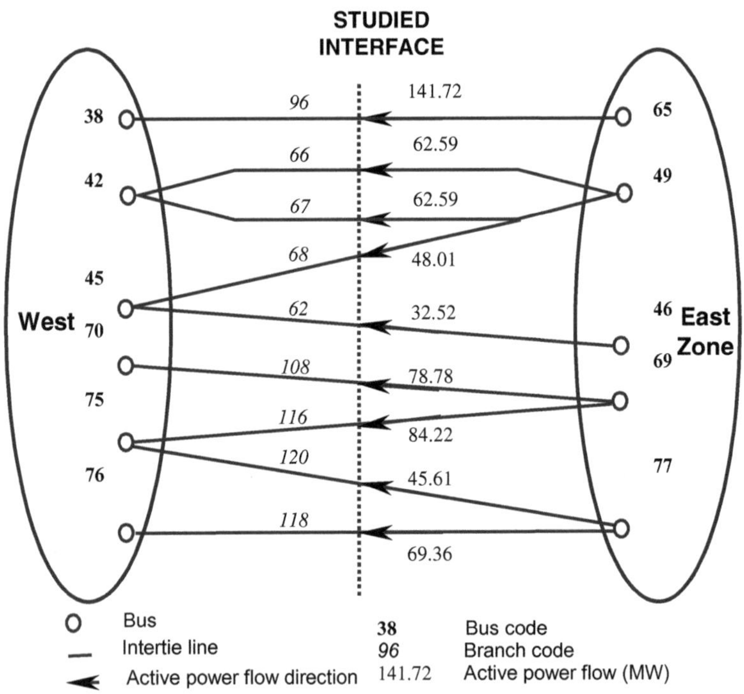

Figure 5.5. Studied interface of IEEE 118 - bus system

Table 5.2. Results of ATC calculation

Case	Description	ATC (MW)	λ_*	CPU time (s)
1	With consideration of the uncertainties	201.92	1.250	103
2	Without consideration of uncertainty	220.84	1.280	19
3	With consideration of the uncertainties	194.41	1.232	73

The proposed method has been coded in Fortran and run on a Pentium II 400MHz PC. CPU times for the ATC evaluations in Table 5.2 show the efficiency of the hybrid method proposed.

Voltage profiles and main branches thermal burden profiles of the system under current and critical situations in case 1 are shown in Figures 5.6 and 5.7 respectively. It is perceivable that the ATC of the studied interface is restricted by the voltage of bus 76, and possibly bus 118 is another relatively weak point that will prevent the ATC value from increasing further. Clearly, reactive power compensation on these weak buses will be effective in the enhancement of the ATC value. Thermal loading factor profiles of the interconnected lines across the studied interface under both situations are depicted in Figure 5.8. Comparing the relevant profiles in Figures 5.7 and 5.8, it is observed that, concerning thermal burden of the studied interface, there is still plenty of space for intensive commitment of the interconnected circuits. Therefore, it is predictable that with appropriate voltage support solutions, the ATC value can be boosted to a great degree.

Finally, apart from the daily peak load, ATC values of other loads are calculated hourly based on the load curve given above. The daily curve of the ATC value achieved is shown in Figure 5.9.

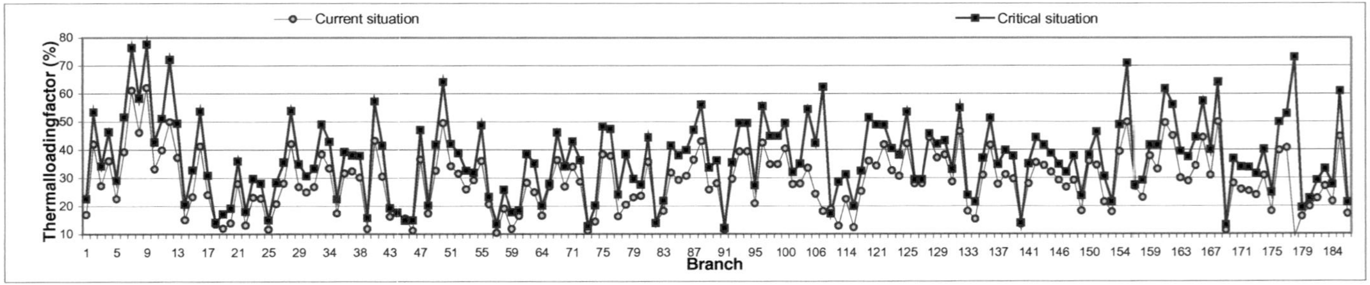

Figure 5.6. Voltage profiles of test system under current and critical situations (note: PV buses and slack bus are omitted here).

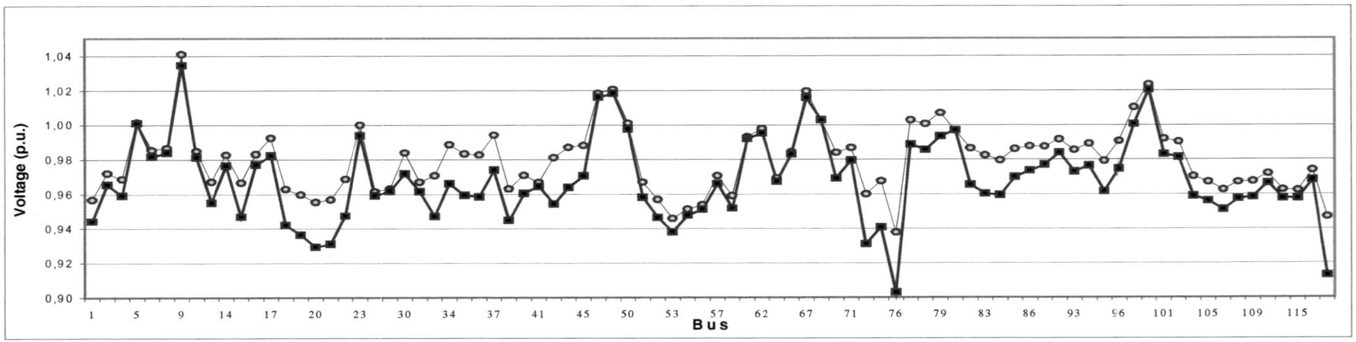

Figure 5.7. Thermal loading factor profiles of test system under current and critical situations.

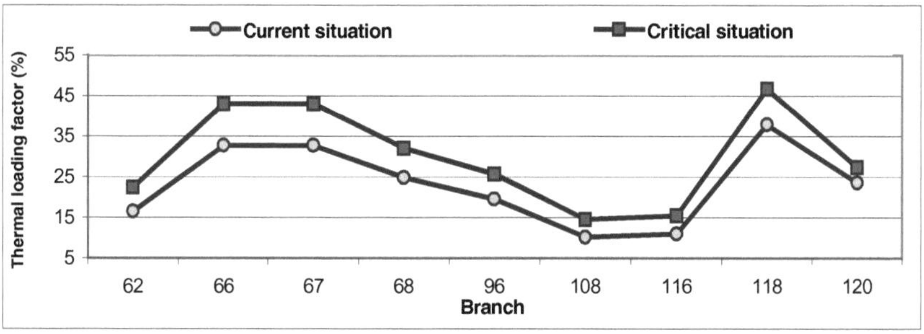

Figure 5.8. Thermal loading factor profiles of interconnected lines across studied interface

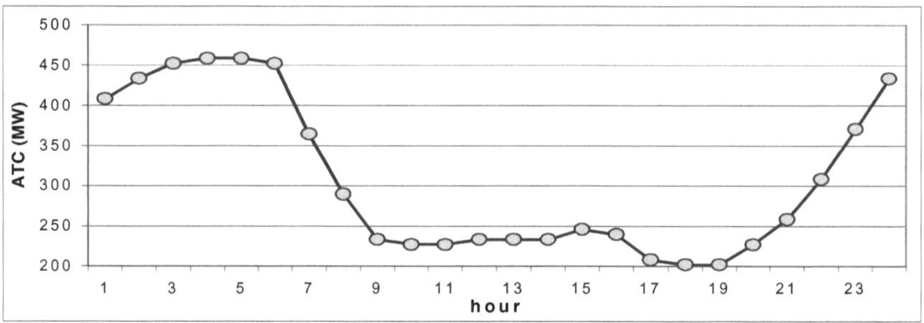

Figure 5.9. Hourly ATC value curve for typical winter day

5.9 Conclusions

With the recent advent of world-wide power industry deregulation, measuring and modelling transmission system transfer capability is assuming greater importance. Transmission capability, gauging the ability of the transmission network to carry through electric power, is a major issue of the electricity market. Realistic and accurate information about the ATC allows ISOs to operate power systems safely, reliably and at higher utilisation levels.

In this chapter, first of all, to give the reader some insight into the terms, the history of standard terms of transfer capability of transmission networks has been reviewed, together with definitions and analysis of their weaknesses. Definitions of the ATC and related terms, including TTC, TRM and CBM and their relationships are presented. Industrial applications of the ATC to transmission services in electricity market have been summarised. On that basis, principles and rules for the calculation and application of the ATC have been identified. Essential criteria for ATC evaluation that should be paid significant attention have also been analysed in detail. Existing methodologies in the literature for ATC evaluation have been reviewed and discussed, together with ATC evaluation software and procedures employed practically in today's power industry. To get accurate and dependable ATC measures, major uncertain factors associated with the ATC estimation have been identified and analysed.

For short-term operational planning purposes, this chapter focuses on the development of a practical approach to evaluate the steady-state ATC of interconnected systems. The ATC value is calculated in respect of current loading conditions and to ensure there are no resultant thermal and voltage limit violations in the interconnected system. To reflect the physical realities of the transmission network, a stochastic model is formulated for ATC calculation, where key uncertainties affecting the level of ATC are modelled explicitly. To deal with the complex problem, with both discrete random variables and continuous random variables involved, efficiently, the solution methodology rests on a novel hybrid stochastic programming formulation, which takes advantages of both two-stage SPR and CCP. From the size comparison of the equivalent problem resulting from the proposed hybrid method and the SPR, it is seen clearly that with substantial size reduction of the transformed deterministic model, the hybrid stochastic programming proposed is more progressive than normal stochastic programming. The results on the IEEE 118-bus system are encouraging, and clearly illustrate the effectiveness and efficiency of the proposed methodology.

Further computational time saving can be accomplished by applying screening techniques to eliminate scenarios with negligible effect on ATC value. Application of decomposition techniques, such as Benders decomposition, is another effective way to improve the execution time by splitting the problem into smaller size independent sub-problems, which can be tackled efficiently using parallel techniques.

It needs to be pointed out that, due to the close relationship between ATC and TTC, the formulated ATC evaluation model can be employed to calculate the corresponding TTC level as well. In addition, the proposed hybrid methodology can be adapted to deal with any other complex stochastic problem with both discrete and continuous random variables involved. In the field of power systems, the method may be employed to solve problems such as optimal bidding strategy, reserve dispatch, unit commitment, hydro-thermal coordination, long-term transmission planning and so forth.

5.10 References

[13] Order No. 888, Promoting wholesale competition through open access, nondiscriminatory transmission services by public utilities, FERC, April 24, 1996.

[14] Notice of Proposed Rulemaking (NOPR), Docket RM95-8-000, Section III-E4f, March 29, 1995.

[15] Transmission transfer capability - A reference document for calculating and reporting the electric power transfer capability of interconnected electric systems, NERC, May 1995.

[16] M. Merrill Hyde, Probabilistic available transfer capability, IEEE PES Winter Meeting, Tampa, FL, Feb. 4, 1998.

[17] North American Electric Reliability Council (NERC), Available transfer capability definition and determination, June 1996.

[18] Determination of ATC within the Western Interconnection, June 2001.

[19] PJM manual for transmission service request, Manual M-02, PJM Interconnection, L.L.C., March 3, 2001.
[20] Y. Xiao, "Operational optimization of FACTS control for improving transmission services", Ph.D. Thesis, Brunel University, London, 2001.
[21] G. C. Ejebe, J. G. Waight, M. Santos-nieto, W. F. Tinney, "Fast calculation of linear available transfer capability", Proceedings of the 21st International Conference on Power Industry Computer Applications, 1999, pp. 226-232.
[22] J. Z. Tong, "Real time transfer limit calculation", Proceedings of the 2000 IEEE/PES Summer Meeting, Seattle, USA, July 2000.
[23] B. S. Gisin, M. V. Obessis, J. V. Mitsche, "Practical methods for transfer limit analysis in the power industry deregulated environment", Proceedings of the 21st International Conference on Power Industry Computer Applications, 1999, pp. 232 - 238.
[24] G. C. Ejebe, J. Tong, J. G. Waight, J. G. Frame, X. Wang, W. F. Tinney, "Available transfer capability calculations", IEEE Trans. on Power Systems, Vol.13, No.4, Nov. 1998, pp. 1521-1527.
[25] X. Luo, A. D. Patton, C. Singh, "Real power transfer capability calculations using multi-layer feed-forward neural networks", IEEE Trans. on Power Systems, Vol.15 No.2, May 2000, pp.903 –908.
[26] D. Gan, R. J. Thomas, R. D. Zimmerman, "Stability-constrained optimal power flow", IEEE Trans. on Power Systems, Vol.15, No.2, May 2000, pp. 535 –540.
[27] R. A. Rosales, D. Ruiz-Vega, D. Ernst, M. Pavella, J. Giri, "On-line transient stability constrained ATC calculations", Power Engineering Society Summer Meeting, 2000. IEEE, Vol. 2, 2000, pp. 1291-1296.
[28] Xia Feng, A. P. Sakis Meliopoulos, "A methodology for probabilistic simultaneous transfer capability analysis", IEEE Trans. on Power Systems, Vol.11, No.3, Aug. 1996, pp. 1296-1278.
[29] J. C. O. Mello, A. C. G. Mello, S. Granville, "Simultaneous transfer capability assessment by combining interior point methods and Monte Carlo simulation", IEEE Trans. on Power Systems, Vol.12, No.2, May 1997, pp. 736-742.
[30] M. H. Gravener, C. Nwankpa, "Available transfer capability and first order sensitivity", IEEE Trans. on Power Systems, Vol.14, No.2, May 1999, pp. 512-518.
[31] EPRI, "Solving the Transfer Capability Puzzle", Grid Operations & Planning News, Sept. 1997.
[32] ATC Calculation/Coordination Procedural Manual, 02-ATCP-60, East Central Area Reliability Coordination Agreement (ECAR), January 2002.
[33] NYISO manual for transmission services, The New York Power Pool/New York Independent System Operator, Sept. 21, 1999
[34] Xiao Ying, Y. H. Song, Y. Z. Sun, "A hybrid stochastic approach to available transfer capability evaluation", IEE Proc.- Generation, Transmission and Distribution, Vol.148, No.5, Sept. 2001.
[35] P. Kall, S. W. Wallace, "Stochastic Programming", John Wiley and Sons, 1994.
[36] G. B. Dantzig, "Linear programming under uncertainty", Management Science, Vol.1, 1955, pp197-206.
[37] A. Charnes, W. W. Cooper, "Chance constrained programming", Management Science, Vol.6, 1959, pp. 73-79.
[38] Billinton Roy, N. Allan Ronald, "Reliable Evaluation of Power Systems", Plenum Press, 2nd edn, 1994.
[39] Xiao Ying, Y. H. Song, "Power flow studies of a large practical power network with embedded FACTS devices using improved optimal multiplier Newton-Raphson method", European Trans. on Electrical Power (ETEP), Vol. 11, No. 4, July/August 2001.

[40] S. Mehrotra, "On the implementation of a primal-dual interior point method", SIAM Journal on Optimization, Vol.2, 1992, pp.575-601.
[41] IEEE 118-bus system, Power systems test case archive, University of Washington, USA, http://www.ee.washington.edu/research/pstca/.

Appendix A: Stochastic Programming with Recourse (SPR)

First, consider a general stochastic linear programming problem:

$$\min \mathbf{c}^T \mathbf{x}$$

$$\text{s.t.:} \quad \mathbf{Ax} = \mathbf{b}$$

$$\mathbf{x} \geq 0 \tag{A.1}$$

where $\boldsymbol{A} \in \Re^{m \times n}; \boldsymbol{c}, \boldsymbol{x} \in \Re^{n}; \boldsymbol{b} \in \Re^{m}$. $\boldsymbol{A}$, $\boldsymbol{b}$ and $\boldsymbol{c}$ can all be treated as random parameters with known probabilities.

The classical SP model uses the two-stage decision process and is known the stochastic linear programming with recourse, which is stated as:

$$\min (\mathbf{c}^T \mathbf{x} + E_{\omega} Q(\mathbf{x}, \boldsymbol{\omega}))$$

$$\text{s.t.:} \ \mathbf{Ax} = \mathbf{b} \tag{A.2}$$

$$\mathbf{x} \geq 0$$

where $Q(\mathbf{x}, \boldsymbol{\omega}) = \min \mathbf{f}(\boldsymbol{\omega})\mathbf{y}$

$$\text{s.t.:} \ \mathbf{D}(\boldsymbol{\omega})\mathbf{y} = \mathbf{d}(\boldsymbol{\omega}) + \mathbf{B}(\boldsymbol{\omega})\mathbf{x}$$

$$\mathbf{y} \geq 0, \boldsymbol{\omega} \in \Omega \tag{A.3}$$

where $\boldsymbol{\omega}$ denotes a realisation of the uncertain parameters with the corresponding probability $p(\boldsymbol{\omega})$ in the probability space $(\boldsymbol{\Omega}, \boldsymbol{P})$. E_{ω} represents the expectation value with respect to $\boldsymbol{\omega}$. The recourse function $Q(\boldsymbol{x}, \boldsymbol{\omega})$ is defined in (A.3), which shows clearly that it depends both on the first-stage decision $\boldsymbol{x}$ and on the random event ω . The technology matrix $D(\omega)$, the right-hand side vector $d(\omega)$, the interstage linking matrix $B(\omega)$ and the objective function coefficients $f(\omega)$ are all random.

The translation of the model formulated above is: as illustrated in (A.2), the first stage decision x is made only with knowledge of the distribution of the random parameters, which is referred to as the "here-and-now" decision, while the second-stage decision $y(\omega)$ is taken to minimise the recourse cost as shown in Equation (A.3), after the random scenarios ω are observed, hence the "wait-and-see" approach. This constraint can be thought of as requiring some action to correct the system after the random event occurs. The decision process of the two-stage SPR is illustrated in Fig. A.1 using the form of a scenario tree.

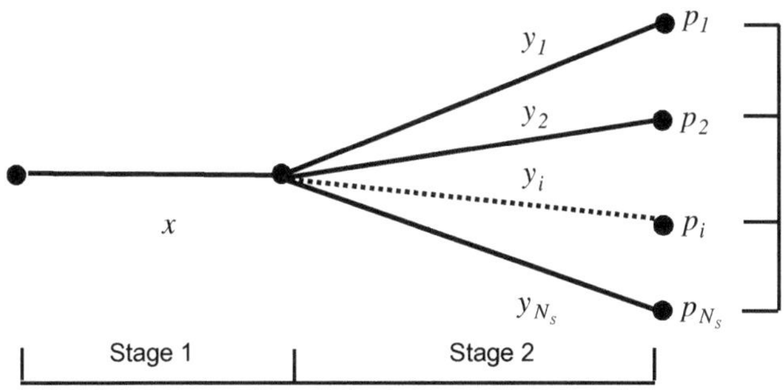

Figure A.1. Scenario tree for two-stage SPR

For simplicity, assuming only discrete random variables involved in the two-stage SPR model, as shown in Equations (A.2) and (A.3), the deterministic equivalent of equation (A.1) can be further formulated as:

$$\min \mathbf{c}^T \mathbf{x} + \sum_{i=1}^{N_S} p_i(\omega) f_i(\omega) y_i$$

$$\text{s.t.: } \mathbf{Ax} = \mathbf{b}$$

$$-B_i(\omega)x + D_i(\omega)y_i = d_i(\omega), \quad i = 1, \ldots, N_S$$

$$\mathbf{x} \geq 0$$

$$y_i \geq 0, \quad i = 1, \ldots, N_S \tag{A.4}$$

where N_S is the number of scenarios and p_i is the probability of the occurrence of scenario i. Following this process, a deterministic model that is equivalent to the original stochastic model can be achieved. Thereafter, conventional optimisation methods can be applied to solve this problem.

Appendix B: Chance-constrained Programming

Assuming the i th constraint of the stochastic linear programming, as illustrated in (A.1), is a chance constraint, which has to been satisfied with a probability of at least β_i, as stated as follows:

$$p\left[\sum_{j=1}^{n} a_{ij}x_j \le b_i\right] \ge \beta_i \tag{B.1}$$

Generally, the CCP can deal with the problem where a_{ij} and b_i are all random variables. For better understanding the basic idea of the CCP, only the special case where a_{ij} is a random variable is introduced here, with further reasonable assumption that the random variable is normally distributed with known mean and standard deviations. The process for equivalent deterministic formation converting is described as follows:

Let $E(a_{ij})$ and $var(a_{ij})$ be the mean and the variance of a_{ij}. Define quantities d_i as:

$$d_i = \sum_{j=1}^{n} a_{ij}x_j \tag{B.2}$$

Since a_{ij} are normally distributed, and x_j are constants to be determined, d_i will also be a normal variable with a mean value of:

$$E(d_i) = \sum_{j=1}^{n} E(a_{ij})x_j \tag{B.3}$$

and a variance of

$$\mathrm{var}(d_i) = \mathbf{x}^T\ \mathrm{var}_i\ \mathbf{x} \tag{B.4}$$

For simplicity, assuming a_{ij} and a_{ik} are unrelated, for any $j \le n, k \le n, j \ne k$, that is the covariance between the any two random variables is 0. Therefore

$$\text{var}_i = \begin{bmatrix} \text{var}(a_{i1}) & 0 & 0 & 0 \\ 0 & \text{var}(a_{i2}) & 0 & 0 \\ 0 & 0 & \dots & 0 \\ 0 & 0 & 0 & \text{var}(a_{i2}) \end{bmatrix} \tag{B.5}$$

The constraint of Equation (B.1) can be expressed as

$$p[d_i \le b_i] \ge \beta_i \tag{B.6}$$

i.e.

$$p\left[\frac{d_i - E(d_i)}{\sqrt{\text{var}(d_i)}} \le \frac{b_i - E(d_i)}{\sqrt{\text{var}(d_i)}}\right] \ge \beta_i \tag{B.7}$$

Therefore, $\frac{d_i - E(d_i)}{\sqrt{\text{var}(d_i)}}$ can be regarded as a standard normal variable. Thus, the probability of realising d_i less than or equal to b_i can be written as:

$$p[d_i \le b_i] = \Phi\left(\frac{b_i - E(d_i)}{\sqrt{\text{var}(d_i)}}\right) \tag{B.8}$$

where $\Phi(z)$ represents the cumulative distribution function of the standard normal variable evaluated at z. If $K_{\beta i}$ denotes the value of the standard normal variable at which

$$\Phi(K_{\beta i}) = \beta_i \tag{B.9}$$

then the constraint can be stated as:

$$\Phi\left(\frac{b_i - E(d_i)}{\sqrt{\text{var}(d_i)}}\right) \ge \Phi(K_{\beta i}) \tag{B.10}$$

This equation will be satisfied if

$$\frac{b_i - E(d_i)}{\sqrt{\text{var}(d_i)}} \ge K_{\beta i} \tag{B.11}$$

By substitute Equations (B.3) and (B.4) into Equation (B.11), we obtain the corresponding deterministic equivalent constraint:

$$\sum_{j=1}^{n} E(a_{ij})x_j + K_{\beta i}\sqrt{\mathbf{x}^T \, \mathrm{var}_i \, \mathbf{x}} - b_i \leq 0 \tag{B.12}$$

Appendix C: Line Thermal Limits of IEEE 118-bus System

Table C.1. Assumed line thermal limits of the IEEE 118-bus system

Line Code	Thermal limits ($\times 10^2$ MVA)
1-6, 10-20, 22-30, 34, 35, 37, 39-49, 52, 53, 55-89, 91, 92, 100, 101, 103-106, 109-115, 117-122, 125, 126, 128-133, 135, 136, 140, 143-162, 164-182, 184-186	2.0
7, 9, 21, 31, 33, 38, 50, 90, 94, 96-99, 108, 116, 123, 124, 137-139, 141, 142, 163, 183	8.0
8, 32, 36, 51, 54, 93, 95, 102, 107, 127, 134	10.0

6. Transmission Congestion Management

X. Wang, Y.H. Song and Q. Lu

The transmission network plays a key role in making the competitive electricity market work and the first important step in any power industry restructuring is open access to transmission. Transmission services have been unbundled from generation to form separate businesses, however, when regarded as a natural monopoly the transmission sector remains more or less regulated to permit a competitive environment for generation and retail services. The operating and planning of the transmission network and the pricing of transmission services are still seen as challenges in the development of electricity markets on both theoretical and practical grounds.

Transmission congestion can be defined as the condition that occurs when there is insufficient transmission capability to simultaneously implement all preferred transactions in electricity markets. Unlike many other commodities, electricity cannot easily be stored and the delivery of electricity is constrained by physical transmission limits, which have to be satisfied at all times to keep the power system operating in a secure fashion. Without transmission limits, deregulation of the power industry would be much simpler. Therefore, transmission congestion management is a major function of any type of Independent System Operator (ISO) in any type of electricity market. It is important to note that if not managed properly, congestion can impose a significant barrier with respect to trading electricity.

This chapter presents a comprehensive review of transmission congestion management methodologies in electricity markets. Covered topics include: the transmission capacity reservation through an auction of Financial Transmission Rights (FTR), which will be discussed in Chapter 8; a framework for real-time system redispatch and congestion management, which has been proposed in Chapter 4, where the different strategies for bilateral contract curtailment are discussed as well. Finally, based on this framework, a Lagrangian Relaxation decomposition approach to active power congestion management across multiple regions is presented in this chapter.

6.1 General Methodologies for Congestion Management

Different market structures and market rules lead to different methods for congestion management [9–10]. Basically, a proper approach to resolving transmission congestion in competitive electricity markets should at least have the following features:

- Fair and non-discriminatory. For the same service, different users should pay the same price and should be treated equally.
- Economically efficient. Individual behaviours of generation, demand and transmission operators should lead to the system optimum through relevant incentives, which could involve cost-reflective charges.
- Transparent and non-ambiguous. The whole congestion management process should be clear to every market participant and should have enough consideration to prevent market gaming. Moreover, simplicity is essential for all the players to understand the rules.
- Feasible. Congestion management processes must always have a feasible solution to maintain system reliability.
- Compatible with various types of contracts. The contracts will include contracts in spot market, real-time balancing market, ancillary services market and contracts in short-term/long-term bilateral markets.

There are at least two main purposes for the transmission congestion management:

- Adjust the preferred transactions to keep the power system operating within its security limits.
- Collect congestion charges from market participants and pay them to transmission grid owners to compensate their investment on the grid.

In this section, three fundamental methods for congestion management will be discussed in detail. They are transaction curtailment, transmission capacity reservation and system redispatch.

6.1.1 Transaction Curtailment

The operation of an electricity market requires sufficient information about generation and loads within each control area as well as between them, in both day-ahead and real-time periods. Without this information the ISO cannot inform market participants of the possible transmission constraints and cannot apply curative actions. The published Total Transfer Capability (TTC) and Available Transfer Capability (ATC) is the minimum information required for market participants to evaluate the risks of seeing their transaction curtailed. ATC summarises complex information and in many cases may provide ambiguous information, even though ATC is based on a clear definition. Therefore, an over-simplified use of this concept could lead to misunderstanding market participants. Furthermore, to help market agents manage the risk of transaction curtailment, ATC publications could include statistical uncertainty of published values and some other dependent information.

This method needs a set of priority rules to curtail transactions when the ATC values are reached. The three common rules are as follows:

- *Pro rata rationing*. In this rule no real priority is defined. All transactions are carried out but the ISO curtails them in the case of congestion, according to the ratio: existing capacity/requested capacity. This rule is transparent but involves the participants in an economically inefficient use of the system.
- *Contribution based on physical flow*. The ISO calculates the contribution of each transaction to the congestion to define its priority. The relative contribution to a transaction is the ratio between the flow induced by the transaction on the congested line and the volume of the transaction. The transactions will be curtailed in accordance with this rank untill congestion disappears. This rule is also transparent, but it is not a market-based method. Its long-term efficiency is not ensured, because this physical contribution factor varies with topology, generation and load patterns.
- *Willing-to-pay*. Transactions submit a price signal to the ISO to show how much they are willing to pay to ameliorate transmission curtailment imposed by the ISO during periods of congestion.
- [11] The ISO picks out some transactions to curtail starting from the one with the lowest will-to-pay bid price. It is a market-based rule for transaction curtailment, because it complies with the principle of "allocating the transmission capacity to the users who value it most highly". However, this method may not be efficient for curtailment against published ATC. More likely, it will be combined with some other congestion management approaches.

With published ATC values and submitted demands for transmission services, the ISO can reject transactions that would cause overloads with a priority rule. One advantage of this curtailment method is that no additional costs are incurred, and therefore there is no cost allocation mechanism. The main drawback is that transaction curtailment based on ATC publication does not convey any economic incentive to the ISO, generators, retailers, or final consumers, and therefore does not promote efficient trade.

6.1.2 Transmission Capacity Reservation

When transmission becomes a scarce resource in the electricity markets, a natural approach to deal with transmission congestion is to allocate the limited transmission capacity in advance to those users who value it best [13]. Auction could be the basic market mechanism for transmission capacity reservation. In a transmission rights auction market, each transmission user submits a price for use of transmission. The bids are selected from the highest one to the lowest one until the capacity is completely used up. The clearing price of the transmission market is calculated and all the participants pay at this price. In some circumstances, the counter-flowing transactions should be paid since they contribute to relieve the congestion.

The auction is efficient regarding competition. The bids exactly reflect the real market value as perceived by participants, which the highest priority for access being granted to the one who is ready to pay the highest price. This method allows the integration of long-term contracts with bilateral or even spot markets. On the other hand, auctions imply additional complexity when a transaction is involved in more than one instance of congestion or when parallel flows are severe. In these cases, transmission users will have to make bids for each bilateral transaction.

An alternative to transmission capacity reservation is the "first come, first served" method. The first reservation made for a given period of time has priority over the following reservations. This method encourages participants to make longer forecasts. Thus, it allows better and faster security assessment for the ISO who knows accurately the volume of exchanges in advance. However, this method may not leave enough room for short-term trading, which is required to ensure successful market dynamics. This method is well suited to bilateral trades, but fails to provide an efficient priority mechanism for day-ahead or real-time pool transactions.

6.1.3 System Redispatch

System redispatch occurs when a central operator directs generation adjustments (incremental or decremental) to relieve congestion and avoid undesired transaction interruptions [11,12]. The cost of these adjustments may be allocated to the responsible participants with either established tariffs or equal shares among all the participants. This allows the transmission users to "buy through" the congestion without the need to enter into contracts with other parties for the redispatch. Financial instruments may be developed to provide transmission users with the opportunity to hedge against the possible high cost of congestion management.

System redispatch is a real-time centralised method for congestion management. It is necessary because the bulk power transmission grid is highly dynamic and predicting constraints well ahead of time is therefore difficult. The main advantage of this method is that due to the centralised nature of the dispatch, no delay occurs between the identification of a constraint and the implementation of redispatch to satisfy the constraint. Furthermore, employing a bid-based auction makes it a market-based system and the congestion management is therefore accomplished based upon market participants' offers and their willingness to buy through congestion to protect their transactions.

Almost all the market participants can contribute to system redispatch.

- Generators. Generators submit incremental/decremental adjustment bids to the ISO. The generators selected by the ISO to relieve transmission congestion will get paid for their contributions.
- Consumers. Some adjustable loads can also provide redispatch service and profit from it. Demand elasticity against spot price signals is an efficient way to alleviate congestion.

- Transmission companies. Many transmission devices, like transformers, FACTS devices, and reactors/capacitors, can be controlled by the ISO to eliminate congestion. Conventionally, these actions are regarded as free resources for system security. However, in a competitive market environment they need to be priced properly to encourage transmission companies to improve the power grid.
- Power marketers. Power markets can adjust their transactions according to the ISO's redispatch commands.

6.1.4 Overall Congestion Management Process

In order to efficiently manage transmission congestion in the real-world market, participants must have freedom to engage in various mechanisms to protect their business. The best solution might always be a combination of several of the basic methods for different time scales. In general, the overall transmission congestion management process, which is shown in Figure 6.1, could consist of three major steps:

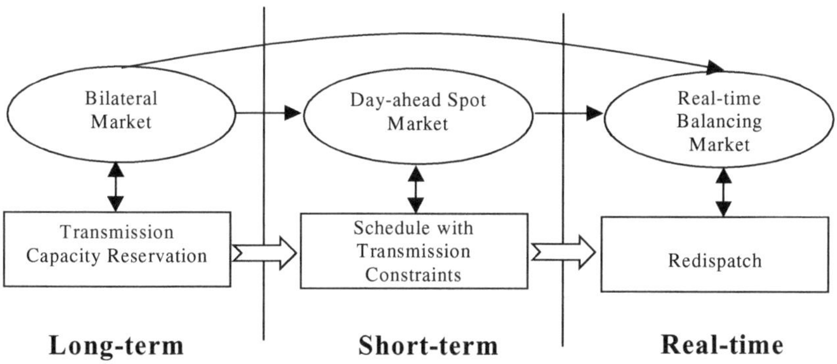

Figure 6.1. Overall congestion management process

1. Long-term transmission capacity reservation - The reservation of transmission capacity can be made yearly, monthly, weekly or even daily. The latest stop time for transmission capacity auction should be no later than the morning of day-ahead. Users can obtain the transmission rights from the ISO through a centralised auction or exchange them through a secondary bilateral market. The transmission rights could be either physical or financial. After obtaining transmission rights, market participants can create new or revise existing bilateral transactions.
2. Short-term scheduling in the day-ahead spot market - The day-ahead schedules are made with consideration of transmission constraints. The input data will include all the signed bilateral contracts and all the generation offers and demand bids in the spot market. Some bilateral contracts might be curtailed if congestion occurs.

3. Real-time redispatch in the real-time balancing market - Because of unpredictable events and fluctuating loads, even after the two previous steps, transmission congestion may still occur during the real-time operation of the market. A centralised balancing mechanism is needed to relieve real-time congestion problems. Although real-time redispatch is also a market-based method, the ISO can take any mandatory actions to maintain system security in emergency cases.

6.2 International Comparison of Congestion Management Approaches

Before any further research of transmission congestion management is discussed, it is very important to note what has been happening in the real world by analysing existing congestion management approaches in some of the major electricity markets across the world. This section will compare congestion management methodologies in five typical electricity markets, England and Wales (UK), PJM (US), California (US), Nord Pool (Norway and Sweden) and New Zealand. Relevant similarities and differences will be highlighted.

6.2.1 UK Market

The UK market [1] had been a pool-based market since 1990 and remained so until the New Energy Trading Arrangement (NETA) [2] went live on 27 March 2001. Congestion management and some related issues will be discussed under both the previous trading arrangement and NETA.

6.2.1.1 Congestion Management in Previous Energy Trading Arrangement

In the previous energy trading arrangement (PETA), almost all the electricity supplied in England and Wales was traded through the Pool [14]. The Pool mechanism set the wholesale price and established the generation merit order to meet the forecast demand (plus a reserve margin) at the day-ahead stage. The National Grid Company (NGC) was responsible for the daily scheduling and dispatch of generation to meet the actual demand. The actual dispatch of plant might not match that anticipated at the day-ahead stage due to transmission constraints, changes in plant availability and differences between actual and forecast demand.

Pool prices were set on the basis of a competitive bidding process for generation. NGC produced a forecast of demand for each half-hour of the following day and then scheduled the generators' bids to meet this demand. This schedule was called the Unconstrained Schedule. Generally, the price of the most expensive unit scheduled to meet forecast demand in each half-hour set the price for energy and was called the System Marginal Price (SMP). To the SMP was added a component called the Capacity Payment, which was provided to give an incentive to generators to maintain an adequate margin of generation over the level of demand for electricity in order to cover for unexpected demand and generator failures on the system. This payment is the product of two factors: the Loss of Load Probability

(LOLP) and the Value of Lost Load (VOLL). Together, SMP and the Capacity Payment constituted the Pool Purchase Price (PPP). PPP values were calculated the day-ahead, i.e. *ex-ante*. Unscheduled Availability Payments (USAV) were made to plant that were available but were not included in the Unconstrained Schedule. Although not generating, these plants contribute to the security of the system since they can be called upon to generate if required.

Constraints on the transmission system can cause the actual half-hourly generation produced by a unit to differ from that anticipated in the Unconstrained Schedule. Units that were scheduled before taking constraints into account might have their output reduced or withdrawn (termed as "constrained-off"). Other units might have their output increased or be dispatched without being included in the Unconstrained Schedule (termed as "constrained-on"). The cost of transmission constraints was embedded in the uplift costs, which was added to the PPP to work out the Pool Selling Price (PSP).

Participants' access rights to the transmission system were not well defined under PETA. The access of generators was limited by their Supplemental Agreements to the Master Connection and Use of System Agreement (MCUSA) to their notified Maximum Export Capacities or Registered Capacities. Suppliers did not have specific access limits, however, the access of distribution network operators was limited by their notified Connection Site Demand Capabilities and this provided an upper bound on the aggregate access limits of all the suppliers within a distribution network. Furthermore, these limits could be reduced if NGC was prevented from transporting electricity due to transmission constraints that could not be avoided. Thus, NGC's connection agreements, i.e. the Supplemental Agreements, did not confer firm access rights. Another issue related to access to the transmission system is the treatment of transmission losses. Suppliers pay for all losses on the transmission system on a uniform basis. Thus, neither generators nor suppliers are exposed to the short-term costs imposed by their choice of location.

6.2.1.2 Congestion Management in NETA

NETA replaces the Pool with voluntary forwards markets for energy trading, a voluntary Balancing Mechanism for resolving energy and system imbalances close to real-time and mandatory imbalance cash-out. Under NETA, there will be no unconstrained schedule and generators will be able to self-dispatch instead. Generators and suppliers will contract bilaterally until the Balancing Mechanism for a half-hour trading period opens and notifications of contract volumes for the period have to be made. At this point, known as "Gate Closure", market participants have to inform NGC, as System Operator, of their intended generation or consumption profiles for the relevant half-hour. Real-time transmission congestion will be mainly managed through the new balancing mechanism [25], which provides a basis whereby NGC can accept offers of electricity (generation increases and demand reductions) and bids for electricity (generation reductions and demand increases) at very short notice. Accepted offers will be paid for at the prices offered (and accepted bids will pay the prices bid). "Gate Closure" is currently three and a half hours ahead, and will reduce to half hour ahead.

In addition, NGC will continue to be able to sign contracts with participants for the provision of specific services to aid in balancing the system. Generators and suppliers whose contract positions do not match their metered volumes will be subject to energy imbalance payments, which are based on the costs of the actions accepted by NGC in the Balancing Mechanism and relevant balancing service costs.

In the latest Ofgem document [24], it is preferred that transmission losses are charged by adjusting participants' metered volumes using estimated zonal loss factors. Ofgem considers that the introduction of a market in firm access rights is likely to be the most effective way of meeting the objective for reform. A regime of firm access rights means that participants must purchase sufficient access rights to match the amount of electricity they wish to transmit across the transmission grid. Auctioning the transmission rights is believed to be an efficient means to allocate these rights in a non-discriminatory way, which allows the value that participants place upon them to be revealed in an efficient manner.

Transmission constraints that can cause congestion on the NGC system, such as thermal, voltage and stability constraints, are studied over time frames from several years to the operational phase. Traditionally, these constraints have been analysed in off-line studies using DC and AC loadflows and transient stability programs. NGC's on-line dispatch program (known as DISPATCH) runs every 5 minutes, and optimises the generation dispatch over the next 2 hours by studying six linked time-steps, while respecting the transmission constraints. A project called CODA (Combined Dispatch Advisor) aims to achieve the on-line constraint management [25]. CODA takes data from the EMS and elsewhere to calculate the values of network constraints for actual and forthcoming conditions up to 2 hours ahead. The intention of this is to reduce transmission service cost by providing control engineers with up-to-the-minute constraint values and recommendations for control actions. The constraint limits calculated by CODA will eventually be passed to the DISPATCH.

6.2.2 PJM Market in the US [16]

PJM started as a Security Coordinator and Control Area Operator in 1927. It was initially a wholesale power exchange in 1997 and finally assumed the formal obligations of an ISO in January, 1998. Though still evolving, it has been an example that represents one of the most stable electricity markets in the US. PJM retains the unit commitment model of daily price clearing and incorporates transmission constraints to the bid and scheduling basis.

PJM includes a spot market coordinated by the ISO, who accepts both bilateral schedules and voluntary bids from the market participants. Using these schedules and bids, the ISO finds an economic, security-constrained dispatch for power flows and the associated Locational Marginal Prices (LMPs). When the transmission system is constrained, the spot prices can differ substantially across locations. Sales through the spot market are settled at the LMPs. The transmission usage charge for bilateral transactions is the difference in the LMPs between origin and destination. An accompanying system of Fixed Transmission Rights provides financial hedges between locations. The Fixed Transmission Rights are the equivalent of perfectly tradable firm transmission rights.

The PJM LMP system was embraced after an experiment during 1997 with an alternative zonal pricing approach that proved to be fundamentally inconsistent with a competitive market and user flexibility. According to PJM's experience, the earlier zonal pricing system allowed market participants the flexibility to choose between bilateral transactions and spot purchases, but did not simultaneously present them with the costs of their choices. The circumstances created a false and artificial impression that savings of $10 per MWh or more could be achieved simply by converting a spot transaction into a bilateral schedule. By contrast, the locational pricing system avoids this perverse incentive. By construction, the LMPs equal system marginal costs. Every generator would be producing at its short-run profit maximising output, given the prices. The market equilibrium would support the necessary dispatch in the presence of the transmission constraints. Spot market transactions and bilateral schedules would be compatible. Flexibility would be allowed and reliability maintained consistent with the choices of the market participants.

In PJM the system experienced transmission constraints, locational prices separated and the opportunity cost of transmission was quite large. The lowest locational prices were sometimes negative, reflecting the value of counterflow in the system where it would be cheaper to pay participants to take power at some locations and so relieve transmission constraints. The highest locational prices were larger than the marginal cost of the most expensive plant running, reflecting the need to simultaneously increase output from expensive plants and decrease output from cheap plants, just to meet an increment of load at a constrained location. Over all hours in April 1998, for example, the low price was -$45 at 1500 hours on April 18 at "JACK PS," and the highest price was $232 at 1100 hours on April 16 at "SADDLEBR," both locations being in the Public Service Gas & Electric territory.

PJM adopts a two-settlement system to enhance the robustness and competitiveness of the market and to provide increased price certainty to market participants. It consists of two markets, with separate accounting for each market:

- Day-ahead market - A forward market in which hourly clearing prices are calculated for each hour of the next operating day based on generation offers, demand bids and bilateral schedules. The day-ahead schedule is developed using a least-cost security-constrained unit commitment and a security-constrained economic dispatch. Day-ahead congestion charges for bilateral transactions are based on the differences of LMPs between the source and sink.
- Real-time market - An energy market in which hourly clearing prices are determined by security-constrained economic dispatch with actual system operating conditions described by state estimation. Transmission customers pay congestion charges for bilateral transaction quantity deviations from day-ahead schedules.

6.2. 3 California Market in the US [3, 15]

On 20 December 1995, the California Public Utility Commission (CPUC) ruled to restructure the electric utility industry in California, to allow for competition in the wholesale and retail electricity markets, in an effort to lower electric rates. The decision required that utilities formed an ISO to begin operation on 1 January 1998, to conduct system dispatch and transmission operation, and created a Power Exchange (PX), and filed plans to voluntarily divest 50% of their fossil generation to mitigate the current utility market power. The new market structure mandated separating the wholesale PX from the ISO. The ISO manages three key markets – competitively procured ancillary services, a real-time energy balancing market and a congestion management market. The PX operates three energy markets, a daily auction for each hour of the next day, an on-the-day market and a block forwards market.

In California, market participants can take part in the process of congestion management through submission of "adjustment bids" to the ISO. The ISO's congestion management process is accomplished in two time scales: day-ahead and hour-ahead. In both time periods, the ISO reschedules to eliminate potential congestion and minimises rescheduling to allow market participants to voluntarily seek their lowest cost of delivered energy. Overall, the congestion management recognises and separates each Schedule Coordinator's (SC) portfolio of generation and load from other SCs while finding the lowest rescheduling cost to maintain system reliability.

To simplify the congestion pricing, California uses zones or geographical locations to define electrical characteristics of the power grid and determine a financial value for the ability to serve its energy needs. Zones are defined as areas where congestion is infrequent and can easily be priced on an average cost basis. The California congestion pricing method uses the term "interzonal" to describe congestion and pricing between zones and the term "intrazonal" to describe congestion and pricing within a zone. Zones can be merged or added if interzonal congestion becomes infrequent and inefficiently priced at marginal cost or intrazonal congestion becomes frequent and inefficiently priced at average cost.

To mitigate the shortcoming of traditional transmission allocation methods, a mix of physical and financial rights was selected for implementation of the Firm Transmission Rights process in California starting 1 February 2000. In this model, Firm Transmission Rights provide scheduling priority as well as financial rights. Firm Transmission Rights that are not sold in the day-ahead market are released in the hour-ahead market, but the original owner retains the financial rights.

However, it is important to note that in 2000 California's electricity market collapsed for a number of reasons [6]. Besides the abnormal climate, high gas prices and strict environmental rules, some fatal flaws in market design caused market power and gaming, which has been severely criticised and the Federal Energy Regulatory Commission (FERC) has proposed remedies for these problems. According to Chandley *et a.l* [7], one of the serious flaws is the poorly structured separation of the ISO from the PX. This separation allowed generators to "game" the market by bidding only a portion of their capacity ahead of time into the PX and then reaping exceptionally high prices when the ISO was forced to buy power in real-time to balance supply and demand. In addition, the frozen retail prices in California meant that rising wholesale prices could in no way be moderated by being passed on to consumers so as to reduce their demand.

6.2.4 Norway and Sweden Market [8-15]

Deregulation in Norway was initiated by the Energy Act of June 1990. Consequently, the market started in May 1992. The principal restructuring was the removal of transmission ownership from Statkraft, the national utility, and the founding of the nationally owned company Statnett to be the transmission owner, market operator and ISO. Sweden passed deregulation legislation in October 1995 and joined the Norwegian market in January 1996. Transmission operation was removed from the national utility Vattenfall, which continues to operate generation, while Svenksa Kraftnät, the national grid owner and ISO, was formed. Meanwhile, the Nord Pool, a market operator owned equally by Statnett and Svenksa Kraftnät, was founded and took over the operation of spot and future markets in both countries.

Norway and Sweden have different philosophies of congestion management. Norway is using the spot market settlement process to prevent congestion efficiently. When transmission congestion is predicted, the ISO declares that the system is split into price zones as predicted congestion bottlenecks. Bidders in the spot market must submit separate bids for each price zone where they have generation or load. If no congestion occurs, the market is settled at one price. Alternatively, if congestion does occur, price zones are settled at different prices caused by binding transmission constraints. Zones with excess generation have lower prices while zones with excess load have higher prices. Revenue from this price difference is paid to the ISO, who uses it to reduce the capacity charge. Bilateral contracts that span price zones must pay for its load at the price in the zone of its load. In this way it is possible to account for its contribution to congestion and to expose the contract to financial consequences of congestion.

In contrast to Norway, Sweden has only one price zone. This is because Sweden does not want the transmission system to affect the market solution. However, Sweden varies the capacity charge portion of its connection point tariff based on geography. Since power flow in Sweden is always from north to south, generation is charged more and load less in the northern part of the country. This affects generation costs and thus the bids submitted to the market, hence deferring some congestion.

Congestion in real time is eliminated by purchase of the adjustment of generation and load from the ISO regulatory markets, which is known as "buyback". Congestion has been infrequent in the Nord Pool market. Despite using different approaches to congestion management, the Norwegian ISO and the Swedish ISO coexist successfully within one market. Because Svenksa Kraftnät does not use price zone congestion management, the Swedish transmission system is always regarded as one price zone. Revenue to market due to congestion between Norway and Sweden is split equally by Statnett and Svenksa Kraftnät. No special fee is charged for energy transfers across national boundaries, and the two ISOs do not charge each other transmission tariffs at their connection points.

6.2.5 New Zealand Market [15-17]

On 1 October 1996, a wholesale electricity market in New Zealand commenced operation under the name Electricity Market Company (EMCO). This market is an *ex post* market and is not mandatory. There is separation between the power exchange, which is EMCO, and the system operator, which is Trans Power. The interconnecting transmission constraint from South Island to North Island has a significant impact upon the operation of the system. Furthermore, transmission losses are also relatively high in New Zealand.

New Zealand's wholesale electricity market is characterised by its adoption of full nodal pricing. Participating generators and purchasers submit offers and bids for each half-hour at the day-ahead stage. Bilateral contract volumes are notified to the system operator, and any deviations are settled at the prices emerging from the spot market. The marginal costs of transmission losses and constraints are reflected in the half-hourly *ex post* market prices calculated for each of the connection points on the network using actual flows. Constrained-on payments are paid to scheduled generators whose offer prices turn out to be higher than the nodal *ex post* price. There are no constrained-off payments to generators who did not run despite bidding below the *ex post* nodal price. The marginal pricing of losses and constraints within the spot market results in a surplus of funds being collected, which is passed on to Trans Power. Bilateral market participants trading outside the spot market pay Trans Power directly for losses and constraints via a charge based upon the volume of power traded and the price differential between the generation and supply nodes. Trans Power uses these funds to lower its use of system charges.

6.3 Real-time Congestion Management across Interconnected Regions

The physical redispatch and real-time balancing mechanisms that have been presented in Chapter 4 have been successfully applied to electricity markets worldwide to relieve the transmission congestion within a separated region.

Coordination of activities among regions is a significant issue in maintaining a reliable bulk transmission system and developing competitive markets. FERC Order 2000 [4] has mandated the formation of Regional Transmission Organisations (RTO), which accelerate interregional transaction and increase the burden of interregional transmission. The Association of European Transmission System Operators (ETSO), founded in July 1999, has been investigating congestion management methods for cross-border transmission between European countries [5]. However, problems remain unsolved with respect to coordinated congestion management across multiple interconnected regions. Since each regional ISO cannot obtain the network operating data of other regions, one of the main difficulties in meeting the requirements presented [4,5] is how to implement coordinated congestion management without a huge amount of information exchange between regions.

In this section, a new approach is proposed to decompose an optimal power flow (OPF) problem by applying augmented Lagrangian Relaxation (LR) in order to implement multi-regional active power congestion management. The LR algorithm has been successfully applied to many aspects of power systems, especially to solving problems of unit commitment, hydro-thermal coordination and OPF. Comparing the proposed method with existing methods for regionally decomposed OPF, neither fictitious buses, generators nor loads are added. Neither is the model of transmission lines modified to decouple interconnecting regions. Using this approach, multi-regional active power congestion management can be implemented as an iterative procedure. The ISO of each region does not need to know any information concerning other regions apart from the corresponding Lagrangian multipliers of the tie-lines between regions, thus the dispatching independence of the ISO is preserved. Optimal transaction prices on all the interconnecting lines are the by-product of multi-regional congestion management. Finally, the IEEE RTS-96 test system with three interconnected regions and 73 buses is studied to illustrate the proposed decomposition approach to congestion management across interconnected regions.

6.3.1 Proposed Method for Regional Decomposition OPF

The original motivation to decompose a large-scale OPF problem into several smaller problems by regions was to improve the computational speed. Over the last decade, a number of mathematical decomposition methods have been successfully applied to regionally decomposed OPF studies. Deeb and Shahidephour [18] applied a Dantzig–Wolfe decomposition method to solve the multi-area reactive power optimisation problem. Kim and Baldick [19] presented a Lagrangian Relaxation based "auxilliary problem principle" approach to parallelising OPF for very large interconnected power systems. A Lagrangian Relaxation decomposition procedure was used by Conejo and Aguado [20] to achieve a multi-area decentralised non-linear DC OPF. Furthermore, Nogales, Prieto and Conejo [21] proposed a decomposition approach, that did not require the solution of sub-problems in each iteration in order to solve a multi-area AC OPF problem.

Since the main purpose of a regionally decomposed OPF is to give regional ISOs the capability to dispatch the interconnected network in a coordinated manner without knowing the operating data of other regions and also to obtain the optimal prices of inter-region transactions, Lagrangian Relaxation based decomposition is applied. The basic procedure of the LR decomposition method is given in Appendix A.

In order to decompose a system into regions by LR-based decomposition, the key point is to find proper coupling constraints between regions and then to relax them into the objective function. To obtain such coupling constraints, a dummy bus is defined at the border for each interconnecting line and then duplicated into two dummy generators [19] while one or two fictitious buses are added per interconnecting line and the model of the interconnecting line is modified [20]. Nogales *et al.* [21] noted that their method is based on decomposition of a set of optimal conditions for a non-linear programming problem without modifying the original problem, but they did not give any further physical explanation about this and the sub-problem for each area. In this research, a new decomposition scheme is presented to achieve the regionally decomposed OPF without any modification to the original network model.

The general OPF problem can be formulated as

$$\text{Minimise: } \sum_{i\in\Omega} C_i(P_i)$$

$$\text{Subject to: } \mathbf{h}(\mathbf{P},\mathbf{Q},\mathbf{V},\boldsymbol{\theta}) = \mathbf{0}$$

$$\mathbf{g}(\mathbf{P},\mathbf{Q},\mathbf{V},\boldsymbol{\theta}) \le \mathbf{0} \tag{6.1}$$

where $C_i(P_i)$ active power cost or bidding function of generator at bus *i*,
$h()$ matrix of equality constraints,
$g()$ matrix of inequality constraints,
$\boldsymbol{V}$ matrix of bus voltage magnitudes,
$\boldsymbol{\theta}$ matrix of bus voltage angles,
$\boldsymbol{P}$ matrix of active power injections,
$\boldsymbol{Q}$ matrix of reactive power injections,
$\boldsymbol{\Omega}$ set of buses in the whole system.

The proposed regional decomposition scheme is shown in Figure 6.2. In the case of a system with two interconnected regions A and B, the tie-line *ij* and bus *j* in region B will be reserved in the computation of region A, while the tie-line *ij* and bus *i* in region A will be reserved in the computation of region B. Therefore, the following coupling constraint equations should be included into the problem:

$$\mathbf{T}^{\mathbf{A}} - \mathbf{T}^{\mathbf{B}} = \mathbf{0} \tag{6.2}$$

where $\mathbf{T}^{\mathbf{A}} = \begin{bmatrix} V_i^A & \theta_i^A & P_{ij}^A & Q_{ij}^A \end{bmatrix}^\dagger$, $\mathbf{T}^{\mathbf{B}} = \begin{bmatrix} V_i^B & \theta_i^B & P_{ij}^B & Q_{ij}^B \end{bmatrix}^\dagger$ are matrices of interconnecting variables in region A and region B, respectively. V_j and θ_j are not included in these matrices because they can be obtained from V_i, θ_i, P_{ij} and Q_{ij}. The nodal power balance equations of bus j^A and bus i^B are not necessary in the regional sub-problems, where the voltages of bus j^A and bus i^B appear only in the nodal power balance equations of bus i^A and bus j^B and are constrained by the branch power flow constraints of P_{ij} and Q_{ij}. Therefore, no dummy generators or loads are added at bus j^A in the region A sub-problem and at bus i^B in the region B sub-problem.

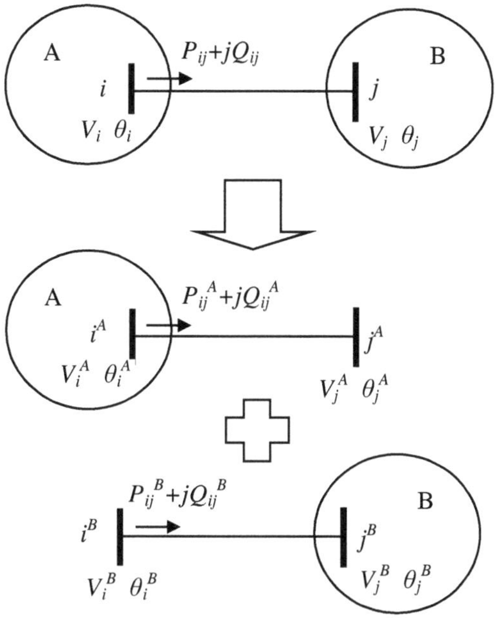

Figure 6.2. Proposed regional decomposition scheme

After relaxing the regional coupling constraints into the objective function, the dual function is obtained as

$$\Phi(\lambda) = \text{Minimise: } \sum_{i\in\Omega} C_i(P_i) + \boldsymbol{\lambda}^{\dagger}(\mathbf{T}^{\mathbf{A}} - \mathbf{T}^{\mathbf{B}}) \tag{6.3a}$$

Subject to: $\mathbf{h}^{\mathbf{A}}(\mathbf{P}^{\mathbf{A}},\mathbf{Q}^{\mathbf{A}},\mathbf{V}^{\mathbf{A}},\boldsymbol{\theta}^{\mathbf{A}}) = \mathbf{0}$

$$\mathbf{h}^{\mathbf{B}}(\mathbf{P}^{\mathbf{B}},\mathbf{Q}^{\mathbf{B}},\mathbf{V}^{\mathbf{B}},\boldsymbol{\theta}^{\mathbf{B}}) = \mathbf{0}$$

$$\mathbf{g}^{\mathbf{A}}(\mathbf{P}^{\mathbf{A}},\mathbf{Q}^{\mathbf{A}},\mathbf{V}^{\mathbf{A}},\boldsymbol{\theta}^{\mathbf{A}}) \le \mathbf{0}$$

$$\mathbf{g}^{\mathbf{B}}(\mathbf{P}^{\mathbf{B}},\mathbf{Q}^{\mathbf{B}},\mathbf{V}^{\mathbf{B}},\boldsymbol{\theta}^{\mathbf{B}}) \le \mathbf{0}$$

where λ is the matrix of coupling constraints multipliers. Then the dual problem is to

$$\text{Maximise}_{\lambda}: \quad \Phi(\boldsymbol{\lambda}) \tag{6.3b}$$

The LR-based solution can be obtained by solving problem (6.3a) at a fixed λ and then updating λ to increase the dual objective function until the dual objective function and λ do not change significantly. Therefore, the decomposed sub-problem of region A can be formulated as follows:

$$\text{Minimise: } \sum_{i \in \Omega^A} C_i(P_i) + \boldsymbol{\lambda}^{\dagger}\mathbf{T}^{\mathbf{A}} \tag{6.4}$$

$$\text{Subject to: } \mathbf{h}^{\mathrm{A}}(\mathbf{P}^{\mathrm{A}}, \mathbf{Q}^{\mathrm{A}}, \mathbf{V}^{\mathrm{A}}, \boldsymbol{\theta}^{\mathrm{A}}) = \mathbf{0}$$

$$\mathbf{g}^{\mathrm{A}}(\mathbf{P}^{\mathrm{A}}, \mathbf{Q}^{\mathrm{A}}, \mathbf{V}^{\mathrm{A}}, \boldsymbol{\theta}^{\mathrm{A}}) \le \mathbf{0}$$

where
- $h^A()$ matrix of equality constraints in Region A,
- $g^A()$ matrix of inequality constraints in Region A,
- V^A matrix of voltage magnitudes of buses in region A and buses directly adjacent to Region A,
- θ^A matrix of voltage angles of buses in region A and buses directly adjacent to Region A,
- P^A matrix of active power injections,
- Q^A matrix of reactive power injections,
- Ω^A set of buses in region A.

In (6.3) and (6.4), it can be seen that the coupling constraints forcing the interconnecting variables to match on both regions of a tie-line are dualised and this decomposes the OPF problem into separate regional OPF problems. The Lagrangian multipliers are updated iteratively as follows by simply applying the sub-gradient method until the interconnecting variables on both interconnected regions match:

$$\lambda^{k+1} = \lambda^{k} + \alpha(\mathbf{T}^{A,k+1} - \mathbf{T}^{B,k+1}) \tag{6.5}$$

where k is the index of iteration. The values of Lagrangian multipliers are the costs to maintain the regional coupling equation constraints. The economic explanations of the multipliers for active and reactive power flows between two interconnected regions are the optimal prices of inter-regional transactions.

Because some of the coupling constraints (P_{ij} and Q_{ij}) are non-linear, relaxing them into the objective function could affect its convexity. To ensure local convexity, augmented Lagrangian decomposition should be used and the augmented Lagrangian function of problem (6.1) can be written as

$$L = \sum_{i \in \Omega} C_i(P_i) + \boldsymbol{\lambda}^{\dagger}(\mathbf{T}^{\mathrm{A}} - \mathbf{T}^{\mathrm{B}}) + \frac{1}{2}\beta\left\|\mathbf{T}^{\mathrm{A}} - \mathbf{T}^{\mathrm{B}}\right\|^2 \tag{6.6}$$

If β is large enough the quadratic term can ensure local convexity of (6.6). However, this new quadratic term makes the relaxed primal problem nondecomposable. To make (6.6) decomposable again, an iterative approach called the Alternating Direction Method [27] is adopted, in which the variables not in the analysing region are fixed to the values of the previous iteration:

$$(\mathbf{T}^{\mathbf{A},\mathbf{k}+1}, \mathbf{T}^{\mathbf{B},\mathbf{k}+1}) = \arg\min \left\{ \begin{array}{l} \sum_{i \in \Omega^A} C_i(P_i) + (\boldsymbol{\lambda}^k)^\dagger \mathbf{T}^\mathbf{A} + \frac{1}{4}\beta \left\| \mathbf{T}^\mathbf{A} - \mathbf{T}^{\mathbf{B},k} \right\|^2 \\ + \sum_{i \in \Omega^B} C_i(P_i) - (\boldsymbol{\lambda}^k)^\dagger \mathbf{T}^\mathbf{B} + \frac{1}{4}\beta \left\| \mathbf{T}^{\mathbf{A},k} - \mathbf{T}^\mathbf{B} \right\|^2 \end{array} \right\} \tag{6.7}$$

The stopping criterion for solving the regionally decomposed OPF is the maximum mismatch between coupling variables is smaller than a preset threshold value:

$$\left\| \mathbf{T}^{A,k} - \mathbf{T}^{B,k} \right\| \le \varepsilon \tag{6.8}$$

It is worth noting that the proposed method can solve all the regional subproblems in parallel. The only information needed to exchange between regions is the LR multipliers of the coupling constraints.

6.3.2 Application of the Proposed Method to Congestion Management across Interconnected Regions

A large electric network with multiple interconnected regions requires coordination of all the regional ISOs to have good use of the grid within its secure capacity. In particular, in the deregulated environment, with the introduction of competition and greater inter-regional trading, a new efficient coordination mechanism should be developed to manage the transmission congestion situation across multiple interconnected regions. The general principles that should coordinating congestion management across interconnected regions are as follows be applied to:

- System reliability is preserved based on market bids.
- Avoid any unnecessary information exchange between regional ISOs and any unnecessary network reduction.
- The whole procedure should be simple, robust and priced correctly.

Cadwalader *et al.*[22] proposed a LR-based approach that decomposed the global congestion management problem into sub-problems corresponding to different regions. However, full information about the whole system is still needed for each of the regional sub-problems. Moreover, some other drawbacks of this method have been pointed out by Oren and Ross [23], such as the convexity problem.

The real-time balancing mechanism has been adopted widely for congestion management in recent years. A framework for congestion management through the real-time balancing market (RBM) has been proposed in Chapter 4. To achieve coordination between regions, the regionally decomposed OPF method is applied under the framework presented in Chapter 4.

6.3.2.1 Mathematical Model

Assume that a preferred schedule has been built in the bilateral contract market and the day-ahead auction market, all the generators and consumers are encouraged to submit their incremental and decremental bids to the ISO in the balancing market for the requirement of system balancing and congestion management. Ignoring the curtailment of bilateral contracts and operating reserves in the model presented in Chapter 4 (see 4.1 to 4.12), the problem of active power congestion management can be simplified as follows:

$$\text{Mimise: } \sum_{i\in\Omega^A} C_i^+(\Delta P_i^+) + \sum_{i\in\Omega^A} C_i^-(\Delta P_i^-) \tag{6.9}$$

$$\text{Subject to: } \Delta P_i^+ - \Delta P_i^- - \sum_{l\in\Omega} B'_{il}\Delta\theta_l - \Delta P_{Loss,i} = 0 \quad \forall i \in \Omega^A$$

$$\Delta P_{mn}^{\min} \le \Delta P_{mn} = B'_{mn}(\Delta\theta_m - \Delta\theta_n) \le \Delta P_{mn}^{\max} \quad \forall m,n \in \Omega; m \ne n$$

After applying the decomposed OPF method, the active power congestion management in region A can be formulated as follows:

$$\text{Minimise: } \begin{aligned} &\sum_{i\in\Omega^A} C_i^+(\Delta P_i^+) + \sum_{i\in\Omega^A} C_i^-(\Delta P_i^-) + \sum_{m\in\Omega^A, n\in\Xi^A} \lambda_{\Delta Pij}^k \Delta P_{mn}^A + \\ &+\frac{1}{4}\beta \sum_{m\in\Omega^A, n\in\Xi^A} \left\| \Delta P_{mn}^A - \Delta P_{mn}^{B,k} \right\|^2 \end{aligned} \tag{6.10}$$

$$\text{Subject to: } \Delta P_i^+ - \Delta P_i^- - \sum_{l\in\Omega^A} B'_{il}\Delta\theta_l - \sum_{j\in\Xi^A} B'_{ij}\Delta\theta_j^A - \Delta P_i^{Loss} = 0 \quad \forall i \in \Omega^A$$

$$\Delta P_{mn}^{\min} \le \Delta P_{mn} = B'_{mn}(\Delta\theta_m - \Delta\theta_n) \le \Delta P_{mn}^{\max}$$
$$\forall m,n \in \{\Omega^A \vee \Xi^A\}; m \ne n$$

$$\Delta P_i^{\min} \le \Delta P_i = \Delta P_i^+ - \Delta P_i^- \le \Delta P_i^{\min} \quad \forall i \in \Omega^A$$

The objective of congestion management is to follow the schedule as closely as possible and to minimise the cost of real-time dispatch. $C_i^+(\Delta P_i^+)$ and $C_i^-(\Delta P_i^-)$ are the cost functions of incremental and decremental adjustment at bus i, respectively. They can be either quadratic functions or linear functions. Ω^A is the set of buses in region A and Ξ^A is the set of buses which have direct connection with buses in region A.

The adjustment of active power flows of tie lines has been relaxed into the objective function by the augmented Lagrangian Relaxation method. The first set of constraints represents the linearised nodal active power flow balance equations constraints represents the linearised nodal active power flow balance equations of all the buses in region A. As described in Chapter 4, ΔP_i^{Loss} is the total change of losses on branches that are connected to bus i and in which flows on the branches are flowing to bus i. The second set of constraints represents the changing ranges of active power flow on branches within region A or connected to region A. The third set of constraints represents the changing ranges of nodal injections in region A.

6.3.2.2 Sequential Solution versus Parallel Solution

According to the analysis in Section 6.3.1, there should be some terms in the objective function of problem (6.10) for the relaxed coupling constraint $\Delta\theta_i^A = \Delta\theta_i^B$. Since the change of voltage angle is a relative value, this coupling constraint can be maintained by adjusting the voltage angle of the slack bus in each region. Therefore items for this constraint have been removed from the objective function.

This algorithm can be implemented either sequentially or in parallel, depending on whether single or multiple slack buses are used in the whole system. If the slack bus in region A is the unique slack bus in the whole system, $\Delta\theta_i^A$, which is the result of the sub-problem of region A, is needed for the solution of the sub-problem of region B in order to maintain the coupling constraint $\Delta\theta_i^A = \Delta\theta_i^B$. In other words, the sub-problem of region B cannot be solved until the solution of region A is obtained. So with a unique slack bus, the algorithm presented can only be implemented sequentially.

When each region has its own slack bus [19], the voltage angles of slack buses are chosen at each iteration so that the average of the changes of border angles in region A equals the average of changes of border angles in region B. As a result, all the regional sub-problems can be solved in parallel.

6.3.2.3 Global Congestion Management versus Two-level Congestion Management

Using this decomposition model, ISOs in all the regions can relieve the transmission congestion in a global way without exchanging a huge amount of information between each other. The global optimal solution can be obtained by solving sub-problem (6.10) for each region iteratively, which is the same as solving the problem of the whole system. However, all regional ISOs must take part in the procedure, no matter whether there is any congestion in their region or not. A regional ISO may prefer to relieve intra-regional congestion management independently rather than to perform the calculation together with all other regional ISOs.

To keep this independence of intra-regional dispatching, another option is required, which is the two-level inter/intra-regional congestion management. To eliminate inter-regional congestion, all the available adjustment bids in the RBMs of all regions are regarded as control variables. Furthermore, all regional ISOs must coordinate their actions and model (6.10) is again adopted. In the event of intra-regional congestion, the control variables include only submitted adjustment bids within the congested region and no additional inter-regional transactions are made. If a regional ISO wants to keep independence for the intra-regional congestion management, it might still need help from the adjacent regions if the internal violation cannot be relieved by local resources only. To model this situation, the objective function of problem (6.10) should be modified to

Minimise:

$$\sum_{i\in\Omega^A} C_i^+(\Delta P_i^+) + \sum_{i\in\Omega^A} C_i^-(\Delta P_i^-) + \sum_{i\in\Omega^A,\, j\in\Xi^A} (\gamma^+_{\Delta Pij}\Delta P_{ij}^{A,+} + \gamma^-_{\Delta Pij}\Delta P_{ij}^{A,-}) \tag{6.11}$$

where $\gamma^+_{\Delta Pij}$ and $\gamma^-_{\Delta Pij}$ are the pre-negotiated coefficients for the active power interchange adjustments, incremental and decremental, respectively. The adjustments should be large enough to keep the additional inter-regional transactions as the last control option to eliminate the intra-regional congestion.

6.3.3 Test Results

The IEEE RTS-96 test system shown in Figure 6.3, which has three interconnected regions, 96 generators, 119 lines and 73 buses, has been used to demonstrate the performance of the proposed algorithm. The complete detail of the power flow data of the IEEE RTS-96 test system is available [26]. To simplify the demonstration of multi-regional congestion management, it is assumed that all the participants in the RBM of each region submit uniform incremental bidding prices and uniform decremental bidding prices, as shown in Table 6.1. Obviously, Region 1 has the most expensive real-time adjustable resources and Region 2 has the cheapest.

To illustrate the application of the proposed regionally decomposed OPF algorithm to relieve inter-regional transmission congestion and intra-regional transmission congestion, two cases have been studied and their results are analysed in the following sections.

6.3.3.1 Case 1: Inter-regional Congestion Management

Assume that there is an unexpected contingency that causes the operational MW limit of tie-line 113-215 (between region 1 and region 2) to reduce from 400MW to 160MW. Since the base MW flow on line 113-215 is 197MW, inter-regional transmission congestion occurs. Using the presented regionally decomposed congestion management algorithm, the cheapest and most efficient resources in the RBM are called upon to eliminate this congestion. Initially all the values of LR multipliers are set to zero, i.e. flat start. The initial values of LR multipliers will not affect convergence, however, good initial values can reduce iteration times significantly.

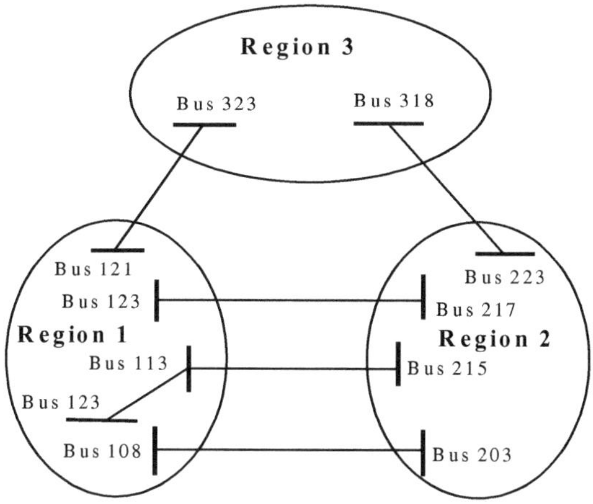

Figure 6.3. IEEE RTS-96 test system

Table 6.1. The incremental and decremental bidding prices in each region

Regions	Incr. Prices (p.u./MW)	Decr. Prices (p.u./MW)
1	1.1	0.5
2	0.5	0.3
3	0.9	0.5

The convergence of the total redispatching cost of inter-regional congestion management, which is the sum of three regional costs, is shown in Figure 6.4. The convergence of MW change in all the other four tie-lines, except for the congested one, are shown in Figure 6.5 (the results from sub-problems of regions on both sides) and Figure 6.6 shows the convergence of the inter-regional transaction prices for all five tie-lines as functions of the number of iterations. From these three figures, it can be seen that overall convergence is reached smoothly within 15 iterations. As the violated tie-line power flow constraint will be fixed at its upper or lower limit if the problem is feasible, the corresponding coupling constraint is satisfied automatically. This is why the Lagrangian multiplier corresponding to tie-line 113-215 is zero all the time in Figure 6.6. As a result, in the objective function of problem (6.10) the relaxed terms, which are related to the violated interconnected constraint, can be removed.

The updating process of the Lagrangian multiplier corresponding to tie-line 123-217 and the convergence of tie-line active power flow change on both sides are shown together in Figure 6.6, which demonstrates how the Lagrangian multiplier is updated in accordance with the MW mismatch of the corresponding inter-regional coupling constraint at every iteration and how the iterative process drives $\Delta P_{123-217}^{1}$ and $\Delta P_{123-217}^{2}$ together.

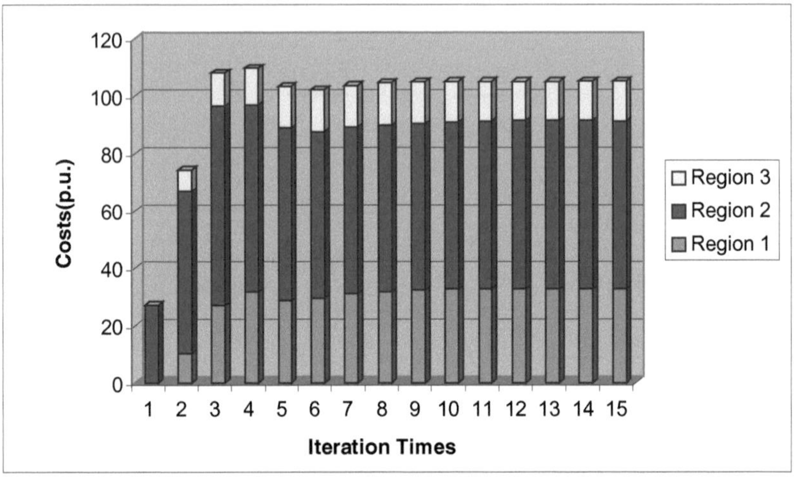

Figure 6.4. Convergence of the redispatching cost for each region in case 1

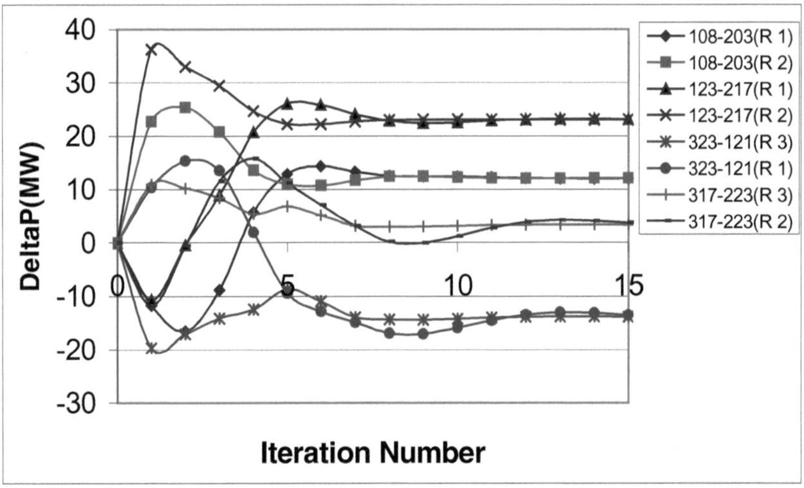

Figure 6.5. Convergence of MW changes of tie-lines obtained from the solutions of different regional sub-problems in case 1

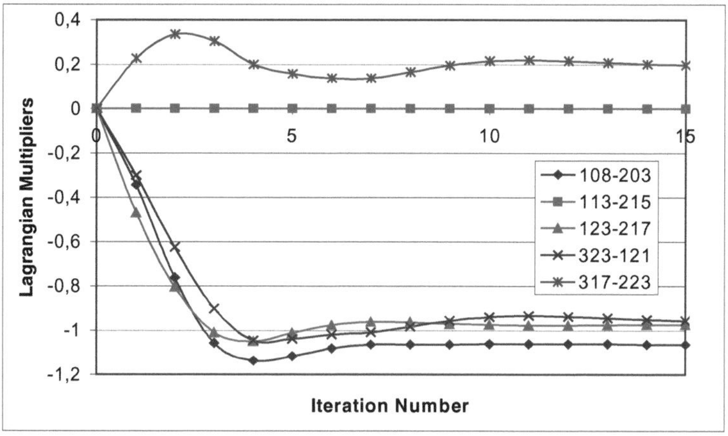

Figure 6.6. Convergence of the Lagrangian multipliers in case 1

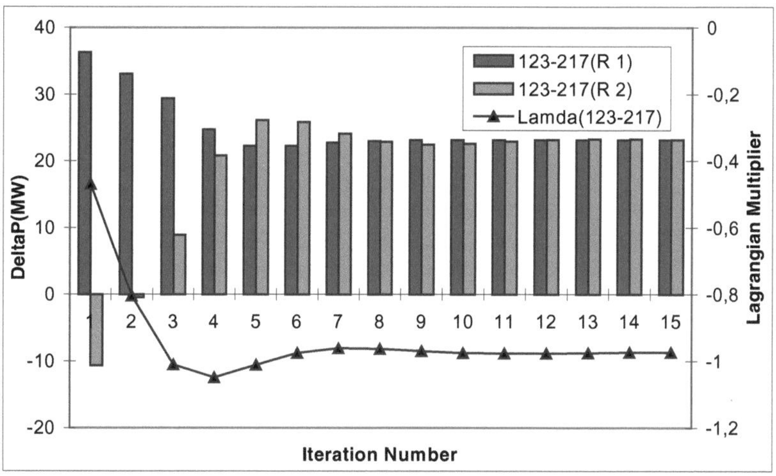

Figure 6.7. Updating of the Lagrangian multiplier corresponding to tie-line 123-217 in accordance with the mismatch of the coupling constraint

6.3.3.2 Case 2: Intra-regional Congestion Management

The base MW flow on line 113-123 is 245MW. Now the reduction of the operational MW limit of line 113-123 from 400MW to 200MW will cause intra-regional transmission congestion in Region 1.

The convergence of MW change of tie-lines 108-203, 113-215 and 123-217 from the results of sub-problems of both Region 1 and Region 2 as functions of iteration numbers are shown in Figure 6.8. Figure 6.9 shows the convergence of the

Lagrangian multipliers of all the three tie-lines between Regions 1 and 2 as functions of iteration number. From these two figures it can be seen that the iterative behaviour is similar to sellers and buyers haggling over the inter-regional transaction prices. The initial prices of additional interchange between regions are zero and the ISO in Region 1 certainly prefers buying as many incremental/decremental resources as possible from other adjacent regions to using its own resources. On the other hand, the ISOs in other regions without any congestion do not want to accept this deal because the prices of inter-regional transactions are too low. With the iterative process, they increase or decrease the amount of interchange in light of the change of prices and finally the optimal solution is reached, which can be accepted by both sides.

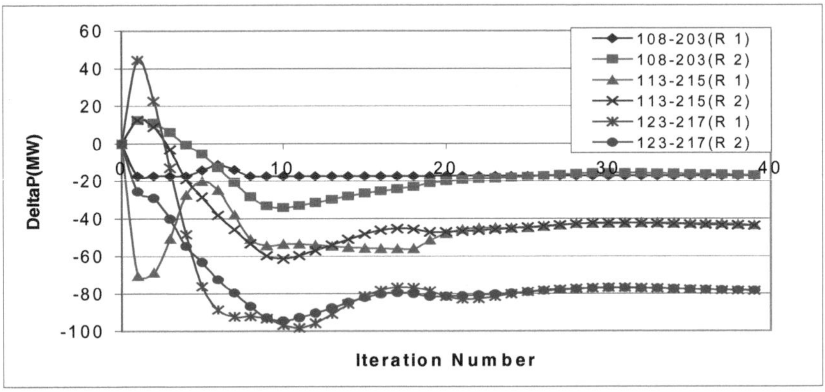

Figure 6.8. Convergence of MW changes of tie-lines between region 1 and region2 from the results sub-problems of both regions 1 and 2 in case 2

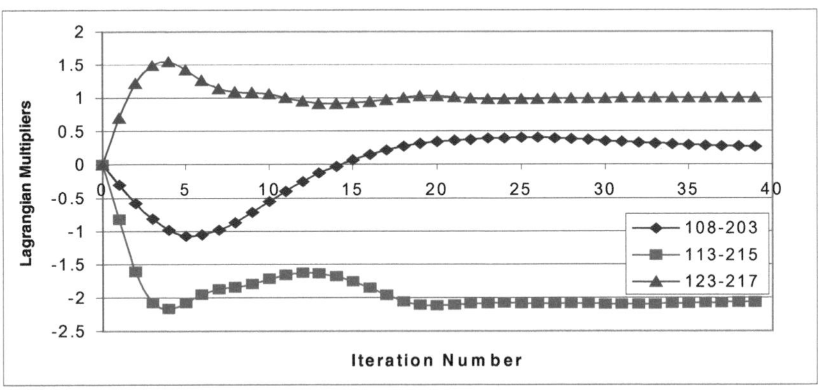

Figure 6.9. Convergence of the Lagrangian multipliers of three tie-lines between regions 1 and 2 in case 2

When the ISO in region 1 prefers using his own regional resources first, although the bidding prices in other regions are much cheaper, none participate in this procedure. Table 6.2 compares the test results of case 2 using the global and the two-level methods. It is easily determined that the total cost of the global method is lower than the two-level method. The reason is that the result using the global method is a global optimal solution and the cheaper incremental resource ΔP_{213} in region 2 has been purchased.

Table 6.2. Comparing test results of case 2 using the global and two-level method

	Global (MW)	Two-level (MW)
$\Delta P_{108\text{-}203}$	-17.40	0.00
$\Delta P_{113\text{-}215}$	-45.28	0.00
$\Delta P_{123\text{-}217}$	-80.07	0.00
$\Delta P_{317\text{-}223}$	4.43	0.00
$\Delta P_{323\text{-}121}$	-4.56	0.00
ΔP_{101}	2.16	84.94
ΔP_{102}	20.00	20.00
ΔP_{107}	0.00	17.40
ΔP_{123}	-165.81	-116.98
ΔP_{213}	148.02	0.00
Total cost	**181.29 p.u.**	**193.06 p.u.**

6.3.3.3 Parameters Selection and Discussion

Two parameters α and β are important in the LR-based decomposition method to keep the objective function convex and they have significant impacts on the total number of iterations. Here they can be experimentally set to $\alpha = 1$ and $\beta = 2$. The impact of β on the total number of iterations is shown in Figure 6.10. The number of iterations in case 2 is more sensitive to the value of β than in case 1. Different systems might need to tune α and β to improve convergence. According to previous analysis in the appendix of Kim and Baldick [19], convergence was reliable with $\beta = 2\alpha$, which was used in this research as a guideline to tune α and β for different systems.

The mismatch tolerance of tie-line power flow is set to $\varepsilon = 0.01$ MW. From the test results of cases 1 and 2, it can be seen that convergence is smoothly attained, but the presented method has poor tail behaviour. The mismatch is reduced significantly in the first several iterations, but it takes many more iterations to make the mismatch smaller than the pre-set tolerance. Although the main purpose of the regionally decomposed OPF method presented here is not to improve the computing speed, it is still very important to reduce iteration time to be of practical use industry. The strategy to update LR multipliers is very important to reduce the tail behaviour. In this chapter, a simple method is used to update the LR multipliers as shown in Figure 6.5, but further research can be done to find a better method. Even an *ad hoc* approach to update LR multipliers more efficiently could be useful.

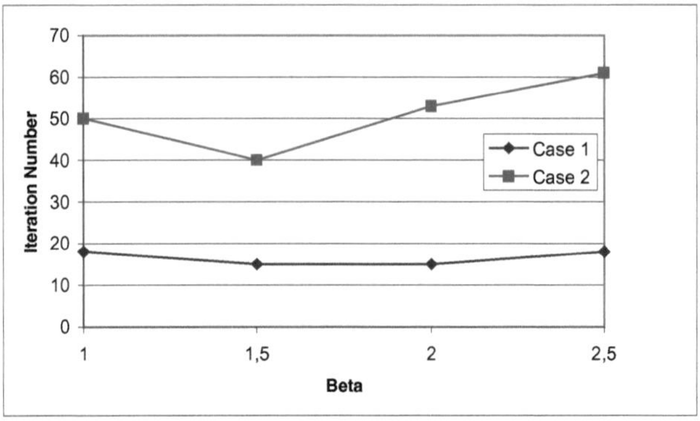

Figure 6.10. Effect of β on the total number of iterations ($\alpha = 1$)

6.4 Conclusions

There are many approaches to congestion management, dependant on the market model, the policy, the technical development and many other factors. Generally speaking, they can be classified into three fundamental categories: transaction curtailment, transmission capacity reservation and system redispatch. However, the best solution might actually be a combination of several of the basic methods for different time scales.

To find out what is going on in the real world, the congestion management approaches of five typical electricity markets in the world were investigated. Every approach had its own advantages and disadvantages, however, none of them is the final version and continuous reforms are still taking place everywhere.

A new scheme for an augmented LR-based regionally decomposed OPF has been presented in this chapter. Compared with existing algorithms, neither fictitious buses, generators or loads are added. Nor is the transmission line model modified to decouple interconnecting lines in the proposed algorithm. Furthermore, all the regional sub-problems can be solved in parallel.

Applying the regionally decomposed OPF algorithm to active power congestion management across interconnected regions through the RBM, presented in Chapter 4, provides an efficient redispatch method to relieve inter/intra-regional congestion without exchanging too much information between regional ISOs. The iteration process is similar to sellers and buyers haggling over prices and when convergence is reached, optimal interchanges and prices can be obtained.

The proposed method is of particular interest to a multi-utility or a multinational interconnected system, such as the USA and Europe, where the independence of regional dispatching should be retained. Case studies based on the three-region IEEE RTS-96 have been presented to illustrate the proposed method.

6.5 References

[1] J. Surrey: "The British Electricity Experiment – Privatization: the Record, the Issues, the Lessons", Earthscan Publications Ltd, London, 1996

[2] OFFER, The New Electricity Trading Arrangements, July 1999, UK

[3] California ISO Scheduling Applications Implementation: Functional Requirements, December 1996, http://www.energyonline.com/wepex/reports/reports2. html#ISO_bus

[4] FERC, 'Order No. 2000, Regional Transmission Organization (RTO)-Final Rule', http://www.ferc.fed.us/news1/rules/pages/rulemake.htm, December 20, 1999

[5] European Transmission System Operators (ETSO), Evaluation of Congestion Management Methods for Cross-border Transmission", November 1999

[6] J. Makansi: "California's Electricity Crisis Rooted in Many Failings", IEEE Spectrum, Feb 2001, pp.24-25

[7] J.D. Chandley, S.M. Harvey, and W.W. Hogan: "Electricity Market Reform in California", http://ksgwww.harvard.edu/people/whogan/, November 2000

[8] R. D. Christie, and I. Wangensteen,: "The Energy Market in Norway and Sweden: Congestion Management", IEEE Power Engineering Review, May 1998, pp.61-63

[9] H. Singh, S. Hao, and A. Papalexopoulos,: "Transmission Congestion Management in Competitive Electricity Markets", IEEE Trans. on Power Systems, Vol. 13, No.2, May 1998, pp. 672-679

[10] R.S. Fang, and A.K. David: "Transmission Congestion Management in an Electricity Market", IEEE Trans. on Power Systems, Vol. 14, No.3, August 1999, pp. 877-883

[11] K. David: "Dispatch Methodologies for Open Access Transmission Systems", IEEE Trans. on Power Systems, Vol.13, No.1, February 1998, pp.46-53

[12] X. Wang, and Y.H. Song: "Advanced Real-Time Congestion Management through Both Pool Balancing Market and Bilateral Market", IEEE Power Engineering Review, Feb 2000, 20(2), pp.47-49

[13] S.M. Harvey, W.W. Hogan, and A.L. Pope: "Transmission Capacity Reservations and Transmission Congestion Contracts", Harvard Electricity Policy Group, Harvard University, June 1996

[14] OFFER, "Review of Electricity Trading Arrangement. Background Paper 1: Electricity Trading Arrangements in England and Wales", February 1998

[15] OFFER, "Review of Electricity Trading Arrangement. Background Paper 2: Electricity Trading Arrangements in Other Countries", February 1998

[16] PJM Interconnection, www.pjm.com

[17] Transpower of New Zealand, www.transpower.co.nz

[18] N. Deeb and S.M. Shahidehpour: "Linear Reactive Power Optimization In A Large Power Network Using The Decomposition Approach", IEEE Transactions on Power Systems, Vol.5, May 1990, pp.428-438

[19] B.H. Kim and R. Baldick: "Coarse-Grained Distributed Optimal Power Flow", IEEE Transactions on Power Systems, May 1997, 12(2), pp.932-939

[20] A.J. Conejo and J.A. Aguado: "Multi-area Coordinated Decentralized DC Optimal Power Flow", IEEE Transactions on Power Systems, November 1998, 13(4), pp.1272-1278

[21] F.J. Nogales, F.J. Prieto, and A.J. Conejo: "Multi-area AC optimal power flow: a new decomposition approach", Proceeding of the 13th Power System Computer Conference, June 28-July 2, 1999, pp.1201-1206

[22] M.D. Cadwalader, S.M. Harvey, W.W. Hogan, and S.L. Pope: "Coordinating congestion relief across multiple regions", http://ksgwww.harvard.edu/people/whogan/, Oct 7 1999

[23] S.S. Oren and A.M. Ross: "Economic Congestion Relief Across Multiple Regions Requires Tradable Physical Flow-gate Rights", http://www.ucei.berkeley.edu/ucei/PDFDown.html, May 2000
[24] OFGEM, "Transmission Access and Losses under NETA: A Consultation Document", May 2001
[25] M.E. Bradley, U. Bryan, D. Waterhouse, and C.E. Fiford: "Optimisation of Transmission System Constraints in Real Time", GIGRE Symposium on the Impact of Open Trading on Power Systems, Tours, 1997
[26] http://www.ee.washington.edu/research/pstca/rts/pg_tcarts.htm
[27] J. Eckstein: "Parallel Alternating Direction Multiplier Decomposition of Convex Programs", Journal of Optimization Theory and Applications, January 1994, 80(1), pp.39-63
[28] X. Wang, Y.H. Song, and Q. Lu: "A Lagrangian Decomposition Approach to Active Power Congestion Management across Interconnected Regions", IEE Proceeding C, Generation, Transmission and Distribution, Vol.148, No.5, September 2001, pp. 497-503

6.6 Appendix A: Lagrangian Relaxation Decomposition Approach

The Lagrangian Relaxation (LR) decomposition procedure, which is used to solve multi-regional congestion management problem, is presented below.
Assume that the primal problem has the structure below:

$$\text{Minimise: } f(x) \tag{A.1}$$

$$\text{Subject to} \quad \begin{array}{l} a(x) = 0 \\ b(x) \le 0 \\ c(x) = 0 \\ d(x) \le 0 \end{array}$$

where $f(x): R^n \to R$, $a(x): R^n \to R^{\overline{a}}$, $b(x): R^n \to R^{\overline{b}}$, $c(x): R^n \to R^{\overline{c}}$, $d(x): R^n \to R^{\overline{d}}$, and $\overline{a}$, $\overline{b}$, $\overline{c}$ and $\overline{d}$ are scalers.

Constraints $c(x) = 0$ and $d(x) \le 0$ are complex constraints, which should be relaxed to simplify the problem (A.1). Therefore, the Lagrangian function is defined as

$$L(x, \lambda, \mu) = f(x) + \lambda^\dagger c(x) + \mu^\dagger d(x) \tag{A.2}$$

where λ and μ are Lagrange multiplier vectors. Under the local convexity assumptions ($\nabla_x^2 L(x^*, \lambda^*) > 0$) the dual function is defined as

$$\begin{aligned} \phi(\lambda, \mu) = \text{minimise}_x \quad & L(\lambda, \mu, x) \\ \text{subject to} \quad & a(x) = 0 \\ & b(x) \leq 0 \end{aligned} \tag{A.3}$$

The dual function is concave and in general non-differentiable. This is a fundamental fact in the algorithm stated below. The dual problem is then defined as

$$\begin{aligned} \text{maximise}_{\lambda,\mu} \quad & \phi(\lambda, \mu) \\ \text{subject to} \quad & \mu \geq 0 \end{aligned} \tag{A.4}$$

The LR decomposition procedure will be attractive if problem (A.3) can be solved with fixed values of λ and μ. The problem to be solved to evaluate the dual function for the given values $\tilde{\lambda}$ and $\tilde{\mu}$ is the so-called relaxed primal problem, i.e.

$$\begin{aligned} \text{minimise}_x \quad & L(\tilde{\lambda}, \tilde{\mu}, x) \\ \text{subject to} \quad & a(x) = 0 \\ & b(x) \leq 0 \end{aligned} \tag{A.5}$$

Problem (A.5) typically decomposes into the following sub-problems, which can be solved in parallel. The decomposition facilitates its solution, and normally allows physical and economical interpretations.

$$\begin{aligned} \text{minimise}_{x_i, \forall i=1,\ldots,n} \quad & \sum_{i=1}^{n} L(\tilde{\lambda}, \tilde{\mu}, x) \\ \text{subject to} \quad & a_i(x_i) = 0, \quad i = 1, \ldots, n \\ & b_i(x_i) \leq 0, \quad i = 1, \ldots, n \end{aligned} \tag{A.6}$$

Under local convexity assumptions, the local duality theorem says that

$$f(x^*) = \phi(\lambda^*, \mu^*) \tag{A.7}$$

where x^* is the optimum for the primal problem and (λ^*, μ^*) is the optimum for the dual problem. In the non-convex case, given a feasible solution for the primal problem x, and a feasible solution for the dual problem (λ, μ.), the weak duality theorem says that

$$f(x) \geq \phi(\lambda, \mu) \tag{A.8}$$

Therefore, in the convex case the solution of the dual problem provides the solution of the primal problem, while in the non-convex case the objective function value at the optimal solution of the dual problem provides a lower bound to the objective function value at the optimal solution of the primal problem. The difference between the objective function values of the primal and dual problems at their optimal solutions is called the duality gap. Once the solution of the dual problem is achieved, its associated primal problem solution could be unfeasible. So feasibility procedures are usually required.

Lagrange multiplier updating is crucial to implementation of the LR decomposition procedure. Existing methods include the subgradient method, cutting plane method, bundle method, dynamically constrained cutting plane method, and so on.

7. Dynamic Congestion Management

J. Ma, Q. Lu and Y.H. Song

Chapter 7 analyses the dynamic security issue in the restructuring power market. A general framework is proposed to manage this issue in combination with the market mechanism and the power system intrinsic characteristics. Under this framework, the Security Management Market has been specially set up and categorised into several markets based on the market nature of the participants. By bidding into the respective market with their offers, different market participants provide their own control measures for utilisation under emergencies. Based on the available resources in this market, ISO would secure the system in the most economic way. Numerical examples are presented to illustrate the proposed framework.

7.1 Stability Analysis and Control of Power Systems

Stability has always been the key issue in power systems operation. During early stages of the utilisation of electric energy, there was no concern about stability because separated DC systems were applied. When AC systems finally won over DC systems and took the dominant position in modern power systems at the end of the 19th century, the stability problem arose. However, the stability of power systems was not recognised as an important problem until 1920 [1]. From then on, the stable operation of power systems has become a major challenge to electrical engineers. Early stability problems were mainly concerned with the transmission of electrical energy from hydropower station to the load centre. As the development of power engineering as well as the rapid advance in computer technology and control theory occurred, the benefits of integrated modern power systems have been realised. Now, power systems have become one of the most complex systems that human beings make. Integrated power systems not only can be operated in a more economic way, but also can provide emergency assistance. However, the complexity and large scale of modern power systems create great challenges to its stable operation. The damage that can be inflicted by loss of stability in these large power systems cannot be overestimated, as evidenced by world-wide grid failures over the last half-century. During the last decade of the 20th century, new concepts have been introduced into traditional power systems. Originating in the United Kingdom, the restructuring of power systems has been undertaken world-wide, and has brought new challenges to power system stability.

There are different classifications of power system stability based on different criteria [1]. Based on the time scale relevant to the researched stability problem, stability can be classified as static stability, transient stability, short-term stability and long-term stability. From the physical nature of the stability problem, it is normally classified as angle stability or voltage stability. Figure 7.1 gives a sketch of the classification of power system stability as well as its major methods of analysis.

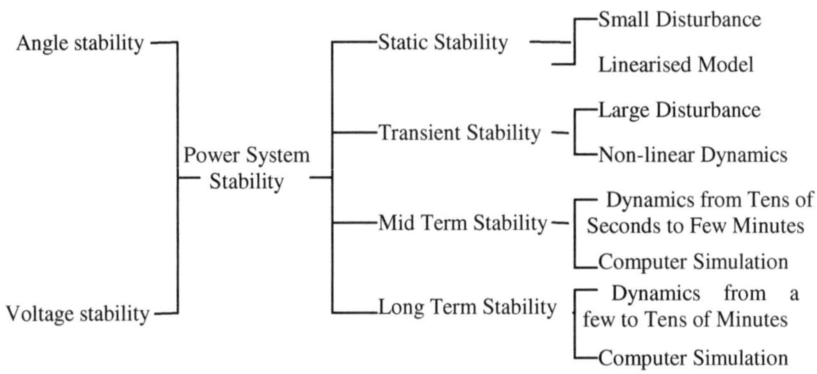

Figure 7.1. Classification of power system stability

Angle stability requires that the power system is operated in synchronism. Voltage stability requires that the voltage profile of the power system lies in the accepted range. During the long history of power system operation, angle stability has been paid more attention than voltage stability. Only in recent years has voltage stability raised increasing concerns. In fact, voltage collapse and losing synchronism are sometimes closely related. Since a power system can experience disturbance at any time, the stability referred to here has two meanings. One is that the system remains at equilibrium under various small disturbances, which is called static stability or small signal stability. To analyse static stability, a linearised power system model can be applied, and eigen value analysis has been the major research method [3]. However, another scenario is that when a system experiences a large disturbance induced often by a utility fault, it can be stabilised at another accepted equilibrium to avoid interruption of the energy supply. Due to the large impact of such a disturbance, system stability may involve dynamics over a much longer time span, which are defined respectively as transient stability, mid-term stability and long-term stability. Under such scenarios, coherent non-linearity of the power system must be considered. Thus, computer simulation becomes the major research method.

The fact that such huge power systems can be operated stably and economically is attributed to the various control facilities distributed around the power system. The dynamic behaviour of the power system is strongly influenced by the control dynamics. The structure of the power system control is depicted in Figure 7.2.

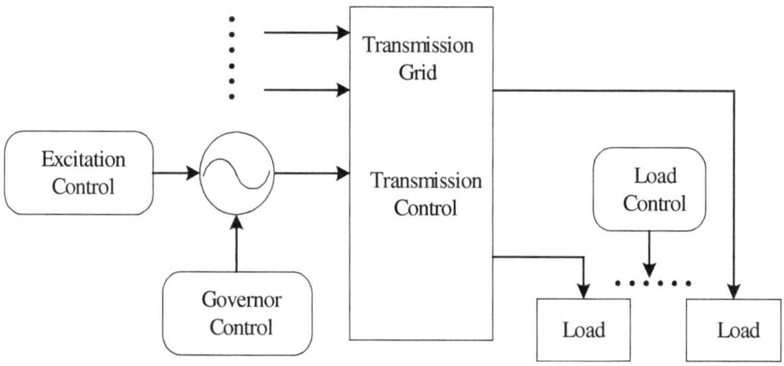

Figure 7.2. Power system control structure

As Figure 7.2 shows, power system control can be classified into three areas, generator control, transmission control and load control. The governor controls the active power generated by the generator, while by adjusting the voltage input of the generator field winding through the excitation control, the generator terminal voltage can be controlled. Governor control and excitation control can be conventional PI controllers [4], or can be non-linear controllers [5]. As far as stability is concerned, transmission control mainly refers to the control of the FACTS device. Relying on power electronic techniques and appropriate control strategies, the FACTS device can maintain the voltage level and adjust the power imbalance under emergencies to enhance the power system stability [6]. Load control measures ensure system stability by controlling the power consumption, which includes under-frequency load-shedding to prevent frequency collapse and under-load tap changing (ULTC) to maintain system voltage level.

Power system controls are so complex that it is impossible to use one diagram to describe them all. Figure 7.2 gives a classification of controls based on their respective control objectives, which is very important in the power market environment. Restructuring in the power market breaks the traditional integrated system structure. Power producer, grid owner as well as energy retailer become independent entities. And with the emergence of the Independent Power Producer (IPP), a nested power system structure has been formed [7]. Prior to the power market, all these control measures belonged to one utility, which fulfilled its responsibility automatically. However, in the market scenario, extra controls from different entities will involve extra individual costs. Naturally, there have been some gaming activities in this respect. How to address this problem has become a major issue in the power market as far as system stability is concerned.

The complexity of power system control in the power market environment not only lies in the different ownership of the control facilities, but also in the different scenarios that may prevail. It is well known that an operating power system may experience the following four states:

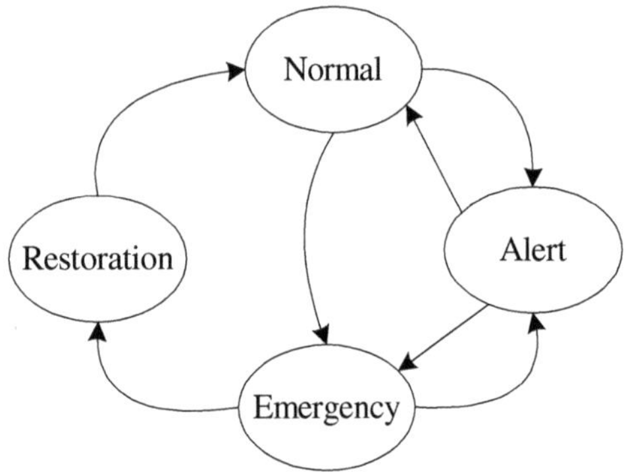

Figure 7.3. Operating states of power system

Due to the unique characteristic of electrical energy, power system operation must meet one equality and several inequalities. The equality is the power balance request, i.e. the load demand must be met by the power generation at any time. The inequalities include the line capacity limit, bus voltage bounds, maximum and minimum power output of the generator, etc. In the normal operating state, the equality and inequalities are all satisfied and the system is operated at a pre-specified equilibrium. When the inequality is not satisfied, such as heavy-loaded transmission lines operated at their maximum transmission capacity, or voltages at some buses below the lowest permitted value, the system enters the alert state. Although in this state, generation can still meet the load demand, the system security level is sharply down. An unexpected disturbance may bring cascading catastrophes to the system. Obviously, proper control at this state may bring the system back to the normal state. Static congestion management presented in other chapters of this book is one such control strategy. If controls in the alert state are not effective, the system may enter an emergency state if the disturbance continues. In the emergency state, the power balance between generation and load does not hold any more. The energy imbalance may lead to the acceleration of some generators and deceleration of others. Without appropriate control facilities, the power system may collapse and lead to power outage over wide areas. By effective control measures, the system may return to the alert state, then, again by appropriate control, the system can return to normal. If the system has collapsed, necessary restoration control measures are needed to restore the system to the normal state. The above analysis clearly shows the importance of the controls in power system operation. On the other hand, it also infers that since different control services corresponding to the different operating states take different responsibilities, they may involve different financial considerations under the market mechanism, which further aggravate the difficulty of stability control in the open power market.

Considering both the different ownership and services of power system control in the power market environment, power system stability analysis may require a completely new approach. The market mechanism brings more competition into power systems, which not only creates opportunities for a more energetic power system, but also imposes potential dangers to system security. How to provide market equilibrium to balance system security as well as individual financial benefits has been debated for a long time and there has been no clear answer until now. Section 7.3 gives a proposal to try to settle this issue in a market way. However, there is still a long way to go to completely solve this problem.

7.2 Stability-constrained Optimal Power Flow

It is well known that the normal operation of a power system can be formulated as an optimisation problem, as has been shown in prior chapters of this book. Mathematically, it takes the following form:

$$\min \quad f(\mathbf{X},\mathbf{U},\mathbf{P}) \tag{7.1}$$

$$\text{s.t.} \quad \mathbf{G}(\mathbf{X},\mathbf{U},\mathbf{P})=0 \tag{7.2}$$

$$\mathbf{H}(\mathbf{X},\mathbf{U},\mathbf{P}) \le 0 \tag{7.3}$$

where f is the objective to be minimised, often corresponding to the total social cost under power market conditions; both vector equality constraints $\mathbf{G}: R^{n\times r\times p} \to R^m$ and inequality constraints $\mathbf{H}: R^{n\times r\times p} \to R^l$ are smooth vector functions. Physically they represent power flow constraints for each node as well as various practical limitations on power system operation, which have been shown in previous chapters. Vectors $\mathbf{X}$, $\mathbf{U}$ and $\mathbf{P}$ with respective dimensions n, r and p correspond to the states, controls and parameters in OPF. Each of them has its physical counterpart in the real situation. States include voltage amplitudes and angles at load nodes; controls can be power generated and the manageable load. The impedance of the transmission line and the fixed load are parameters. Techniques in both linear and non-linear programming have been successfully applied to solve this problem [8].

With the increased demand on system dynamic security, especially after restructuring of the power market, there has been a wide investigation into stability constrained OPF [9–13]. The conventional OPF problem can only give a static generation schedule; however, it can't ensure stability of the power system under possible contingencies. Furthermore, in the power market environment, the system is often operated under stress, which creates potential threats to system security. Sharply decreased stability levels in contrast with increased demand on energy quality require, naturally, the incorporation of stability constraints into the conventional OPF. On the other hand, as stated in [10], 'the traditional trial and error method can produce a discrimination among market players in stressed power systems', so it can't meet the new power market requests.

As far as the stability is concerned, system dynamics must be considered. Different from the equality constraint (7.2) in conventional OPF, a power system has to be represented by a set of differential algebraic equations [9]:

$$\dot{\mathbf{x}}(t)=\mathbf{f}(\mathbf{x}(t),\mathbf{V}(t),\mathbf{u}) \tag{7.4}$$

$$\mathbf{g}(\mathbf{x}(t),\mathbf{V}(t))=0 \tag{7.5}$$

where $\mathbf{x}(t)$ is a vector of states, which consists of the rotor angle, rotor frequency, q-axis transient voltage, etc; $\mathbf{V}(t)$ is a vector of nodal voltages, $\mathbf{u}$ is a vector of control variables, which can be power generation and the voltage level at the PV generator node.

There are several attempts to incorporate stability constraints into the OPF. In [13], Artificial Neural Networks (ANN) are applied to give preventive control measures to assure system transient stability. In [11], the Single Machine Equivalent (SIME) method is used to find the critical machines and respective control measures. Then an iterative algorithm is implemented to find the solution to OPF. The following method is taken from [9-10].

In general, (7.4) and (7.5) can be discretized to numerically equivalent algebraic equations as follows:

$$\mathbf{H}_i(\mathbf{y}_i,\mathbf{y}_{i,r},\mathbf{u})=0 \qquad i=0,1,2,\cdots,n_T \tag{7.6}$$

where

$$\mathbf{y}_i=[\mathbf{x}_i^T,\mathbf{V}_i^T]^T,\quad \mathbf{y}_{i,r}=[\mathbf{y}_{i-1}^T,\mathbf{y}_{i-2}^T,\cdots,\mathbf{y}_{i-r}^T]^T$$

n_T is the total number of time steps with respect to the integration interval [0,T] and r is the number of steps of the adopted implicit discretization rule (where $r \le i$).

Inequality constraints in conventional OPF, such as the thermal limit, voltage limit, generation capacity limit, etc., can also be discretised as:

$$\mathbf{l}_i(\mathbf{y}_i,\mathbf{u})\le 0 \quad i=0,1,2,\cdots,n_T \tag{7.7}$$

In addition to the above inequalities, the stability constraint has to be considered to ensure the transient stability of the power system. In [9], a stability test variable is defined as:

$$P^i=\sum_{j=1}^{n_g}P_{aj}^i(\mathbf{y}_i)(\theta_j^i-\theta_j^s) \tag{7.8}$$

where P_{aj} denotes the accelerating power of the *j*th machine, θ are rotor angles with respect to the centre of inertia (COI) and *i* and *s* refers to the *i*th time step and SEP respectively, n_g is the total number of the generators.

The criterion to judge system stability is as follows:

As long as the defined variable P^i is less than zero at each *i*th time step of the discretised trajectory, the system is stable. Namely, the set of inequality

$$P^i < 0 \quad i = n_{cl}, n_{cl+1}, \cdots, n_T \tag{7.9}$$

must be satisfied, where n_{cl} is the time step corresponding to the fault-clearing time.

If $C_1(\mathbf{u})$ is defined as a cost function, the stability-constrained OPF can be formulated as a problem to minimise $C_1(\mathbf{u})$ constrained by (7.6)–(7.9). Thus, the OPF problem with transient stability constraint has become a static optimisation problem and all developed algorithms in conventional OPF can be applied. Detail of the algorithm can be found in [9].

In [10], a more intuitive stability test is carried out. Instead of judging the inequality (7.9), the following criterion must be satisfied at each time step:

$$\bar{\delta}_i^j = \delta_i^j - (\sum_{k=1}^{n_g} H_k \delta_i^k) \Big/ (\sum_{k=1}^{n_g} H_k) \leq 100^\circ \quad i = n_{cl}, \cdots, n_T \tag{7.10}$$

This can be physically interpreted as the deviation of the individual rotor angle with respect to the COI at each time step cannot be larger than 100 degrees.

The solution of the stability-constrained OPF gives preventive control measures in the case of predefined contingencies. However, the following concerns still need to be further investigated. First, the optimal generation pattern is given under the specified contingency. For a different contingency, the optimal solution becomes meaningless. Second, not all contingencies can be prevented simply by generation redispatching. Special control measures often have to be used to rapidly stabilise the disturbed system. Under some extremes, equilibrium of the post-fault system doesn't exist under the assumption that generation is constant during the optimisation. Third, various control facilities are not considered in the aforementioned stability constrained OPF. In the following section, stability constrained OPF will be formulated as a dynamic optimisation problem instead of a static optimisation problem. The solution to this dynamic optimisation problem will be taken to manage dynamic congestion in the power market environment.

7.3 Market-based Dynamic Congestion Management [14–16]

As stated in previuos sections, the power industry has been undergoing great changes during the last decade. Although the restructured power market encourages full competition among participants, it is widely recognised that there is still a need for regulated monopoly in the operation of transmission systems. Therefore an independent System Operator (ISO) is established to coordinate system operations in light of the criteria of security, economy, and reliability. In the traditional power industry, the obligation to maintain the dynamic security of the power system is distributed to designated power plants. Load curtailment under the contingency is compulsory. However, this mechanism may not be appropriate in a competitive market environment. Although the importance of maintaining the system dynamic security is well known to every market participant, driven by self-interest, none of them is willing to take the responsibility for the system's benefit without being paid. Such self-oriented economic behaviour may finally jeopardise the healthy operation of the power markets.

Although power market security has become the top issue during recent years, much research work has been centred on static security management [17–19], typically, static congestion management and reserve scheduling. The basic idea behind such research work is to ensure real-time balance between power generation and consumption through market mechanisms. It should be noted that all such management can only be effective when the power system is operated in a stable state, namely, situations in which system dynamics has little effect on power system operation. However, such request may not always be met, especially under contingencies. Disturbances may raise oscillation or even infringe system stability. Extra dynamic control measures may be necessary to stabilise the power system on such occasions. Caramanis *et al.* [20] first researched security control and its pricing; then, Berger *et al.* [21] investigated dynamics in Automatic Generation Control (AGC) in a free market framework; following the same line, Baughman *et al.* [22] described a general pricing mechanism for power markets with dynamics involved. However, they didn't give mathematical solutions to the proposed theoretical framework. Although not addressing dynamic security in power markets directly, La Scala *et al.* [9] and Gan *et al.* [10] try to incorporate system dynamics into conventional Optimal Power Flow (OPF) to develop the Stability Constrained OPF tool. Nevertheless, dynamics in the power system is not a snapshot, but a procedure, so formulating it as a static optimisation problem may not be quite appropriate with respect to the nature of dynamics. Singh and David's work [12] provides a good way to redispatch generation by use of an energy function method when system stability is in danger, however, the market nature of energy transactions has not been fully reflected with their proposed method.

This section proposes a market-based dynamic security management framework, which has been sketched in Figure7.4. In this framework, the service market for security management can be categorised according to the nature of the market participants. Market individuals participate in security management by entering this market with their offers and bids. Control measures available in this market can be utilised later by ISO when system security is seriously endangered. Furthermore, based on the proposed framework, system security will not only be ensured, but also be done in an economic way. All this would be shown by Dynamic Congestion Management, a very important issue in market based Dynamic Security Management.

The British new electricity trading arrangement that went live on 27 March 2001 has classified the power market into Multilateral/Bilateral Contract Market (MCM), Pool Auction Market (PAM) and Real-time Balancing Market (RBM).

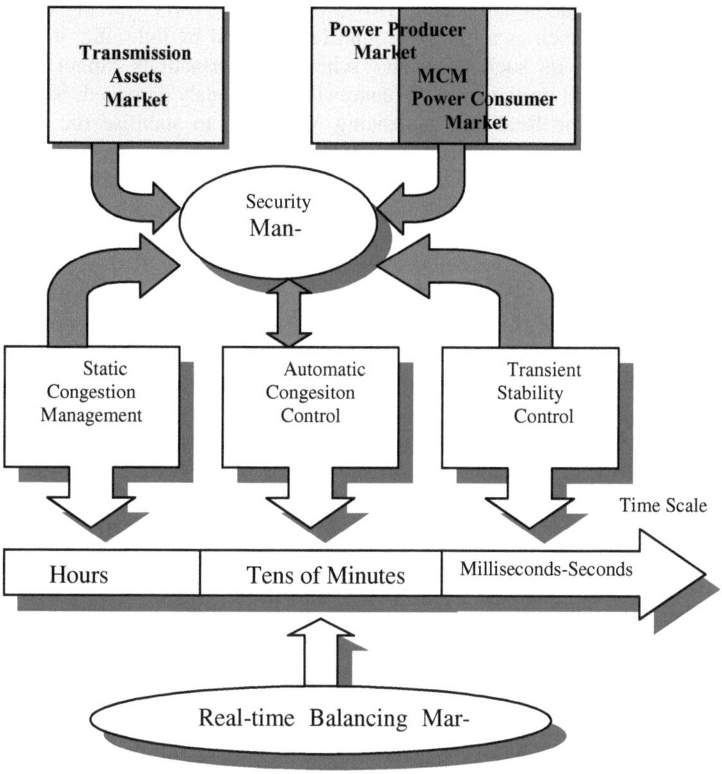

Figure 7.4. Market mechanism for dynamic security management

The real-time Balancing Market has three major tasks to do. First, real-time power production and consumption may exceed the thermal and static stability limit of the transmission line. Redispatching generation and load to eliminate overload of the transmission line is called Static Congestion Management, which is carried out on an hourly time scale. By satisfying the N and (N-1) security criteria [23], Static Congestion Management guards the steady-state security of the power market against contingencies. However, even when the transmission grid is operated within its capacity, undefined power consumption and generation may cause frequency fluctuation, which is called frequency dynamics. Different from transmission congestion, which is mainly due to bottlenecks in the transmission grid, frequency dynamics can be attributed to imbalance between generation and load. Therefore, the second task of the real-time balancing market is to eliminate frequency dynamics by adjusting generator power output, generally called Automatic Generation Control (AGC). AGC is a kind of slow dynamics, the time scale for which can be several minutes to half an hour. NERC operation guides require that the Area Control Error returns to zero within 10 minutes of previously reaching zero [24]. Under AGC control, the effects on power systems of frequent but small energy imbalances can be attenuated. However, when a power system experiences large disturbances, such as a line to ground fault caused by lightning, it may lose stability. Generally, on such occasions, scheduled transactions cannot be maintained and the market participant's revenue will be seriously damaged. So the third important task of the Real-time Balancing Market is to stabilise the post-fault power system and to maintain the scheduled transactions as much as possible. Compared with AGC, transient stability control involves a sudden and large energy imbalance that may lead to system collapse if dealt with improperly. The time scale for transient stability control is quite short, usually from tens of milliseconds to several seconds.

Driven by economic interest, market individuals in a competitive power market always try to maximise their own profit. As a natural result, the responsibility to maintain system dynamic security is often neglected. To promote market participants' involvement in Dynamic Security Management, a proper market mechanism must be set up. In Figure 7.4, a special Security Management Market is shown, which consists of Transmission Assets Market (TAM), Power Producer Market and the Power Consumer Market. Characterised by its identity, different market participants enter the respective market with available dynamic security services together with bids. The transmission Assets Market provides grid owned dynamic control services, such as FACTS devices, energy storage devices, etc. The Power Producer Market provides controls from the generator side, such as pilot node voltage control, steam turbine fast-valving or generator tripping etc., while the Power Consumer Market provides a load-shedding service. Since the originally scheduled energy transactions will inevitably be affected during dynamic security management, the bid corresponding to the service would comprise two cost parts. One is the variable cost due to implementation of the controls, while the other is the compensation for the financial loss due to altering energy transactions.

Generally speaking, the control measures in the Transmission Assets Market have the lowest bid due to the following facts. First, fully utilising offers in this market will not affect the scheduled transactions, which minimises the financial loss relevant to altering those transactions; second, services available in this market are often specially invested for system security management. Under normal operating conditions they are usually inactive. Only when the system suffers serious disturbances do they take effect. Third, offers in this market financially are long-term investments and lower bids can make it easier to enter the market and obtain their reimbursement. In contrast to TAM, offers in MCM, shown in Figure 7.4 as the intersection between the Power Producer Market and Power Consumer Market, have the highest bids due to the fact that curtailing transactions in this market would affect more than one side's benefits as well as long-term energy transactions. Naturally, transactions signed between market participants and the ISO in the Pool Auction Market will be curtailed prior to multi-lateral transactions since they usually have lower bids in the Security Management Market. It should be noted that, different from the conventional ancillary market, control strategies have also been introduced into the market under the proposed framework. Since maintaining system security needs more effort than conventional control, the extra cost incurred during Dynamic Security Management will be reimbursed according to the respective bids.

For power system security, sufficient control measures must be available in this market. A compromise between system security desired by ISO and the individual profit required by market participants has to be reached. Technically, such a compromise should be based on solving off-line security-constrained optimal power flow by ISO hourly; financially, the hierarchical bids described above not only reimburse market participants for their services, but also provide incentives to bring new control technologies into this service market with higher bids.

Transient stability control shown in Figure 7.4 is one of the special tasks for ISO. It is well known that power systems may undergo discrete changes in system configuration due to outage and contingencies, which will affect system dynamic performance and might threaten system stability. In this respect, the ISO must utilise available resources to maintain system security and reliability. This process is called dynamic congestion management (DCM).
The market position of Dynamic Congestion Management is illustrated in Figure 7.5.

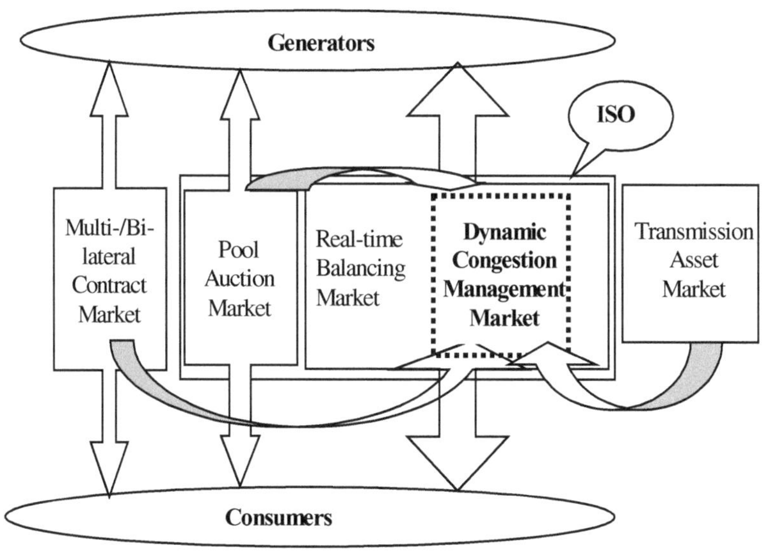

Figure 7.5. Dynamic congestion management in the power market

A market for Dynamic Congestion Management has been set up in the Real-time Balancing Market. Since DCM is only part of the Dynamic Security Management previously described, market participants in MCM, PAM and TAM would bid into this market following the same philosophy analysed in previous paragraphs.

The objective of Dynamic Congestion Management is twofold: to stabilise the power system and improve its dynamic performance; to minimise management cost. In other words, ISO would try to secure the system in the most economical way based on the available control resources in the respective market. To achieve such an objective, appropriate Total Dynamic Congestion Management Cost (TDCMC) has to be defined, which should consist of the System Dynamic Disturbance Cost (SDDC) and Congestion Management Cost (CMC). The SDDC is stipulated according to the ISO's stability requirement on the post-fault power system. It would reflect not only the severity of the disturbance on power system stability, but also market participants' requests as well as tolerances on it. Obviously, a competitive market environment provides a chance for market participants themselves to choose security criteria. The other part of the TDCMC consists of the extra control cost incurred by the DCM, which certainly is based on the bids submitted by involved market participants, namely, pay as he bids. Under the proposed market framework, ISO's responsibility is to minimise the Total Dynamic Congestion Management Cost by fully utilising available resources in the respective markets.

According to different market participants in DCMM, there exist two models for DCM. One is DCM with only RBM generators, the other is coordination between RBM generators and MCM generators. Since the participation of MCM generators in DCM will affect contracts signed day ahead and more than one side of the market participants will be involved, those MCM generators may, normally, enter DCMM with higher bids than those from RBM generators. As a natural result, the coordination model may lead to higher CMC, however, with more generators available in DCMM, ISO has more alternatives and higher manageable capabilities to mitigate the disturbing effects of the fault on the power markets.

Based on the framework of DCM proposed above, the following dynamic optimisation problem is formulated:

$$\begin{aligned} Min: J &= \Delta\boldsymbol{\omega}^T(t_f)\mathbf{Q}\Delta\boldsymbol{\omega}(t_f) \\ &+ \int_0^{t_f} \left[\Delta\boldsymbol{\delta}^T(t)\mathbf{R}\,\Delta\boldsymbol{\delta}(t) + \Delta\boldsymbol{\omega}^T(t)\mathbf{Q}\,\Delta\boldsymbol{\omega}(t)\right]dt \\ &+ \int_0^{t_f} (\Delta\mathbf{U}_V^T\mathbf{B}_1\,\Delta\mathbf{U}_V + \Delta\mathbf{U}_E^T\mathbf{B}_2\,\Delta\mathbf{U}_E)dt \end{aligned} \tag{7.11}$$

The objective consists of three parts. The first one is the criterion that requires the frequency oscillation to be quenched on the post-fault system; the summation of the two integrals corresponds to the TDCMC with the second item of the objective to SDDC and the third one to CMC. The choice of the respective cost is based on the principle stated before and may not be unique. The weighting matrices $\mathbf{R}$ and $\mathbf{Q}$ are based on the participants' tolerance of the disturbance induced by the fault, while $\mathbf{B}_1$ and $\mathbf{B}_2$ reflect the bids for available resources in DCMM with $\mathbf{B}_1$ corresponding to fast valve control and $\mathbf{B}_2$ to pilot voltage control, in the example here.

Under the market mechanism, the dynamics of different market participants must be modelled individually. The *i*th generator who enters the DCM with the fast valve control will be described by [5]:

$$\frac{d\delta_i}{dt} = \omega_0 \Delta\omega_i \tag{7.12}$$

$$H_i \frac{d\Delta\omega_i}{dt} = P_{Hi} + C_{ml} P_{m0i} - P_{ei} - D_i \Delta\omega_i \tag{7.13}$$

$$T_{Hi} \frac{dP_{Hi}}{dt} = -P_{Hi} + C_H P_{m0i} + C_H \Delta u_i \tag{7.14}$$

where Equations(7.12) and (7.13) are generator swing equations, while (7.14) describes turbine dynamics. For these generators, turbine control is provided with bids in the DCMM and can be utilised when needed.

For generators who show no interests in dynamic congestion management, in addition to Equations (7.12)–(7.14), the local governor dynamics will be included as:

$$T_{Govj} \frac{d\Delta u_j}{dt} = -\Delta u_j - \alpha \cdot K_A \Delta\omega_j \tag{7.15}$$

Here the valve is controlled by the local governor and is available in the DCMM. However, this local control can be bypassed because of economic considerations or gaming, which has been observed in AGC implementation. The coefficient α is a local participating factor. If it is zero, then the local controller is not in effect (since the initial value of Δu_j is zero).

By controlling the generator field voltage input, the excitation system maintains the voltage of the generator node at the prespecified level. The dynamics due to the excitation system can be described by [7]:

$$T_{d0i} \frac{dE'_{qi}}{dt} = -E'_{qi} - (x_{di} - x'_{di}) I_{di} + E_{fdi} \tag{7.16}$$

$$T_{Ei} \frac{dE_{fdi}}{dt} = -(K_{Ei} + S_{Ei}) E_{fdi} + V_{Ri} \tag{7.17}$$

$$T_{Ai} \frac{dV_{Ri}}{dt} = -V_{Ri} - K_{FRi} (E'_{qi} - E_{refi}) \tag{7.18}$$

Equations (7.16), (7.17) and (7.18) correspond to, respectively, the generator field dynamics, the typical exciter dynamics as well as the voltage regulator dynamics.

Most of the voltage control from the generator is local behaviour, which takes into effect automatically responding to voltage fluctuations. However, completely distributed voltage control driven by individual financial motivation in the power market may not be enough to cope with all the operating situations, especially when a contingency emerges. The voltage adjustment from the pilot node will enter DCMM as the available control resource to stabilise the system. Instead of Equation (7.18), the dynamics of voltage control from the pilot node in DCMM will be:

$$T_{Ai}\frac{dV_{Ri}}{dt} = -V_{Ri} - K_{FRi}(E'_{qi} - E_{refi} - \Delta E_{refi}) \tag{7.19}$$

Other constraints include the active and reactive power balance equation:

$$P_{ei} = E'^2_{qi}G_{ii} + E'_{qi}\sum_{\substack{k=1\\k\neq i}}^{N} E'_{qk}(G_{ik}\cos\delta_{ik} + B_{ik}\sin\delta_{ik}) \tag{7.20}$$

$$Q_{ei} = -E'^2_{qi}B_{ii} + E'_{qi}\sum_{\substack{k=1\\k\neq i}}^{N} E'_{qk}(G_{ik}\sin\delta_{ik} - B_{ik}\cos\delta_{ik}) \tag{7.21}$$

and the physical or contractual limits of the offers imposed on the control variables in DCM:

$$\mathbf{S}(\Delta\mathbf{U}_V, \Delta\mathbf{U}_E) \leq \mathbf{0} \tag{7.22}$$

Neglecting the effects of damper winding, the following equation holds [7]:

$$Q_{ei} = E'_{qi}I_{di} \tag{7.23}$$

The load dynamics can also be considered. To include the load dynamics, the structure of Bergen–Hill's model [25] used in stability analysis is adopted here:

$$-P_{D0l} - D_l\frac{d\delta_l}{dt} - P_{el} = 0 \tag{7.24}$$

where P_{D0l} is the pre-fault load contracted in the power pool market or bilateral market.

By optimizing the objective (7.11) under the dynamic as well as static constraints (7.11)–(7.24), the available measures in DCMM are utilised in an optimal way to eliminate the congestion raised by the stability concern.

To solve the dynamic optimisation problem presented, the Hamiltonian function is formed as follows:

$$
\begin{aligned}
\widetilde{H} &= \Delta\boldsymbol{\delta}^T(t)\mathbf{R}\Delta\boldsymbol{\delta}(t) + \Delta\boldsymbol{\omega}^T(t)\mathbf{Q}\Delta\boldsymbol{\omega}(t) \\
&+ \Delta\mathbf{U}_V^T\mathbf{B}_1\,\Delta\mathbf{U}_V + \Delta\mathbf{U}_E^T\mathbf{B}_2\,\Delta\mathbf{U}_E \\
&+ \sum_{i=1}^{N}\alpha_i\Delta\omega_i\omega_0 \\
&+ \sum_{i=1}^{N}\frac{\beta_i}{H_i}(P_{Hi} + C_{mi}P_{m0i} - P_{ei} - D_i\Delta\omega_i) \\
&+ \sum_{i=1}^{N}\frac{\nu_i}{T_{Hi}}(-P_{Hi} + C_H P_{m0i} + C_H\Delta u_i) \\
&+ \sum_{j=1}^{N-N_0}\frac{\xi_j}{T_{GOVi}}(-\Delta u_j - \alpha\cdot K_A\Delta\omega_j) \\
&+ \sum_{i=1}^{N}\frac{\gamma_i}{T_{d0i}}[-E'_{qi} - (x_{di} - x'_{di})I_{di} + E_{fdi}] \\
&+ \sum_{i=1}^{N}\frac{\rho_i}{T_{Ei}}[-(K_{Ei} + S_{Ei})E_{fdi} + V_{Ri}] \\
&+ \sum_{i=1}^{N_1}\frac{\mu_i}{T_{Ai}}[-V_{Ri} - K_{FRi}(E'_{qi} - E_{refi} - \Delta E_{refii})] \\
&+ \sum_{i=N_1+1}^{N}\frac{\mu_i}{T_{Ai}}[-V_{Ri} - K_{FRi}(E'_{qi} - E_{refi})] \\
&+ \sum_{l=1}^{N_L}\frac{\varepsilon_l}{D_l}[-P_{D0l} - P_{el}] \qquad (7.25) \\
&+ \sum_{i=1}^{N}\eta_i[E'^2_{qi}G_{ii} + E'_{qi}\sum_{\substack{k=1\\k\neq i}}^{N}E'_{qk}(G_{ik}\cos\delta_{ik} + B_{ik}\sin\delta_{ik}) - P_{ei}] \\
&+ \sum_{i=1}^{N}\sigma_i[-E'^2_{qi}B_{ii} + E'_{qi}\sum_{\substack{k=1\\k\neq i}}^{N}E'_{qk}(G_{ik}\sin\delta_{ik} - B_{ik}\cos\delta_{ik}) - Q_{ei}] \\
&+ \sum_{i=1}^{N}\tau_i(E'_{qi}I_{di} - Q_{ei}) \\
&+ \sum_{l=1}^{M}\kappa_l S_l(\Delta u_i, \Delta E_{refi})
\end{aligned}
$$

Define

$$\mathbf{X} = [\boldsymbol{\delta}, \Delta\boldsymbol{\omega}, \mathbf{P_H}, \Delta\mathbf{u}, \mathbf{E'_q}, \mathbf{E_{fd}}, \mathbf{V_R}]$$

$$\boldsymbol{\lambda} = [\overline{\partial}, \overline{\beta}, \overline{\nu}, \overline{\xi}, \overline{\gamma}, \overline{\rho}, \overline{\mu}]$$

$$\Phi(t, \mathbf{X})_{t_f} = \Delta\boldsymbol{\omega}^T(t_f)\mathbf{Q}\Delta\boldsymbol{\omega}(t_f)$$

The necessary condition to achieve the optimum [26] will be:

$$\dot{\mathbf{X}} = \frac{\partial\widetilde{H}}{\partial\boldsymbol{\lambda}} \qquad (7.26)$$

$$\dot{\boldsymbol{\lambda}} = -\frac{\partial \tilde{H}}{\partial \mathbf{X}} \tag{7.27}$$

$$\mathbf{X}(0) = \mathbf{X}_0 \tag{7.28}$$

$$\boldsymbol{\lambda}(t_f) = \left.\frac{\partial \Phi(t, \mathbf{X})}{\partial \mathbf{X}}\right|_{t_f} \tag{7.29}$$

$$\frac{\partial \tilde{H}}{\partial \Delta \mathbf{U}_V} = 0 \tag{7.30}$$

$$\frac{\partial \tilde{H}}{\partial \Delta \mathbf{U}_E} = 0 \tag{7.31}$$

In dynamic optimisation theory, Equations (7.26) and (7.27) are called state and co-state dynamic equations, respectively, with their boundary conditions as Equations (7.28) and (7.29). It should be noted that these boundary conditions limit the initial conditions on the state variables, and the final conditions on the co-state variables. Mathematically, it is infeasible to obtain an analytical solution to Equations (7.26)–(7.31) due to the complexity of the system. Thus, many numerical algorithms have been proposed to solve such dynamic optimisation problems [26,27]. Here a revised Quasi-Newton Method is used to find the optimal solution. The flowchart for the algorithm is illustrated in Figure 7.6.

In Figure 7.6, $u^{(0)}$ is the initial control guess, while $u^{(n)}$ is the control at the beginning of each iteration. The first step taken in this method is a Cauchy method. Following iterations calculate $\bar{h}$ based on the equation given in the algorithm until it is less than the predefined error ε. For every iteration a line search algorithm [27] is used to find α^* that minimises the objective J. The evaluation of objective J consists of two stages. First, Equations (7.26) and (7.28) are integrated to find the state variables under the control law $u^{(n)}$; second, the objective (7.11) is evaluated using the controls and states in the current iteration; the ∇J with respect to the control can be computed using a finite difference approximation, however, this is computationally expensive. A more efficient method is to find the co-state variables by integrating Equations (7.27) and (7.29), then ∇J can be evaluated based on Equations (7.30) and (7.31). The details of the algorithm can be found in [27].
The optimal solution presents an economic way to stabilise the disturbed power system in a competitive power market environment. A compromise is reached between system security and the economic benefit of individual market participants, which will be shown in detail by case studies in the following section.

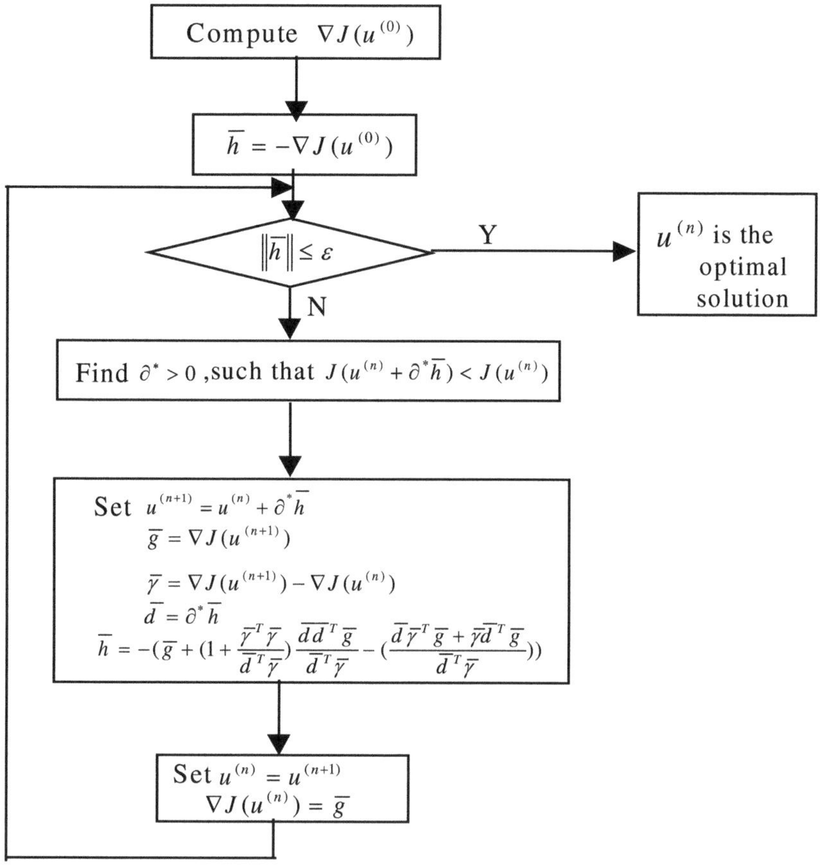

Figure 7.6. Computer algorithm flowchart

7.4 Case Studies and Analysis

The IEEE 30-bus system, shown in Figure 7.7, is used here to demonstrate the proposed framework of the DCM. The network structure and the initial power injections can be found in [28]. The market structure is shown in Table 7.1. There are three generating companies. The GenCo A has generators G1 and G8 with the former arranged in PAM and the latter contracted in the MCM. The power pool generator G13 and the MCM generator G2 belong to GenCo B. All generators owned by GenCo C, namely G11 and G5, are contracted in the MCM.

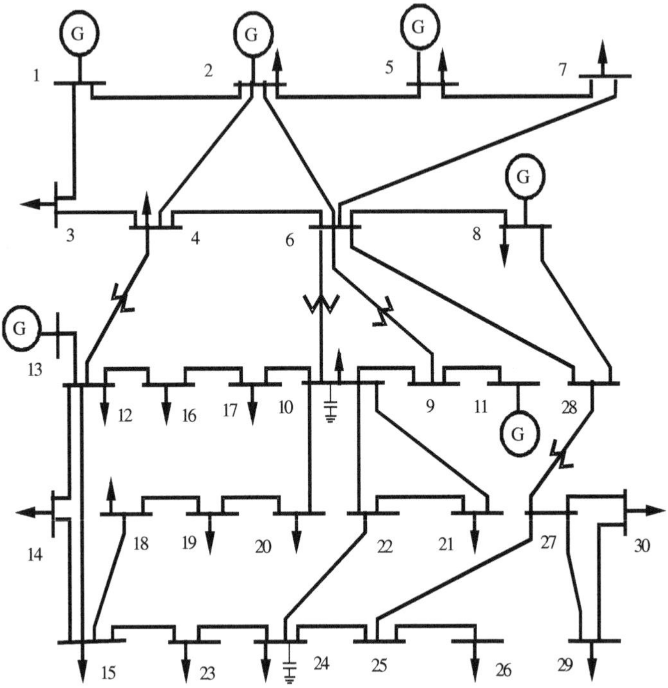

Figure 7.7. Base case of the IEEE 30-bus test system

Table 7.1. The market structure

Generating Companies	PAM	MCM
GenCo A	G1	G8
GenCo B	G13	G2
GenCo C		G11 & G5

The power consumption at bus 10 and bus 17 are increased, respectively, to 108 MW and 14 MW to aggravate system loading. Then, the following contingency is imposed on this test system. At t=0.5s, a three-phase short circuit occurs at one end of line 6-28 near bus 6. At t=0.55s, the circuit breaker at the near fault end of the line disconnects the faulty line from bus 6 and at t=0.6s, the circuit breaker at the remote end of line 6-28 disconnects the faulty line from bus 28. After the fault is cleared, the breakers at both ends of the line are reclosed at t=1s to restore the pre-fault network structure. Individual bids, adjustable offers as well as the relevant cost of the following cases are summarised in Table 7.2.

No DCM is involved in Case I. The system loses its stability and the scheduled market transactions cannot be maintained. The rotor dynamics of this case is shown in Figure 7.8.

Cases II, III as well as IV correspond to the RBM management model when only RBM generators enter DCMM in these cases. In Case II, fast valve control is the only available resource in DCMM, however, it is not enough to eliminate the dynamic congestion. After voltage control is available in DCMM, pre-contingency energy contracts and system stability can be maintained in Cases III and IV. The slight difference between these two cases is that the local governor controls from the MCM generators are bypassed in Case IV, which leads to a smaller increase in SDDC and CMC compared with Case IV. The effects of the local governor controls from MCM generators are clearly shown here. The rotor angle as well as the frequency dynamics in Case III are shown respectively in Figures 7.8 and 7.9. The fast valve control in DCM of this case is shown in Figure 7.10.

Case V and Case VI illustrate the coordination between RBM generators and MCM generators in DCM since GenCo C enters the DCMM with the MCM generator G11. Compared to DCM with RBM generators only, the SDDC decreases. This clearly indicates that the coordination model is more beneficial to the stability of the power system. Although the CMC increases due to the higher bid from the MCM generator in DCMM, the TDCMC still decreases.

Figure 7.11 shows the fast valve control in DCM in Case V while Figure 7.12 corresponds to that in Case VI. Due to the much higher bid from the MCM generator G11 in Case VI, the control from G11 in Case VI is smaller and the time spent at the upper limit is shorter than in Case V, which can clearly be seen from Figures 7.11 and 7.12. Compared with Figure 7.10, both Figures 7.11 and 7.12 show that the involvement of the MCM generator G11 shortens the time interval over which RBM generators stay at their upper limits. The respective excitation controls in DCM of Case III, V together with VI are shown in Figure 7.9. The amplitudes of the excitation control are different corresponding to different valve control strategies in different cases. It reaches the highest amplitude in Case III since in this case only RBM generators participate in the DCM and the available resources in DCMM are limited. The excitation control reaches its lowest value in Case V compared with other cases due to the co-ordinated DCM model and the fact that the bid from the MCM generator is not too high. The differences among these control strategies under different scenarios demonstrate the optimal utilisation of the available resources in DCMM to eliminate the dynamic congestion.

Table 7.2. Optimal dynamic congestion management

Cases	Offers		Bids	SDDC ($)	CMC ($)	TDCMC ($)
Case I	No Dynamic Congestion Management			The system loses stability under the scheduled contracts.		
Case II	G1 in GenCo A as well as G13 in GenCo B offers fast valve control	$\lvert\Delta u_1\rvert \le 1$ $\lvert\Delta u_{13}\rvert \le 1$	$B_1 = 10\$/\Delta u$ $B_{13} = 12\$/\Delta u$	No feasible control solution to eliminate the congestion		
Case III	G1 in GenCo A and G13 in GenCo B offer fast valve control; G1 also offers the voltage control	$\lvert\Delta u_1\rvert \le 1$ $\lvert\Delta u_{13}\rvert \le 1$ $\lvert\Delta EQ_{ref1}\rvert \le 0.05$	$B_1 = 10\$/\Delta u$ $B_{13} = 12\$/\Delta u$ $B_{EQ} = 20\$/\Delta u_E$	161.5000	27.7454	189.2454
Case IV	G1 offers voltage and fast valve control; G13 offers fast valve control, multilateral gen-erators' local governor controllers are not in effect	$\lvert\Delta u_1\rvert \le 1$ $\lvert\Delta u_{13}\rvert \le 1$ $\lvert\Delta EQ_{ref1}\rvert \le 0.05$	$B_1 = 10\$/\Delta u$ $B_{13} = 12\$/\Delta u$ $B_{EQ} = 20\$/\Delta u_E$	161.9738	27.7971	189.7709

Case V	G1 offers voltage and fast valve control; G13 offers fast valve control; G11 offers fast valve control	$\lvert\Delta u_1\rvert \le 1$ $\lvert\Delta u_{13}\rvert \le 1$ $\lvert\Delta u_{11}\rvert \le 1$ $\lvert\Delta EQ_{ref1}\rvert \le 0.05$	$B_1 = 10\$/\Delta u$ $B_{13} = 12\$/\Delta u$ $B_{11} = 25\$/\Delta u$ $B_{EQ} = 20\$/\Delta u_E$	124.8258	34.7657	159.5915
Case VI	G1 offers voltage and fast valve control; G13 offers fast valve control; G11 offers fast valve control	$\lvert\Delta u_1\rvert \le 1$ $\lvert\Delta u_{13}\rvert \le 1$ $\lvert\Delta u_{11}\rvert \le 1$ $\lvert\Delta EQ_{ref1}\rvert \le 0.05$	$B_1 = 10\$/\Delta u$ $B_{13} = 12\$/\Delta u$ $B_{11} = 50\$/\Delta u$ $B_{EQ} = 20\$/\Delta u_E$	131.1819	38.2641	169.4460

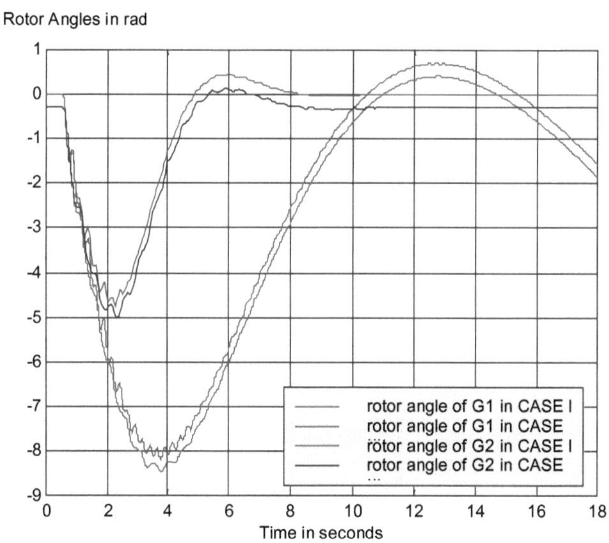

Figure 7.8. The dynamics of rotor angles in Case I and Case III

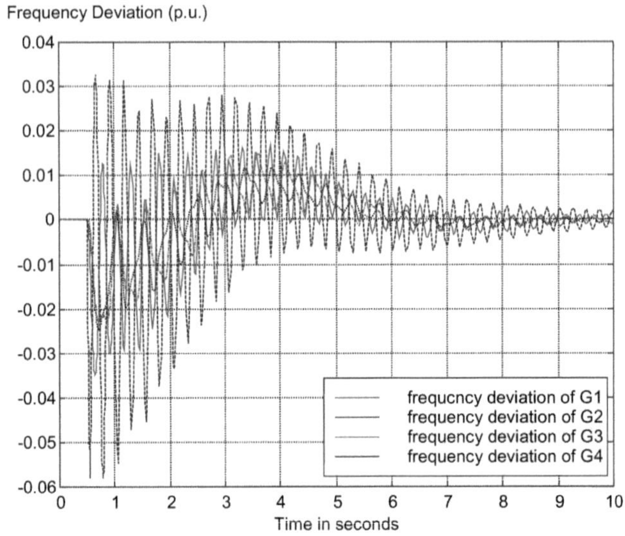

Figure 7.9. The dynamics of frequency in Case III

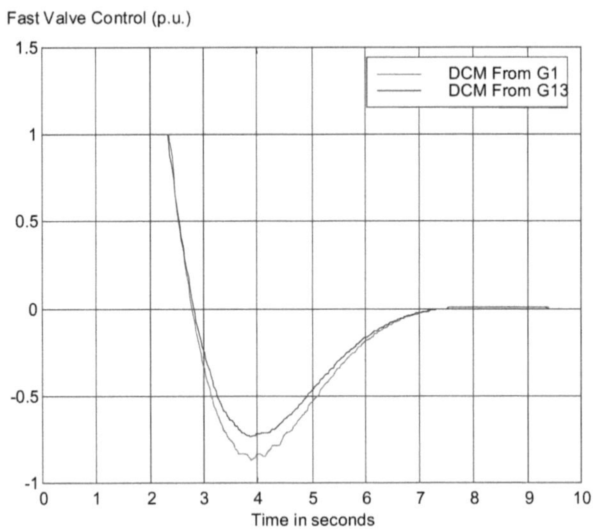

Figure 7.10. Fast valve control in Case III

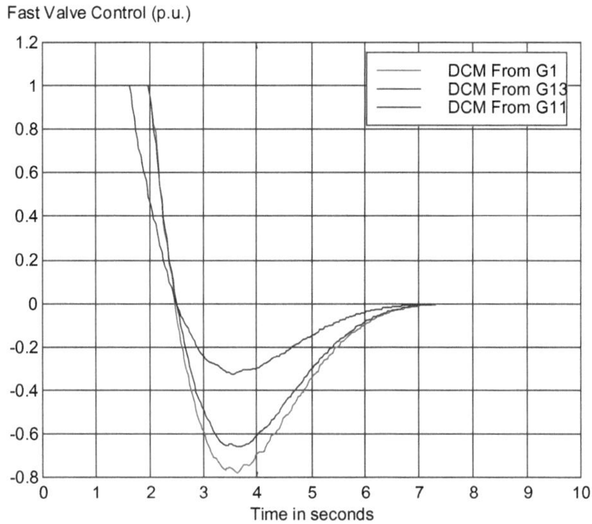

Figure 7.11. Fast valve control in Case V

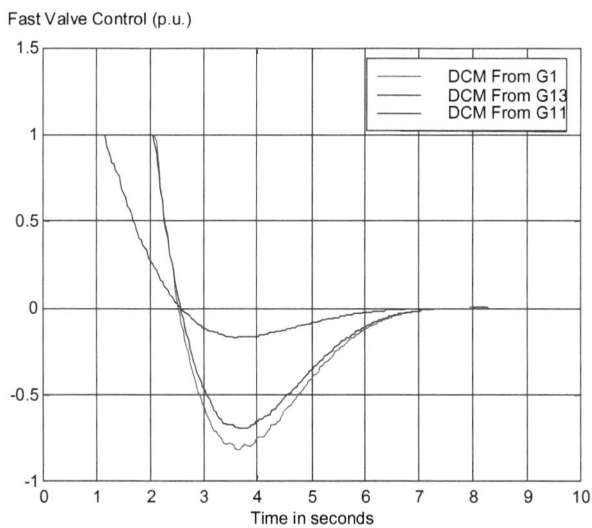

Figure 7.12. Fast valve control in Case VI

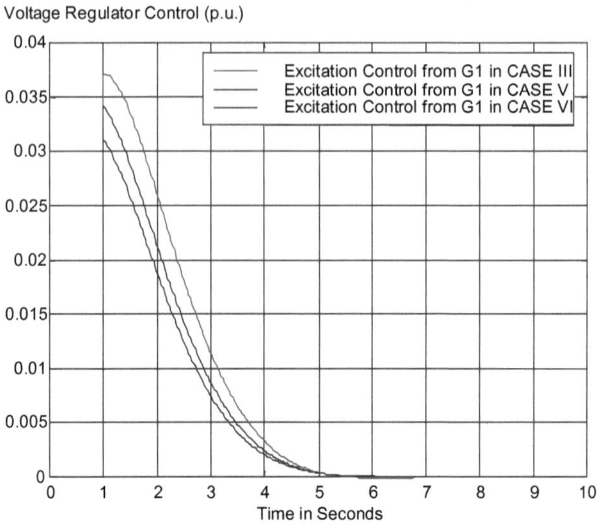

Figure 7.13. Excitation control in Case III, V and VI

Now we assume that the transmission owners also enter the DCM market with control facilities. The assumed scenario could be Superconducting Magnetic Energy Storage (SMES) installed on Bus 8 and also bids into the DCM market. The dynamics of SMES can be found in [29] and could be conveniently incorporated into the dynamic optimisation problem as proposed in the previous section. Table 7.3 presents two cases with SMES involved in DCM.

Compared with Case II and Case V respectively, Case VII and VIII clearly show that involvement of SMES control from the Transmission Asset Market sharply reduces the SDDC and ensures better dynamic response during the large disturbance. This confirms the analysis in the previous section. Although the CMC is higher than that without the participation of SMES, the Total Dynamic Congestion Management Cost is reduced and security still plays a more important role in this example system. It should be noted that the increase of the CMC due to the participation in multi-lateral generator is smaller in Cases VII and VIII than that in Cases III and V. This fact demonstrates that the participation of cheaper control facilities from TAM reduces the utilisation of the expensive generator unit so that the Total Dynamic Congestion Management Cost can be minimised.

Table 7.3. Dynamic congestion management with SMES involved

Cases	Offers		Bids	SDDC($)	CMC($)	TDCMC ($)
Case VII	SMES control is provided in TAM; G1 in GenCo A as well as G13 in GenCo B offers turbine control; G1 also offers the voltage control	$\lvert\Delta u_1\rvert \le 1$ $\lvert\Delta u_{13}\rvert \le 1$ $\lvert\Delta EQ_{ref1}\rvert \le 0.05$	$B_{SMES} = 5\$/\Delta u_{SMES}$ $B_1 = 10\$/\Delta u$ $B_{13} = 12\$/\Delta u$ $B_{EQ} = 20\$/\Delta u_E$	145.2258	33.8237	179.0495
Case VIII	SMES control is provided in TAM; G1 in GenCo A as well as G13 in GenCo B offers turbine control; G1 also offers the voltage control; Multilateral generator G11 also enters DSMM by providing turbine control.	$\lvert\Delta u_1\rvert \le 1$ $\lvert\Delta u_{13}\rvert \le 1$ $\lvert\Delta u_{11}\rvert \le 1$ $\lvert\Delta EQ_{ref1}\rvert \le 0.05$	$B_{SMES} = 5\$/\Delta u_{SMES}$ $B_1 = 10\$/\Delta u$ $B_{13} = 12\$/\Delta u$ $B_{11} = 25\$/\Delta u$ $B_{EQ} = 20\$/\Delta u_E$	116.6715	37.9318	154.6033

7.5 Conclusions and Future Work

This chapter analyses dynamic security in the restructuring power market. A general framework is proposed to manage this issue in combination with the market mechanism and the power system intrinsic characteristics. In this framework, the Security Management Market has been specially set up and categorised into three markets based on the market nature of the participants.

By bidding into the respective market with their offers, different market participants provide their own control measures for utilisation in emergencies. Based on the available resources in this market, ISO would secure the system in the most economic way. Thus, power system security can be ensured under the market mechanism.

Dynamic security in the power market environment is a completely new issue. Although much research work has been dedicated to the incorporation of dynamic security into the routine market operation, the solution so far has been far from satisfactory. How to fully play the role of the market in this special field is still an open question. Proper market rules should be stipulated to stimulate, but also regulate the involvement of market participants with different market objectives. Technically, efficient algorithms for on-line dynamic security analysis must be developed to ease the intensive computation burden of current versions. The authors hope that this chapter may stimulate new ideas in this field to promote a secure and efficient power market.

7.6 References

[1] P. Kundur, "Power System Stability and Control", McGraw-Hill, Inc., 1993
[2] E. Mariani and S.S. Murthy, "Control of Modern Integrated Power Systems", Springer-Verlag, 1997
[3] G. Rogers, "Power System Oscillations", Kluwer Academic, 2000
[4] Yao-nan Yu, "Electric Power System Dynamics", Academic Press, 1983
[5] Qiang Lu, Yuanzhang Sun, Shengwei Mei, "Nonlinear Control Systems and Power System Dynamics", Kluwer Academic Publishers, 2001
[6] Narain G. Hingorani and Laszlo Gyugyi "Understanding FACTS - Concepts and Technology of Flexible AC Transmission System", Institute of Electrical and Electronics Engineers, Inc., New York, 1999
[7] M. D. Ilic, S. Liu, "Hierarchical Power Systems Control: Its Value in a Changing Industry", Springer, 1996
[8] Rao S.S. "Optimization: theory and applications", Halsted Press, New York, 1984
[9] M. La Scala, M. Trovato, C. Antonelli, "On-Line Dynamic Preventive Control: an Algorithm for Transient Security Dispatch", IEEE Trans. on Power Systems, Vol. 13, No.2, pp. 601-608, May 1998
[10] D. Gan, R. J. Thomas, D. Zimmerman, "Stability-Constrained Optimal Power Flow", IEEE Trans. on Power Systems, Vol. 15, No.2, pp. 535-540, May 2000
[11] A. L. Bettiol, D. Ruiz-Vega, D. Ernst, L. Wehenkel, M. Pavella, "Transient Stability Constrained Optimal Power Flow", International Conference on Electric Power Engineering, 1999, pp. 62
[12] S. N. Singh, A. K. David, "Towards Dynamic Security-Constrained Congestion Management in Open Power Market", IEEE Power Engineering Review, pp. 45-47, August 2000
[13] V. Miranda, J. N. Fidalgo, J. A. Pecas Lopes, L. B. Almeida, "Real-time Preventive Actions for Transient Stability Enhancement with a Hybrid Neural Network-Optimization Approach", IEEE Trans. on Power Systems, vol.10, No.2, May, 1995
[14] Jin Ma, Y.H.Song, Qiang Lu, Shengwei Mei, "Market-Based Dynamic Congestion Management", IEEE Power Engineering Review, May, 2002
[15] Jin Ma, Y.H. Song, Qiang Lu, Shengwei Mei, "Framework for Dynamic Congestion Management in Open Power Market", IEE Proc.-Gener. Transm. Distrib., Vol.149, No.2, March 2002
[16] Jin Ma, Qiang Lu, Shengwei Mei, Y.H. Song, "Dynamic Security Management in Open Power Market", Proceedings of IEEE CRIS'2002, Beijing, China, Sept. 23-27, 2002

[17] R. D. Christie, B. F. Wollenberg, Ivar Wangensteen, "Transmission Management in the Deregulated Environment", Proceedings of the IEEE, Vol. 88, No.2, pp. 170-195, Feb. 2000
[18] Christie, R. D., Wangensteen, I., "The Energy Market in Norway and Sweden: Congestion Management", IEEE Power Engineering Review, May 1998, pp.61-63
[19] Singh, H., Hao, S., Papalexopoulos, A., "Transmission Congestion Management in Competitive Electricity Markets", *IEEE Trans. on Power Systems*, Vol. 13, No.2, May 1998, pp. 672-679
[20] Caramanis, M. C., Bohn, B. E., Schweppe, F.C., "System Security Control and Optimal Pricing of Electricity", Electric Power and Energy Systems, Vol.9, No.4, Oct. 1987, pp.217-224
[21] Berger, Authur, W., Schweppe, F.C., "Real Time Pricing to Assist in Load Frequency Control", IEEE Trans. on Power Systems, Vol.4, No.3, Aug., 1989, pp.920-926
[22] M.L.Baughman, S.N.Siddiqi, J.W.Zarnikau, "Advanced Pricing in Electrical Power Systems", IEEE Trans. on Power Systems, Vol.12, No.1, pp.489-502, 1997
[23] A. J. Wood, B. F. Wollenberg, "Power Generation, Operation and Control", Wiley and Sons, 1996
[24] NERC Operating Guide I: Systems Control, February 27,1991
[25] P. Varaiya, F. F. Wu, R-L. Chen, "Direct Methods for Transient Stability Analysis of Power Systems: Recent Results", Proceedings of the IEEE, Vol. 73, No.12, pp. 1703-1716, Dec.1985
[26] Hsu, Meyer, "Modern Control Principles and Applications", McGraw-Hill, 1968
[27] S. K. Agrawal, B. C. Fabien, "Optimization of Dynamic Systems", Kluwer Academic Publishers, 1999
[28] http://www.ee.washington.edu/research/pstca/pf30/pg_tca30bus.htm
[29] Yoke Lin Tan, Youyi Wang, "Stability Enhancement Using SMES and Robust Nonlinear Control", Proceedings of International Conference on Energy Management and Power Delivery'98, Vol.1, pp. 171-176, 1998

8. Financial Instruments and Their Role in Market Dispatch and Congestion Management

X. Wang, Y.H. Song and M. Eremia

In Chapters 4 and 6, congestion management in the real-time operation of electricity markets has been discussed. A framework was proposed and implemented, in which the Independent System Operator (ISO) can balance the system and relieve transmission congestion coordinatedly and efficiently through a real-time balancing market. This chapter is about how to avoid and manage transmission congestion during short-term (day-ahead to hour-ahead) scheduling of electricity markets.

Traditional approaches to transmission access and pricing focus on "contract path" and cost-recovery-based transmission tariffs, which ignore the economic and physical realities of the power grid. Locational Marginal Pricing (LMP), developed by Schweppe in 1988 [1], provides a more economic method of transmission pricing and congestion management.

Despite the advantages of LMP, it can also create temporal and locational price risks. However, these risks can be hedged through some purely financial instruments such as Contract for Differences (CfDs) and Financial Transmission Rights (FTRs). Bushnell and Stoft explained how it works in a long-run electric grid investment [2].

In this chapter, the concepts of two typical financial instruments, CfDs and FTRs, and how they have been used to hedge against price risks are introduced in Section 8.1. In Section 8.2, the basic model of optimal dispatch in the spot market and the fundamentals of LMP theory are presented. In particular, the impact of bus generation and load limits on nodal prices are emphasized from the analysis of different forms of nodal price. After that, on the basis of the typical spot market dispatch model, some new individual revenue adequacy constraints are added to produce a more reasonable result for bilateral contract delivery in transmission congestion situations. An iterative procedure is employed to solve the formulated complex problem with dual variables in constraints. A 5-bus system and the IEEE 30-bus system are analysed to illustrate the proposed approach.

8.1 CfDs and FTRs

In a spot market, both the seller and the buyer of a bilateral trade face two types of price uncertainties: temporal uncertainty and locational uncertainty. In spite of the fact that the two parties are forced to trade directly with the grid at fluctuating spot prices, they can completely insulate themselves from these fluctuations by the use of a CfD provided that they face the same spot price. If spot prices are different locationally due to transmission congestion, new price risk arises. This locational price risk can be eliminated by an FTR. In this section, we will present mathematical models of these two financial contracts, and then describe how they work together to hedge against market price risks.

8.1.1 CFDs

Bilateral contracts take many forms, and in theory can include any provisions agreed to by the parties involved. A typical bilateral contract in which the parties can make direct physical trades is defined as a Physical Bilateral Contract (PBC). Two major characteristics of such a contract are price and quantity. Financial penalties are generally specified along with the target quantity. These penalties are to enforce "physical performance" of the contract. In summary, a PBC specifies requirements for both financial performance and physical performance.

In the presence of a pool-based spot market, CfD is a form of long-term financial bilateral power supply contract to hedge against temporal price risks. In describing CfDs, we assume a uniform locational spot price for market participants. Imagine a generator at node *i* and a consumer at node *j* wishing to trade P_{ij}^{CfD} units of power at a certain time in the future when the system spot price in the market will be ρ. However, traders wish to trade a negotiated strike price ρ_{ij}^{C}. This can be achieved indirectly by signing a CfD, which can be defined as:

Under a CfD, the consumer will pay the generator $(\rho_{ij}^{C} - \rho)P_{ij}^{CfD}$*, where* ρ_{ij}^{C} *is the contract price,* P_{ij}^{CfD} *is the contract quantity and* ρ *is the market spot price.*

The importance of specifying only financial performance cannot be seen as long as both parties actually do perform physically in line with the contract's nominal quantity. It is only when traders fail to supply or consume the contracted quantity that the potential benefit of CfD becomes apparent.

Generally if the spot price is very low, the generator will find it more profitable to stop generating, while if the spot price is very high, the demander will find it beneficial to stop demanding. This behaviour is consistent with economic rationality and the parties capture the benefits of this rational behaviour. Under a PBC, the participants' benefits from fluctuations in the spot price may be more limited. For this reason, combining a CfD with the spot market produces a synthetic bilateral contract which can offer short-term advantages over PBC. The scale of these advantages will depend on the degree to which performance penalties and transactions costs limit the ability of parties holding bilateral contracts to take advantage of favourable spot market prices.

8.1.2 FTRs

Before analysing FTRs and their effect on spot market dispatch and congestion management, we need to describe the necessity for such an instrument. An FTR is essentially an indirect way of conferring a property right for transmission. Generally speaking there are two approaches to defining property rights to the transmission network: physical rights and financial rights. The physical right is obviously the strongest form of ownership, which can affect usage on the entire system. The function of financial rights is to allocate the economic rents that should accrue to portions of the network.

A physical transmission right means the control of its usage: to be able to transmit electricity along that transmission link whenever one wants to. However, exercising (or not exercising) control of a link can affect the ability of others to exercise the control of their links. As a matter of fact in a meshed transmission network, power transmitted between any two nodes flows on every link. The rigidity introduced by defining transmission property in this way will further limit the ability of the ISO to adjust to fluctuating demand and generation conditions in an efficient manner.

Financial rights of transmission can have two varieties: Link-based Transmission Rights (LTR) and FTR. LTR associates ownership with the right to collect rent accrued by the link in the network. For example, the owner of P_{ij}^{LTR}, the right to a link connecting node *i* and node *j*, would collect the price difference between those two nodes times the actual power flow on that link. However, the most telling criticism of this approach is the possibility of providing the wrong signal to grid construction. A classic example of this is the construction of a line from *i* to *j* with low capacity and high admittance relative to an existing path from *i* to *j*. Such an addition to the network can easily reduce the total capacity from *i* to *j*. Thus, rewarding an expansion with a LTR can encourage extremely harmful improvements.

FTR, also known as Transmission Congestion Contracts (TCC), was developed by W. Hogan in 1992 [3]. Like LTR, FTR pays the right holder the price difference between the two nodes specified by that right. The two approaches differ in that the quantity that is multiplied by this price difference is defined by the right itself, rather than by the actual flow on a specific link. FTR can be defined as:

FTR is a financial instrument that entitles holder to receive compensation for Transmission Congestion Charges that arise when the transmission grid is congested in the spot market, leading to different LMPs at different locations.

In the spot market, given the LMPs ρ_i and ρ_j, under an FTR with magnitude P_{ij}^{FTR} from *i* to *j*, would pay its owner $(\rho_j - \rho_i)P_{ij}^{FTR}$, which can be shared equally by both parties, no matter how much power flows between node *i* and *j*. This is exactly the marginal loss the transaction could suffer under congestion. One very important implication of this fact is that FTRs, unlike LTRs, need not be limited to existing physical links. This allows FTRs to be applied to any bilateral transaction between two nodes anywhere on the network.

Basically, FTR is defined as a point-to-point contract. However, it can have a form of network that is from multiple points to multiple points. FTR could be option (one-sided) or obligation (two-sided). In this chapter we consider FTR as an obligation.

- Option FTR: income = max {0, $(\rho_j - \rho_i)P_{ij}^{FTR}$ } >= 0.
- Obligation FTR: income = $(\rho_j - \rho_i)P_{ij}^{FTR}$. When $\rho_j > \rho_i$, income > 0; when $\rho_j < \rho_i$, income < 0.

FTRs provide long-term transmission rights that can be different from the actual dispatch of the system. Although it is impossible to maintain a perfect match between long-term rights and the actual dispatch, it is possible to guarantee the financial payments to the FTR holders as long as the outstanding FTRs continue to pass the simultaneous feasibility test.

With well-defined FTRs, it is natural to allocate part or all of the FTRs to provide open access to the network through a market mechanism, such as an auction. An Optimal Power Flow (OPF) dispatch model can be adapted to provide a formulation of an FTR auction model for selecting the long-term capacity awards based on the willing-to-pay principle. The power flow equations embedded into the FTR auction make it straightforward to identify which FTRs are available by characterizing all possible rights and selecting a set of feasible rights that provide the highest valued use of the network. The auction enables market participants to purchase FTRs that would not be able to get through bilateral transactions in a secondary market. A secondary market provides a contractual mechanism for long-term pricing of a transmission network. FTR's purchase and its role in congestion management are shown in Figure 8.1.

Here, the objective of the ISO is to maximize the profit from FTR auction while keeping all existing FTRs simultaneously feasible without violating any operating limits. Each FTR can be either from single bus to single bus or from multiple buses to multiple buses. According to [4–6], with a DC model the optimal FTR auction can be formulated as the following linear programming problem:

$$Max\,(\mathbf{b}_{FTR})^{\mathrm{T}}\,\mathbf{P}_{FTR} \tag{8.1}$$

Subject to:

$$\mathbf{B}'\boldsymbol{\theta} - \mathbf{M}\ \mathbf{P}_{FTR} - \mathbf{M}_B\mathbf{P}_B = \mathbf{0}$$

$$\mathbf{P}_{FTR}^{\min} \le \mathbf{P}_{FTR} \le \mathbf{P}_{FTR}^{\max}$$

$$\mathbf{P}_l^{\min} \le \mathbf{H}\boldsymbol{\theta} \le \mathbf{P}_l^{\max}$$

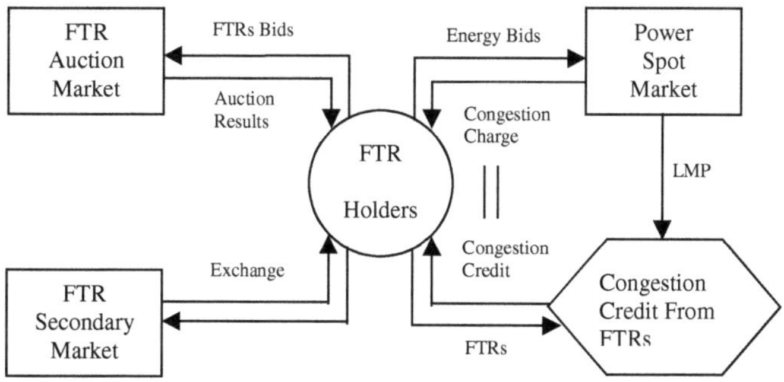

Figure 8.1. FTR's purchase and its role in congestion management

where $\boldsymbol{P_{FTR}}$ is the matrix of winning FTR bids (MW), $\boldsymbol{b}^{FTR}$ is the matrix of bidding prices of FTR bidders, $\boldsymbol{P_B}$ is the matrix of FTR injections in the base case, $\boldsymbol{B}$` is the linearised active power Jacobian matrix, $\boldsymbol{H}$ is the matrix of branch power flow constraint coefficients, $\boldsymbol{\theta}$ is the matrix of bus voltage angles, $\mathbf{P}_l^{\min}$ and $\mathbf{P}_l^{\max}$ are matrices of lower and upper MW flow limits of branches, $\mathbf{P}_{FTR}^{\min}$ and $P_{FTR}^{\min}$ are matrices of minimum and maximum amount of FTR MW value the bidders are willing to pay for, $\boldsymbol{M_B}$ is the nodal injection mapping matrix of FTRs in the base case, $\boldsymbol{M}$ is the nodal injection mapping matrix of FTRs in the auction. $\boldsymbol{M}$ is an $N \times N\ bidder$ Matrix. N is the total number of buses while $N\ bidder$ is the total number of bidders in the auction.

$$\mathbf{M} = \begin{array}{c} \\ bus\ 1 \\ bus\ 2 \\ \vdots \\ bus\ N \end{array} \begin{array}{c} \begin{array}{cccc} FTR^1 & FTR^2 & \cdots & FTR^{Nbidder} \end{array} \\ \begin{bmatrix} m_{1,1} & m_{2,1} & \cdots & m_{Nbidder,1} \\ m_{1,2} & m_{2,2} & \cdots & m_{Nbidder,2} \\ \vdots & \vdots & \ddots & \vdots \\ m_{1,N} & m_{2,N} & \cdots & m_{Nbidder,N} \end{bmatrix} \end{array} \tag{8.2}$$

The control variables in problem (8.1) are the nodal injections associated with the MW values of FTRs submitted to the auction. An FTR can have either positive or negative value depending on whether it is an FTR to be purchased or an FTR to be sold. To ensure the final FTRs set is simultaneously feasible, the FTRs not entering the auction should be treated as base case loads and generations in nodal power balancing equality constraints.

To enter the auction, bidders should submit their data to the ISO. These data include the bidding prices, the points of injections and extractions, and the minimum and maximum MW values of FTRs. Results of the FTR auction include an optimal set of winning bids of FTRs and the Market Clearing Prices (MCPs) of winning FTRs. The MCP of an FTR is defined by the opportunity cost of that FTR in the auction, i.e. the difference in market prices between both sides of the FTR. The MCP is the price that the purchaser of the FTR will pay, and that the seller of the FTR will be paid. The revenue from the FTR auction will be allocated to the Transmission Owners (TO) to compensate their investment in improving the power network.

Case studies on the FTR optimal auction model will be given in Chapter 15.

8.1.3 How CfDs and FTRs Hedge Price Risks

Assume that a generator at node *i* who signs a CfD with a consumer at node *j* agrees to pay the consumer the difference between a negotiated strike price and the true spot price, in exchange for a fixed payment. The consumer thereby locks in a constant price for power even in the event of transmission congestion. A CfD can be modelled by a power supply amount P_{ij}^{CfD} between the two nodes *i* and *j* at the negotiated strike price ρ_{ij}^{C}.

However, if the spot prices at nodes *i* and *j* differ because of transmission congestion, the whole transaction could still be exposed to the locational price risk. Given the nodal spot prices ρ_i and ρ_j, the payments of this CfD to the generator at node *i* and consumer at node *j* are $(\rho_{ij}^{C} - \overline{\rho}_{ij})P_{ij}^{CfD}$ and $-(\rho_{ij}^{C} - \overline{\rho}_{ij})P_{ij}^{CfD}$, respectively, where $\overline{\rho}_{ij} = (\rho_i + \rho_j)/2$. If the spot price is much higher at the consumer's node than at the generator's node and the average price between these two nodes is above the strike price at the consumer's node, the transaction will suffer a marginal congestion charge which equals the difference between the nodal prices. So another financial instrument, FTR, is needed to hedge the locational price risk.

FTR is a purely financial contract, according to which the holder is paid the spot prices difference between nodes times a quantity specified in the contract. For an FTR with magnitude P_{ij}^{FTR} from *i* to *j*, where the nodal spot prices are ρ_i and ρ_j respectively, the owner would be paid $(\rho_j - \rho_i)P_{ij}^{FTR}$, which is shared equally by both parties. This is exactly the marginal loss the transaction could suffer under congestion.

Through the combined application of a CfD and a matching FTR, both consumer and supplier can hedge against temporal and locational price uncertainties. Even in the absence of CFDs, FTRs can remove spot risk at an aggregate level when the system injections are matched by consolidation of all FTRs. The payment relationship between the ISO, generators and consumers is given in Figure 8.2.

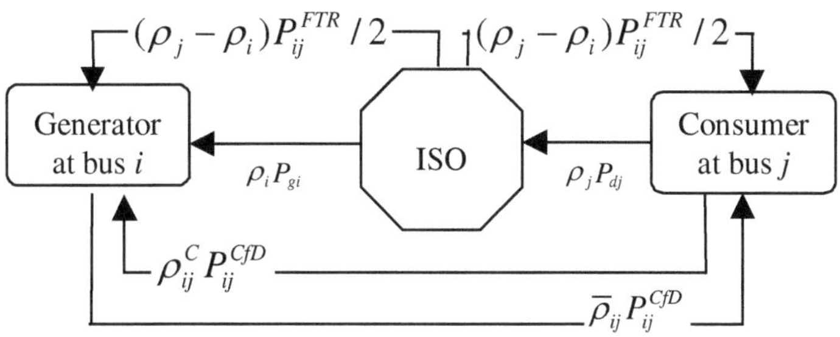

Figure 8.2. Cashflow between ISO and market participants

8.2 Spot Market Dispatch and Congestion Management with Individual Revenue Adequacy Constraints

A fully open electricity market should encourage more bilateral contracts and give market participants more freedom to arrange their own transactions. On the other hand, due to the special characteristics of the power energy commodity, a bid-based spot market is still needed to balance the system and eliminate potential transmission congestion. How to redispatch bilateral contracts when required has always been one of the main problems facing the ISO who runs the spot market. Some of the physical approaches to redispatch bilateral contracts have been presented by Fang and David [7], and Wang and Song [8].

Bilateral contracts are encouraged to buy FTRs so as to reserve necessary transmission capacity and maintain the revenue adequacy of related parties. Otherwise, without the protection, they may be exposed to the risks of unprofitability, as a result of serious congestion charge. Therefore, managing bilateral contracts from the perspective of their revenue adequacy is a very rational and straightforward means of congestion management in spot markets.

8.2.1 Impact of Operating Limits on Locational Marginal Prices

The spot market dispatch problem can be formulated as a bid-based DC OPF model, whose objective is to maximise the social welfare.

$$\underset{\mathbf{P_g},\mathbf{P_d}}{Max} \quad \sum_D \mathbf{B}(\mathbf{P_d}) - \sum_G \mathbf{C}(\mathbf{P_g}) \tag{8.3}$$

Subject to:

$$\mathbf{P}_{\mathrm{g}} - \mathbf{P}_{\mathrm{d}} - \mathbf{B}'\boldsymbol{\theta} = 0: \quad \boldsymbol{\lambda}$$

$$\mathbf{P}_{\mathrm{l,min}} \le \mathbf{H}\boldsymbol{\theta} \le \mathbf{P}_{\mathrm{l,max}}: \quad \boldsymbol{\mu}_l$$

$$\mathbf{P}_{\mathrm{g,min}} \le \mathbf{P}_{\mathrm{g}} \le \mathbf{P}_{\mathrm{g,max}}: \quad \boldsymbol{\mu}_{\mathrm{g,min}}, \quad \boldsymbol{\mu}_{\mathrm{g,max}}$$

$$\mathbf{P}_{\mathrm{d,min}} \le \mathbf{P}_{\mathrm{d}} \le \mathbf{P}_{\mathrm{d,max}}: \quad \boldsymbol{\mu}_{\mathrm{d,min}}, \quad \boldsymbol{\mu}_{\mathrm{d,max}}$$

Where:

D the set of consumers,
G the set of generators,
θ bus voltage angle matrix,
P_g generators' active power output vector,
P_d consumers' active power load vector,
P_l branches' active power flow vector,
λ vector of shadow prices of bus active power balancing equality constraints,
$\boldsymbol{\mu}_l$ vector of shadow prices of branch active power flow operational constraints,
$\mu_{g,min}$, $\mu_{g,max}$ vector of shadow prices of generators' operational constraints,
$\mu_{d,min}$, $\mu_{d,max}$ vector of shadow prices of consumers' load constraints,
B' linearised active power Jacobian matrix,
H matrix of branch power flow constraint coefficients,
$C(\bullet)$ active power offer functions of generations in the spot market,
$B(\bullet)$ active power bid functions of consumers in the spot market.

The variables following colons are the shadow prices of the corresponding constraints. According to the marginal price theory, the marginal price at bus i is

$$\rho_i = \lambda_i = \lambda_s - \sum_l^L \mu_l \frac{\partial P_l}{\partial P_i} \tag{8.4}$$

where

i index of network buses,
l index of branches,
L the set of network branches,
ρ_i nodal marginal price at bus i,
λ_s shadow price of bus power balancing equation at slack bus s.

When losses are ignored, nodal marginal price is decided by the system lambda (i.e. the nodal price at the slack bus) and transmission congestion charge. Baughman and Siddiqi pointed out the impact of generation limits on nodal prices [9]. Here, both generation and demand are treated as control variables (i.e. the demand elasticity is fully taken into account by consumers' submitted bidding curves to the ISO), so alternatively, the locational marginal price of generation buses and demand buses can have the following forms:

$$\rho_i = -\frac{\partial B_i(P_{d,i})}{\partial P_{d,i}} + \mu_{di,\min} - \mu_{di,\max} \qquad i \in N_D \tag{8.5}$$

$$\rho_i = \frac{\partial C_i(P_{g,i})}{\partial P_{g,i}} - \mu_{gi,\min} + \mu_{gi,\max} \qquad i \in N_G \tag{8.6}$$

where

N_D set of demand buses,

N_G set of generation buses.

Here, it can be noted that not only transmission congestion, but also limits of the output or demand of market participants can have significant impacts on LMP. In fact, if we ignore the last two inequality constraints in problem (8.3) and assume all the consumers submit second-order benefit curves to the ISO, there will never be any real risk of high spot prices caused by congestion. Without minimum load level limit, consumers can reduce their loads to zero in accordance with their bidding curves when they suffer from a high congestion charge.

Without considering losses, a physical bilateral contract can be modelled as one generation at the supply bus and the same amount of load at the consumption bus. Bilateral contracts are traded in the bilateral market instead of the spot market, so injections resulting from bilateral contracts cannot be regarded as control variables in problem (8.3). They should be treated as base loads and expressed by lower limits of the offers and bids in the spot market. These lower limits may bring prohibitively high nodal spot price to consumers in the case of transmission congestion.

8.2.2 Formulation of Individual Revenue Adequacy Constraints

With full consideration of risk hedging financial instruments, a consumer's profit can be formulated as below:

$$R_{di} = B_i(P_{di}) - \rho_i P_{di} + \sum_j ((\rho_{ij}^C - \bar{\rho}_{ij}) P_{ij}^{CfD}) + \sum_j ((\rho_j - \rho_i) P_{ij}^{FTR} / 2) \tag{8.7}$$

where the first item is the value of the power purchased, the second is the cost of power, the third is the payment from signed CfDs, and the last one is the payment from FTRs. Similarly, a generator's profit can be formulated as below:

$$R_{gi} = \rho_i P_{gi} - C_i(P_{gi}) + \sum_j ((\rho_{ij}^C - \overline{\rho}_{ij}) P_{ij}^{CfD}) + \sum_j ((\rho_j - \rho_i) P_{ij}^{FTR} / 2) \tag{8.8}$$

To maintain the incentive of a market participant to implement its transactions, its profit should be higher than a minimum profit level, which may consist of operating costs and the cost for purchase of the FTRs. In this chapter it is represented by an individual revenue adequacy constraint:

$$R_{di} \geq R_{di,\min} \tag{8.9}$$

$$R_{gi} \geq R_{gi,\min} \tag{8.10}$$

where R is the profit made by an individual market participant.

8.2.3 Implementation

Although the financial instruments like CfDs and FTRs should be separated from the delivery of physical transactions and their owners should not have any scheduling priority, they may still be able to provide some economic incentives to individual market participants for congestion management. The reason is that the original purpose for purchasing FTRs is to reserve enough transmission capacity for physical delivery while the matching of financial rights with network usage can mitigate the risk of congestion charge most economically.

In order to utilise the economic incentives of individual participants to transmission congestion management, individual revenue constraints should be embedded into the spot market dispatch problem (8.3). In other words, participants will be willing to curtail some of their transactions if they are losing money due to high congestion prices. However, the nodal prices in (8.10) are functions of dual variables of the primary problem (8.3), so it is impossible to add these new constraints directly into problem (8.3). To avoid this obstacle, an iterative procedure, which is similar to the "two-level" approach used in [9,10], is applied to solve the problem of optimal dispatch with individual revenue adequacy constraints.

This iterative procedure is shown in Figure 8.3. First set the lower MW limits of all the generators and consumers according to their physical operation limits and the existing bilateral transactions. Then solve the problem (8.3) by the use of an interior point primal-dual quadratic programming method to obtain optimal generation, demand and nodal prices. Check individual revenue adequacy constraints of all the participants. If any of these constraints cannot be satisfied, reduce the corresponding participants' lower MW limits (i.e. curtail its bilateral contracts) and then solve the problem (8.3) again. Keep doing the iteration until all constraints are satisfied.

There are at least two ways to adjust the lower MW limits of participants. The first is to reduce the lower MW limit step by step at each iteration. In this way convergence can be reached smoothly but a bit slowly. The second way is to obtain the minimum value of P_{di} or P_{gi} that satisfies (8.9) or (8.10), then update the lower MW limits with these values. Using this method, the procedure can reach convergence faster, but some transactions can be over-curtailed. If the minimum profit of a participant is so high that (8.9) or (8.10) cannot be satisfied even when $P_{di}^{(k)} \geq P_{di,\min}^{(k)}$ or $P_{gi}^{(k)} \geq P_{gi,\min}^{(k)}$ (i.e. $\mu_{di,\min}^{(k)} = 0$ or $\mu_{gi,\min}^{(k)} = 0$), the revenue adequacy constraint of this participant will be removed from the iterative procedure.

Without unexpectable contingencies, the minimum feasible solution set of the optimal dispatch of the spot market will be equal to the set of sold FTRs, which has passed the simultaneous feasible test during the FTR auction. Therefore, those bilateral contracts that hold matching FTRs will be guaranteed to deliver.

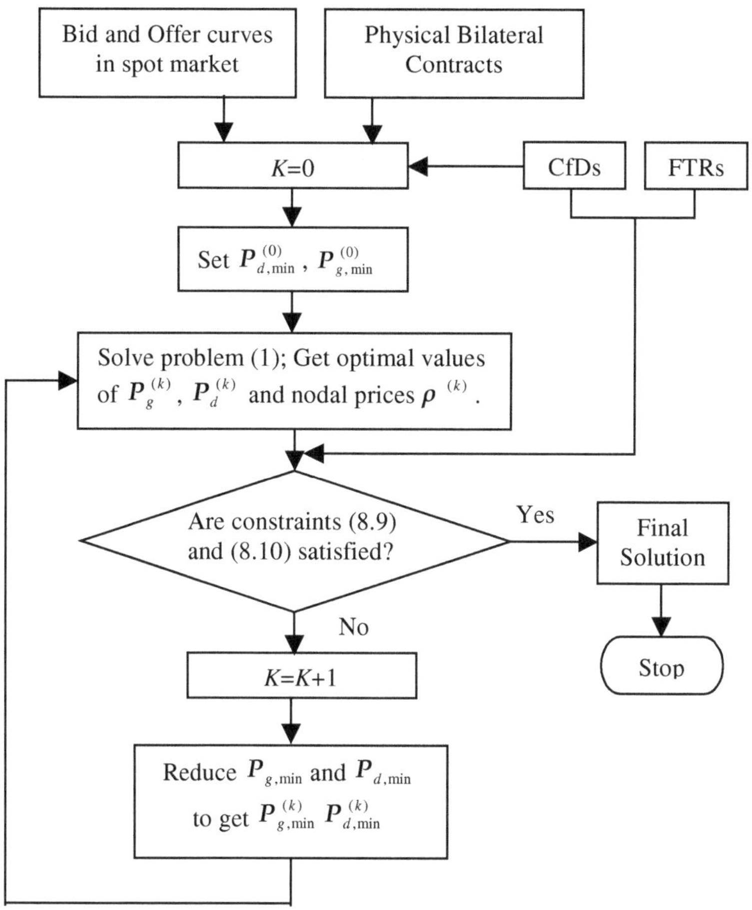

Figure 8.3. Iterative procedure to solve optimal dispatch problem with individual revenue adequacy constraints

8.2.4 Test Results

The proposed approach will be applied to two test systems: a 5-bus system and the IEEE 30-bus system. In the former system, a simple case is studied in detail to illustrate how the proposed approach works. In the latter system, a more complicated case is investigated to show how the nodal prices are affected by an inappropriate transaction and how to eliminate this bad impact.

8.2.4.1 System I: 5-bus System

A 5-bus test system is shown in Figure 8.4, in which generators' offer curves and consumers' bid curves in the spot market are given at the left-bottom corner. Bilateral contracts traded in the bilateral market and their associated financial instruments are listed in Table 1. Based on these bilateral contracts, the lower MW limits for all the participants entering the spot market are set at the right-bottom corner of Figure 8.4. The congested branch is Line 1-3, whose thermal limit is assumed to be 80MW. Under the given conditions, the optimal solution of problem (8.3) and the nodal spot prices are also shown in Figure 8.4. In this case, the nodal price at bus 1 is very high due to the congestion of Line 1-3, while the nodal prices at buses 3 and 5 are very low since D3 must increase its load level to absorb some part of the minimum output of G2. It is perceivable that the ISO won't be happy with this result, as it may bring about serious market power.

Given the minimum profit levels of D1 and G2 as $1,800 and $2,000 respectively, their individual revenue adequacy constraints cannot be satisfied in this situation because they have transactions exposed directly to locational price risk. So they are willing to reduce their transactions to make more profits. With the iterative procedure presented in the previous section, the lower MW limits of D1 and G2 are decreased in steps of 10MW per iteration. At each iteration a new set of nodal prices are produced by solving problem (8.3) with updated $P_{D1,min}$ and $P_{G2,min}$. This iterative procedure is repeated until all the individual revenue adequacy constraints have been satisfied.

The change in nodal prices and total profits of D1 and G2 are plotted against the reduction in their lower MW limit in Figure 8.5, where the impact of the lower MW limits on nodal prices is demonstrated clearly. When the lower MW limit of D1 is reduced to 160MW and the lower MW limit of G2 is reduced to 360 MW, the nodal price at bus 1 goes down to 20.51$/MW and the nodal prices at buses 3 and 5 rise to 9.4$/MW. Therefore, their revenue adequacy constraints can be satisfied. Meanwhile, the spot market produces much more reasonable nodal prices.

8.2.4.2 System II: IEEE 30-bus System

The standard IEEE 30-bus test system is modified here to test the proposed method. Most of the data can be found in [4]: the network branch parameters and thermal limits are in Table 21 of [4]; a feasible FTRs set can be found in Table 16 of [4]; the system one-line diagram is shown as Figure 7 in [4]. It is assumed here that each defined FTR has a matching CfD. In other words all participants have fully hedged against both temporal and locational price risks. The lower MW limits of all the generators and consumers are set according to the CfDs. To simplify the description of the results, in the spot market all the generators' offer functions are set as $C(P_g) = 0.02P_g^2 + 5P_g$ and all the consumers' bid functions are set as $B(P_d) = -0.01P_d^2 + 10P_d$. As result of optimal dispatch of the spot market, the limit binding branches include lines 6–10, 9–10, 15–18, 18–19, 21–22 and 25–27. The nodal prices of this case are shown in Figure 8.6 as the curve for "Base Case".

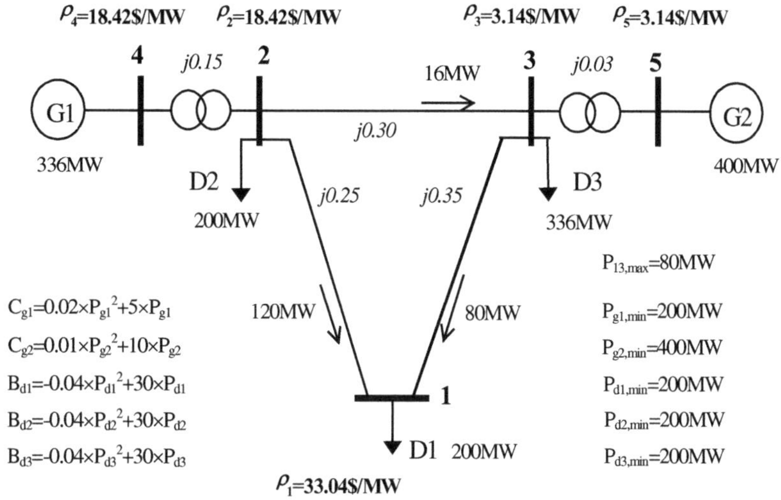

Figure 8.4. 5-bus test system

Table 8.1. Bilateral contracts and associated financial instruments

Transaction Participants	Transaction Amount (MW)	Transaction From/To Buses	Amount of CfD (MW)	Strike Price of CfD ($/MW)	Amount of FTR (MW)
G2→D2	200	5→2	200	17	200
G2→D1	100	5→1	100	17	100
G1→D3	100	4→3	100	17	100
G1→D1	100	4→1	0	0	0
G2→D3	100	5→3	0	0	0

On the basis of the base case, an additional transaction $T_{13\text{-}19}$ is added with an amount 11MW. This transaction has not brought any financial instrument to hedge against price risks. The nodal prices for this case are given in Figure 8.6 as the curve K=0, in which the prices at buses 10, 17, 19, 20, 21 are very high due to the congestion charge. These abnormal high prices may prevent participants from trading energy in the spot market and may cause serious market power. On the other hand, the consumer at bus 19 will not be willing to continue the transaction of T_{13-19} at such a high price, because its revenue adequacy constraints cannot be satisfied without the protection of the corresponding financial instruments. The proposed iterative procedure is applied by reducing the lower MW limit of this consumer in steps of 1MW per iteration. The decrease in nodal prices with each iteration is also shown in Figure 8.6. When T_{13-19} is curtailed to 5MW after six iterations, the nodal price at bus 19 has reduced sufficiently to meet the requirement of the local consumer's revenue adequacy. Then the iterative procedure stops. Not only the consumer at bus 19 but also all other participants including the ISO can benefit from this dispatch result.

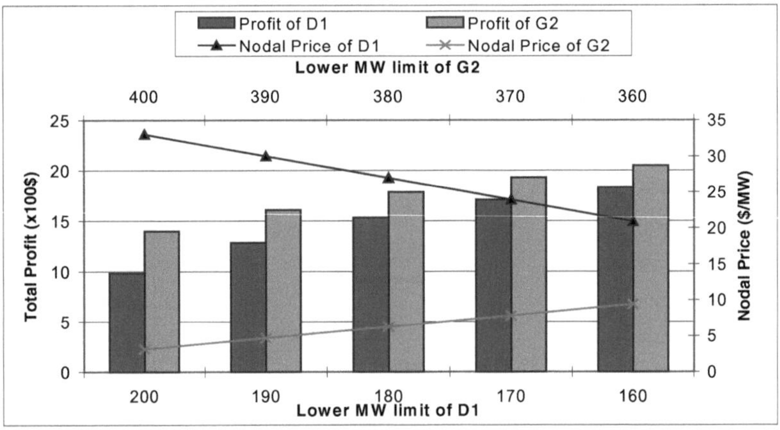

Figure 8.5. D1and G2's feedback to nodal spot prices on total profits and lower MW limits

The evolution of profit components of the consumer at bus 19 and the consumer at bus 21 against number of iterations are illustrated in Figure 8.7 and Figure 8.8 respectively. As analysed in Section 8.2.2, the total profit of a market participant, either a generator or a consumer, can be divided into two parts, the profit earned in the spot market and the profit earned from financial instruments. From Figure 8.7 it can be seen that the total profit is negative at k=0, which means that its profit from financial instruments cannot cover the huge congestion charge in the spot market. With each successive iteration, the profit from financial instruments decreases while the profit in the spot market increases, and eventually the total profit increases to reach the minimum profit requirement. The reason is that the nodal price at bus 19 is decreasing dramatically with curtailment of bilateral transaction T_{13-19}. In Figure 8.8, the total profit of the consumer at bus 21, who is also suffering from a very high nodal spot price, can remain constant, although its two components both change a lot during the iteration process. The explanation for this phenomenon is that the consumer at bus 21 has bought enough financial instruments to fully hedge against price risks. Consequently, its income from CfDs and FTRs can always cover its loss of profit in the spot market exactly, no matter how much its nodal price changes.

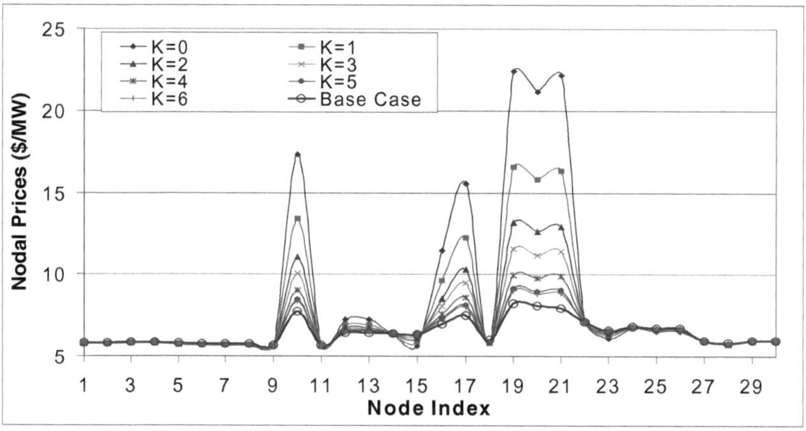

Figure 8.6. Change of nodal prices in IEEE 30-bus system during the iterative procedure of the proposed approach

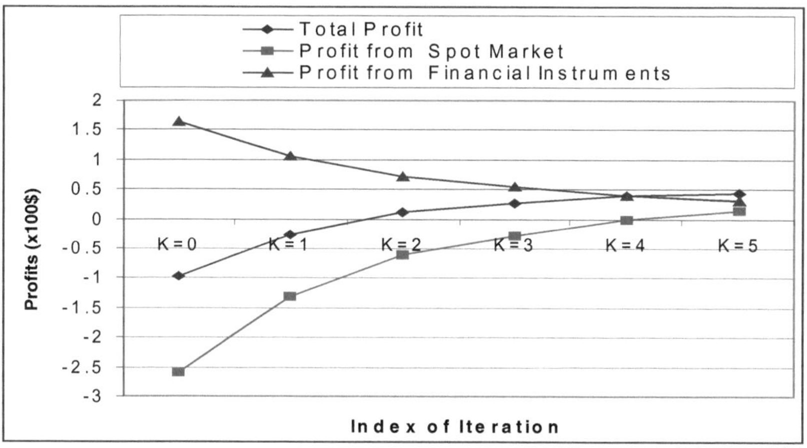

Figure 8.7. Evolution of profit components of consumer at bus 19 against number of iterations

In the proposed approach, the ISO needs to know the information about price and amount of CfDs and FTRs to mitigate congestion while in some previous methods bilateral contracts must submit price signals like "willing to pay" to the ISO for curtailment in the spot market. It is possible and causes no harm for the ISO to knowing the price information about bilateral financial instruments like CfDs, if the ISO is a just and non-profit organization. The formulae presented can also be applied to model the problem of optimal individual bidding strategy.

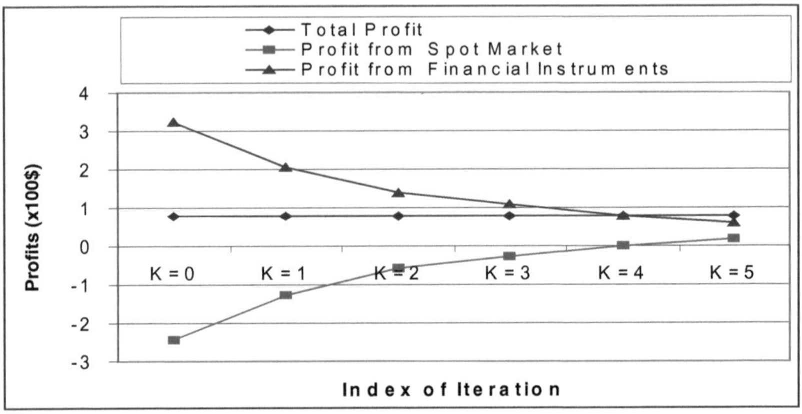

Figure 8.8. Evolution of profit components of consumer at bus 21 against number of iterations

8.3 Conclusions

To hedge against temporal and locational price risks, two purely financial instruments, CfDs and FTRs, are described. Cash flows between the ISO, generators, and consumers are analysed when CfDs and FTRs are combined to manage congestion and price risks. This chapter presents a new optimal dispatch method for the spot market with bilateral contracts, which takes into account the total revenue of an individual participant not only from the spot market but also from its financial instruments. The aim of this approach is to eliminate transmission congestion and to avoid prohibitively high nodal spot prices by utilising the natural profit incentive of individual market participants. To implement this approach, individual revenue adequacy constraints should be included in the original dispatch problem. An iterative procedure is applied to solve this special optimisation problem with both primal and dual variables in its constraints.

8.4 References

[1] F. C. Schweppe, M. C. Caramanis, R.D. Tabors, and R. E. Bohn: " Spot pricing of electricity" (1988), Kluwer Acadamic Publishers, Norwell, USA

[2] J. Bushnell, S. Stoft: "Electric Grid Investment Under a Contract Network Regime" (1995), POWER Working Papers, University of California Energy Institute, **PWP-034**

[3] W. W. Hogan: "Contract networks for Electric power transmission"(1992), Journal of Regulatory Economics, 4, No. 3, pp. 211-242

[4] M. I. Alomoush, S. M. Shahidehpour: "Generalized model for fixed transmission rights auction"(2000), Electric Power Systems Research, 54 (3), pp.207-220

[5] R. S. Fang, A. K. David: "Optimal dispatch under transmission contracts"(1999), IEEE Trans. PWRS-14, (2), pp. 732-737

[6] M. I. Alomoush, S. M. Shahidehpour: "Fixed transmission rights for zonal congestion management"(1999), IEE Proceedings on Generation, Transmission and Distribution., 146, (5), pp. 471-476

[7] Wang Xing, Y. H. Song, Q. Lu, and Y. Z. Sun: "Optimal Allocation of Transmission Rights in a Network with FACTS Devices"(2002), IEE Proceedings on Generation, Transmission and Distribution, 149, No.3, May pp.359-366

[8] R. S. Fang, A. K. David: "Optimal dispatch under transmission contracts"(1999), IEEE Trans., PWRS-14, (2), pp. 732-737

[9] Wang Xing and Y. H. Song: "Advanced Real-Time Congestion Management through Both Pool Balancing Market and Bilateral Market" (2000), IEEE Power Engineering Review, 20(2), pp.47-49

[10] R. Rajaraman, J. V. Sarlashkar, and F. L. Alvarado: "The effect of demand elasticity on security prices for the poolco and multi-lateral contract model"(1997), IEEE Trans., PWRS-12, (3), pp. 1177-1184

[11] M. L. Baughman, and S. N. Siddiqi: "Real-time pricing of reactive power: theory and case study results" (1991), IEEE Trans., PWRS-6, (1), pp. 23-29

9. Ancillary Services (I): Pricing and Procurement of Reserves

M. Rashidinejad, Y.H. Song and M.H. Javidi

This chapter introduces the concept of ancillary services and the various ancillary service markets. In particular, we will discuss the procurement and pricing of reserves in electric power markets. A joint dispatch model for energy and reserve with regards to the real-time and day-ahead market is presented. Finally, development of an option pricing model is described.

9.1 Ancillary Services in the Electricity Industry

The increasingly competitive electricity industry and deregulated power market is resulting in greater interest in efficient utilisation of generation and ancillary services. In most electricity industries around the world, the System Operators (SO) may face serious problems with regard to acquisition, pricing and cost of ancillary service. Problems arise because the importance of ancillary services may not yet be recognised, either technically or economically. Ancillary services are necessary to provide security and reliability for dispatching and scheduling. Furthermore, ancillary services are required to maintain system reliability and to affect commercial transactions.

Since bulk-power systems must maintain a continuous and near-instantaneous balance between generation and load, operating reserve is a backup capability in these systems to prevent any unwanted imbalances. In a power market where payments are based on energy supplied, generators prefer to have their units on-line and loaded close to the high economic dispatch limit. Because most ancillary services are provided by the same pieces of equipment that produce energy, energy markets and ancillary services markets are tightly coupled. Any problem in one market may cause problems in other markets. A real-time market may provide a balance of energy and ancillary services between generation and consumption (balancing mechanism). Contingency Reserves (CR) are the ancillary services that maintain this balance when a major generation or transmission line unexpectedly fails.

In the electricity industry, ancillary services are complementary services that interact the production of energy. Specifically, ancillary services are those functions performed by power systems with regards to generation, control, transmission and distribution of electric power to facilitate technical and commercial electricity transactions. These services are provided by the same equipments that generate and transmit electricity in support of the main services of electric energy and power delivery, collectively called ancillary services. Despite considerable costs of ancillary services, which is about roughly 10 percent of the costs of the energy commodity, these services are important for bulk-power reliability as well as for the support of commercial transactions. In power markets, the availability of sufficient ancillary services make power systems reliable and transactions deliverable.

9.1.1 Types of Ancillary Services

Ancillary services are defined as all those required for the reliable delivery of electricity. According to the NERC, an ancillary service is an interconnected operation service that is necessary to effect a transfer of electricity between purchasing and selling entities, and which a transmission provider must include in an open access transmission tariff. In England and Wales, ancillary services are required for the security and stability of the transmission system and these services are economically contracted from a range of different providers and enable the maintenance of satisfactory voltage and frequency, as well as the restoration of power supplies after system failure.

Generally, ancillary services include, but are not limited to, frequency control, Automatic Generation Control AGC, spinning reserve, non-spinning reserve (dispatchable load and generation) and black-start capability, which are generation-based ancillary services. Reactive power support and voltage control is another type of ancillary service, which relies on generators and requirements and installation of compensation devices, and may be probably better provided by contribution of generating units as well as transmission providers. It can clearly be seen that different power systems have different requirements for ancillary services. The definitions of major types of ancillary services are presented in this section.

- Regulation (Reg.) or Automatic Generation Control (AGC) is the regulating capability under automatic generation control, and provided by generating units to match real-time demand and supply. AGC as the regulating capacity can be converted to such ancillary service commodity, which is a weighted-average of MWs of response available in terms of few seconds. This service is required to continuously balance the ISO control area's supply resource instantaneous demand variations in order to meet the reliability and control performance standards.
- Ten-minute Spinning Reserve (TSR) is an on-line generation capacity synchronised to the system, that should be able to supply energy to the grid immediately and should be fully available within ten minutes. TSR need to be sustained for a period of about thirty minutes to facilitate first contingency protection.

- Ten-minute Non-spinning Reserve (TNSR) is a resource capacity that is not necessarily synchronised to the system, but must be available on ten minutes notice and should be able to supply energy as well. First and second contingency can be provided by TNSR with the aim of contingency protection with regard to the reliability satisfaction.
- Thirty-minute Operating Reserve (TOR) is a generating capacity, non-synchronised to the system that is fully available within thirty minutes. This type of Ancillary System (AS) must supply energy for up to one hour to cover first and second contingencies.
- Voltage Support (VS) is the reactive power required to maintain adequate voltage levels in the case of normal as well as abnormal conditions.
- Black Start (BS) is a self-starting generating capability available following localised and/or system blackouts.

Although the details and definitions of ancillary services remain vague and different from system to system, the key concepts and purposes of these services are now widely understood and appreciated [1, 2].

Based on FERC (Federal Energy Regulatory Commission) order 888, ancillary services are necessary services to support the transmission of electric energy. FERC-AS and the arrangements made by the transmission customer are:

- Scheduling, System Control and Dispatch
- Reactive Supply and Voltage Control from Generation Sources
- Regulation and Frequency Response
- Energy Imbalance
- Operating Reserve (Spinning)
- Operating Reserve (Supplemental)

NERC (North American Electric Reliability Council) Policy 10 includes different sets of Interconnected Operations Services (IOS); these services are defined as the fundamental physical capabilities supplied by generation or load resources needed to maintain the reliability of interconnected power systems in North America. NERC-IOS are defined as:

- System Control
- Reactive Supply from Generation Sources
- Regulation
- Frequency Response
- Contingency Reserve- Spinning
- Contingency Reserves- Supplemental
- Load Following
- System Black Start Capability

In England and Wales, the major ones include:

- Reactive power - This is required to maintain voltage profile.
- Frequency control - This is required to maintain system frequency.
- Reserve - This is required to counter the effects of generation failure/shortfall or demand forecast inaccuracy.

- Black start - This is the ability of a generating set equipped with auxiliary plant to start up and provide electricity to the transmission system without an external power supply.
- Constraints/congestion management - This is occasionally required to mitigate restrictions that occur on the Transmission System from time to time. Table 9.1 lists the major ancillary services in different countries.

Table 9.1. Major ancillary services in a number of markets

	California	New England	New York	PJM	REE (Spain)	UK
Fast Reserves	Spinning Reserves	Ten Minute Spinning Reserves (TMSR)	Ten Minute Synchronous Reserves	Spinning Reserves	Secondary Control	Fast Reserves
Cold Reserves	Non-spinning Reserves	Ten Minute Non-Spinning Reserves (TMNSR)	Ten Minute Non-Synchronous Reserves (TMNSR)	Supplemental Reserves	---------- -------------- -------------- -	Fast Start Capability
Slow Reserves	Replacement Reserves	Thirty Minute Operating Reserves (TMOR)	Thirty Minute Operating Reserves (TMOR)	Load Following	Tertiary Control	Standing Reserves
Spot Reserves	Regulation	AGC	AGC	Regulation	Primary Control	Frequency Response
	Voltage Support	Voltage Support	Voltage Support	Reactive Power	Voltage Support	Reactive Power

9.1.2 Market for Ancillary Services

Two important issues with regards to ancillary services are the cost of providing these services and the value of these services to the power system. Generally, in the electricity industry, ancillary services can be procured either by the ISO or may be purchased through contracts. Irrespective of the fact that ancillary services are the same components that produce energy, charging ancillary services using a bundled rate may not be equitable to users. To reflect the actual usage of these services, a competitive market-based ancillary service is an appropriate scheme regarding procurement and charging users [3]. According to the FERC, ancillary services interconnected operating services required to effect a transfer of electricity between purchasing and selling entities, and which a transmission provider must include in open access transmission tariffs. In England and Wales, ancillary services are considered to be required for the security and stability of the transmission system. These services are economically contracted from a range of different providers to enable the provision of satisfactory voltage and frequency, as well as the restoration of power supplies after system failure [4, 5]. The operation of a real-time market and day-ahead market for these services is one of the responsibilities of the system operator (SO). The SO must ensure that there are sufficient capacities among generators to participate in the balancing market. In order to have optimal operation within the new restructured power market, the SO needs to incorporate all market participant facilities regarding contingency reserves support capabilities, to satisfy security and reliability of energy supplies [6]. In the following section, a brief review of several electricity industries with regard to ancillary service markets is made.

In the design of an ancillary market there are a number of choices that can be considered. The following list presents some of the common and key ones:

- The time frame for the market can be long-term or short term.
- Procurement can be sequential or simultaneous.
- The market can be bilateral or competitive bidding.
- Settlement rules can be: price-based on bid type; price-based on usage type; marginal price or pay as bid.
- Recovery (charging) can be: energy uplift; use of the system charge or other methods.

Table 9.2 lists the major ancillary services market differences in different countries. The cost of ancillary services across 12 major utilities in the USA was surveyed by Hirst Ital. and the averaged cost of ancillary services was about 0.41 cents/KWh, which is 10% of the generation and transmission costs. The distribution of different costs is illustrated in Table 9.3. For the Pool-based England and Wales market, Table 9.4 summarises the cost of several main ancillary services.

Table 9.2. Differences in ancillary services markets

	California	PJM	New England	New York	UK	Spain
Hourly	Yes	Regulation only	Yes	Yes	No	Yes
Combined Operator/Syste Operator	No	Yes	Yes	Yes	Yes	No
Locational in AS prices	Yes	Yes	No	No	Some	No
Self -	Yes	Yes	No	No	Some	No
Load	Yes	No	No	No	Yes	No
AS Imports	Yes	No	No	Allowed	Rare	No
AS Exports	No	No	No	No	No	No

Table 9.3. Distribution of different ancillary services costs in 12 USA utilities

Loss es	Energy im- balance	Supplemental operating	Voltage control	Reliabil- ity	Load fol- lowing	Schedule dispatch
30%	11%	18%	12%	16%	9%	4%

Table 9.4. Ancillary services cost in England and Wales (1996 1998)

Reactive	Response	Reserve	Black start
£50m	£35m	£15m	£10m

9.1.3 General Considerations in England and Wales Ancillary Services Markets

In England and Wales, after privatisation (1990), contracts for ancillary services were made with all the large generators, including those generators in Scotland. Since then, changes in the structure of the Electricity Supply Industry (ESI) have increased the number and variety of potential providers, while the volume of services required has also developed. Three different ranges of services that may be required to support secure and reliable power supplies can be categorised as:

- Mandatory Services (MS): these are fundamental to the satisfactory operation of the electricity supply system.
- Necessary Services (NS): these can assist in the restoration of electricity supplies in the event that all or part of the transmission system become de-energised as a result of plant failure or other unexpected occurrences such as black start capability.
- Commercial Services (CS): these are complementary and additional services, occasionally required to mitigate restrictions that occur on the Transmission System (TS).

Under the conditions of connection, all large generators are obliged to provide the mandatory services; some, but not all, generators need to provide the necessary services; generators are also encouraged to enter into Commercial Ancillary Service (CAS).

These policies facilitate competition in the provision of ancillary services and ensure that there are contracts for all the required services. The main ancillary services are reactive power, frequency control, reserve, and black start. England and Wales have particular geographical conditions that heighten the benefits of placing appropriate incentives on the interdependent roles of network management and system control, which is reflected in the management of ancillary services that are essential for ensuring security and quality of supply. Immediately after privatisation in the UK in 1990, other than for ancillary service contracts, there was no incentive for any party within the electricity industry to minimise the overall cost of these services. As a result the costs rose dramatically until, from 1994, the National Grid assumed responsibility for them through the Transmission Services Scheme (TSS). NGC carried out a variety of functions including System Operator, Settlement System Administrator (SSA) and Ancillary Services Provider (ASP), as well as being the owner of the transmission assets. NGC as a system operator scheduled and dispatched generation and ancillary services including reserve, reactive power, voltage support, frequency support and black start. Under the New Electricity Trading Arrangement (NETA), generating units and NGC agreed to enter into a Supplemental Ancillary Services Agreement (SASA) with regard to the provision of certain types of ancillary services to be provided at each of the generator's power stations. Since the year 1993/94, NGC has carried out a tender process for competitive reserve procurement. In fact, the aim of this process is to supply sufficient standing reserve services to satisfy the operating reserve requirements for England and Wales. Since 1 April 1998, a Power Market has been established, which provides for the payment arrangements for services provided by generators.

9.2 Reserve Provision and Pricing in Power Markets

Electric utilities have traditionally provided reserves as part of the bundled product they sold to their customers. With regards to maintaining a required level of reliability through reserve provision, utilities implicitly agree to carry out the desired amount of reserves. On the other hand, reliability criteria may have substantial effects on the profits from energy generation and sales as well as earnings regarding the provision of reserves. In the following section, contingency reserves are defined and how they are used to restore the electric power system to equilibrium after an outage is described.

9.2.1 Contingency Reserves

By considering a general definition of three types of ancillary services, AGC, TSR, and TNSR, the so-called contingency reserves, they can be substituted based on a simple hierarchy as illustrated in Figure 9.1. AGC capacity is a service provided by generating units that respond in a fraction of a minute to restore the frequency deficit in the presence of frequency disturbances. Once the frequency drop exceeds the deadband of the generator governors, the governors sense the frequency drop and open valves on the steam turbines, which rapidly increases generator output and then increases the frequency. After a few more seconds, generator output declines slightly because the higher steam flow through is not matched by the steam flow from the boiler to the turbine. At this point, the operating reserves, in response to AGC signals from the control centre, kick in. More fuel is added to the boiler, leading to a higher rate of steam production, which leads to higher power output. In this process, the system is restored to its pre-contingency conditions within the required ten minutes.

Considering reliability criteria, the FERC definition, supplemental reserves as "generating capacity that can be used to respond to contingency situations " are suggested. In fact, supplemental reserves are not available instantaneously, they can be provided within a short time period on order of ten minutes. Supplemental reserves, as part of contingency reserves, are provided by on-line generating units but unloaded, quick-start generators and by interruptible loads. These types of reserves can be categorised as spinning and non-spinning reserves.

It can be argued that there may be some one-way hierarchical substitution among these three types of contingency reserves. AGC capacity can be transferred to TSR as an additional capacity, particularly in the case that there is extra AGC capacity. In the case of having extra capacity of TSR, it is also transferable to TNSR to increase the capability of power systems as well as off-line generation capacities and/or interruptible load demand [7].

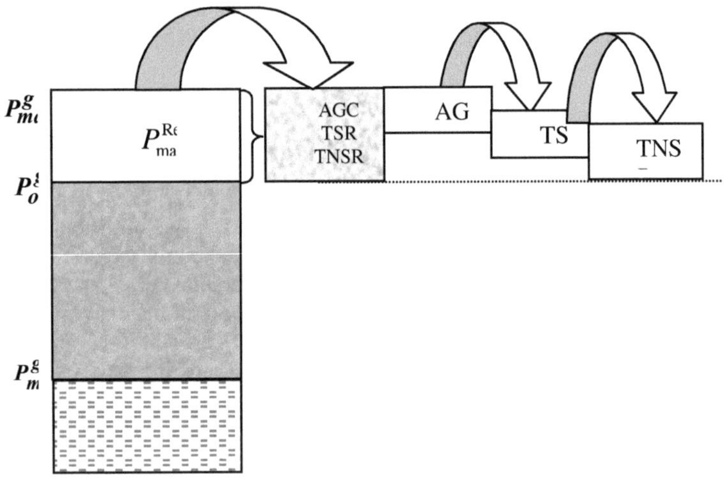

Figure 9.1. Hierarchical substitution of contingency reserves pattern for a unit

Figure 9.1 shows a schematic of a single generating unit for energy and ancillary service procurement, which can be used effectively in the restructured electric power markets. In fact, to procure sufficient reserves in power systems, the ISO deals with several tasks including reserve provision, contribution, dispatching and pricing.

9.2.2 Reserve Procurement Mechanism

In order to procure economical reserves in power systems, the ISO needs to use all facilities to prepare sufficient reserves. These are associated with the least possible price to supply secure and reliable electric. To achieve these goals, the ISO must consider system reserve capability with regard to all participants' potentials, including generators, curtailable loads and their maximum contribution. The ISO should also apply advanced techniques to dispatch energy and reserves considering either cost-based and/or market-based pricing methodologies to clear the optimum price for energy and reserves. As shown in Figure 9.2, in the reserve procurement mechanism the ISO is dealing with several different issues including physical and economical considerations.

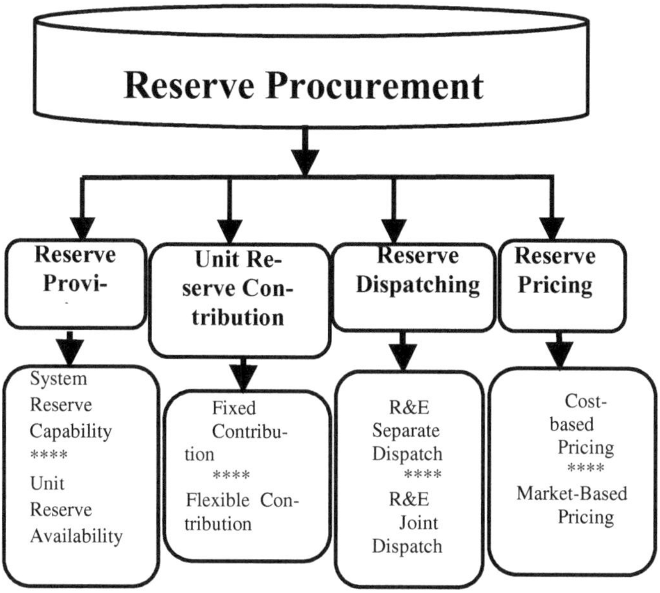

Figure 9.2. Structure of reserve procurement

System reserve capability, unit availability and unit maximum reserve contribution are physical aspects, while dispatching methods and pricing mechanisms are both physical and economical aspects. The main idea of using ED or OPF for reserves provision is to capture the most economic reserve capacity dispatching, incorporating all possible physical and economical restrictions. Such restrictions are reliability and security constraints.

9.2.3 Reserve Markets in Several Power Markets

9.2.3.1 Reserve Markets in England and Wales

In any system, there is a requirement that system frequency be maintained within an acceptable range. Reserves are required in situations when demand is greater than forecast or where plant breakdowns occur.

9.2.3.1.1 Mandatory Frequency Response

The minute-by-minute system frequency control is managed through the "self-balancing actions" of generators and suppliers and the "offer/bid acceptances" by the NGC. All large generators must be capable of contributing to second-by-second frequency control, which is required by the Grid Code and the conditions for connection to the transmission network. Each connected unit has to have a 4% governor droop characteristic. Prices are "cost-reflective" to reflect inefficiency, and "wear & tear" costs associated with service provision. Primary response is defined as the additional active power delivered by automatic governor action from a generating unit (or decrease in active power demand initiated by low frequency relay) that will be available 10 seconds after an event and can be sustained for a further 20 seconds. Secondary response is defined as the additional active power delivered by automatic governor action from a generating unit (or decrease in active power demand initiated by low frequency relay) that will be available 30 seconds after an event and can be sustained for a further 30 minutes. High frequency response is defined as the reduction in active power delivered by automatic governor action from a generating unit that will be available 10 seconds after an event and can be sustained indefinitely.

9.2.3.1.2 Commercial Ancillary Services

Commercial Frequency Response arrangements are entered into when both the System Operator and the provider see a technical or commercial advantage in doing so and are freely negotiated between the parties both in terms of contract form and price. While generators, through mandatory services, provide continuous system frequency control, occasional frequency control requirements are largely met by changing demand through commercial contracts. NGC is always keen to contract for commercial services by volume to obviate the need to use the generally more expensive "obligatory" services. The demand can participate if it is prepared to be interrupted for 30 minutes several times a week; triggered by LF relay within ½ second for primary response and 30 seconds for secondary response; payments are based on the relay setting in £MWh at contract price. Fast Reserve is the rapid and reliable delivery of active power provided as increased output from generation or a reduction in consumption from demand sources, following receipt of an electronic dispatch instruction from the National Grid. Active power delivery must start within 2 minutes of the dispatch instruction at a delivery rate in excess of 25MW/minute, and the reserve energy should be sustainable for a minimum of 15 minutes. Standing Reserve: at certain times of the day NGC needs extra power in the form of either generation or demand reduction to be able to deal with actual demand being greater than forecasted demand and plant breakdowns. This requirement is met from synchronised and non-synchronised sources. NGC procures the non-synchronised requirement by contracting for Standing Reserve, provided by a range of providers including generating units, demand reduction and independent generating plant.

9.2.3.2 Reserve Markets in the USA

9.2.3.2.1 California Markets

In California the Independent System Operator (CAISO) is responsible for ancillary service procurement to ensure that sufficient ancillary services are available to maintain the reliability of CAISO grid. Ancillary service requirements, established by CAISO, may be self-provided by each Scheduling Coordinator (SC); those not self-provided must be competitively procured by contracts for long-term, day-ahead, hour-ahead, or real-time basis. The CAISO is responsible for acquiring or monitoring the acquisition of six ancillary service products, where they are: regulation service, spinning reserve, non-spinning reserve, replacement reserve, reactive power support, and black-start generation capability. Of these six ancillary services, the last two, voltage support and black-start, are not acquired through day-ahead and hourly market-clearing processes, they are instead procured through a longer-term contracting process and this could be an option for some of the other services as well. Usually, for the rest, on the day before these services are required, the ISO holds an auction for each individual ancillary service. The ISO determines a Market Clearing Capacity Price (MCCP) for each of AS, where successful bidders into each of these markets may or may not in fact be called upon to provide energy. If these units are called upon to provide real-time energy, suppliers to the spin, non-spin, and replacement reserve markets are paid the imbalance energy price for any energy that they may provide. This payment is in addition to the capacity payment they receive for making their capacity available to the ISO. Due to metering and software limitations, suppliers of energy into the Regulation market cannot set or earn the imbalance energy price for any energy that they provide. Suppliers of Regulation energy instead earn the Regulation Energy Payment Adjustment (REPA), a $/MW of Regulation capacity bid payment, which is set according to an estimate of the energy provided by each supplier. Suppliers bidding to provide any of these four ancillary services must satisfy various technical operating characteristics for each of these markets. Bidding in each market consists of a capacity price ($/MW) and a schedule of energy price ($/MWh) and quantity (MW and/or MWh) bids. The market clearing ancillary service price paid to all successful bidders (not subject to cost-based price caps) is set at the capacity price bid of the last bidder whose capacity is accepted. Those bidders subject to cost-based price caps are currently paid their bid price for the quantity of each ancillary service that they supply to the market. An important feature of the operation of these markets is that although firms submit their bids to all of these ancillary services markets simultaneously, the markets clear sequentially, with regulation first, followed by spin, non-spin, and finally, replacement. All four markets are cleared and bidders are told the market clearing prices and how much of their capacity was accepted for each market. In this sequential market-clearing process, capacity that is won in a previous auction is subtracted from the capacity that is bid into subsequent markets.

9.2.3.2.2 New York Markets

In July 1998, FERC conditionally approved many of the critical aspects of New York's proposal for creation of an Independent System Operator and a Reliability Council to develop reliability standards to govern the operations of the New York Independent System Operator (NYISO). FERC recognised the Public Service Commission's (PSC's) role in ensuring system reliability by requiring the NYISO to implement local reliability rules written by the PSC. The proposed NYISO would oversee markets for energy, capacity, and ancillary services. The New York ISO (NYISO) assumed the responsibility of ancillary service procurement through an ancillary service market in November 1999. The markets administered by the NYISO include installed capacity, energy, ancillary services, and transmission congestion contracts. NYISO's ancillary service markets include ten-minute spinning reserves, ten-minute non-spinning reserves, thirty-minute non-spinning reserves, and regulation. To ensure the reliability of the electric system, the NYISO obtains reserves (generation capability above that needed to meet forecast demand) through the NYISO administered ancillary services market. The operating reserves required in New York are both ten-minute reserves and thirty-minute reserves. The total ten-minute reserve capability has to be equal to the largest contingency that can occur in the system; half of that amount must be spinning reserve and half can be non-spinning. Thirty-minute reserves have to be available in an amount equal to half of the total ten-minute reserves. The total reserves thus required, ten-minute plus thirty-minute, equals 1.5 times the largest single contingency that may occur on the system. The NYISO has been authorised by a commission to operate day-ahead markets and real-time markets for energy and four ancillary services including regulation and frequency response, ten-minute spinning reserve, ten-minute non-spinning reserve, and thirty-minute non-spinning reserve.

9.2.3.2.3 New England Markets

There have been many problems in the New England ancillary services market under the current structure. New England began operations with four daily ancillary service markets: ten-minute spinning reserve, ten-minute non-spinning reserve, thirty minute operating reserve, and automatic generation control. New England also has several ancillary service products, which are not priced through markets, including replacement reserves, voltage support and black-start capability. The market rules for these products in some cases simply have not reflected general economic assumptions; the rules are going to change to remove inefficiencies. The New England ISO began market-based pricing for energy, ancillary services and capacity products on 1 May 1999. Following a quiet start in May 1999, the New England markets encountered their first problems in June and July 1999, when summer heat waves tested the efficiency of the market design. During the summer 1999, NEISO corrected numerous hourly market prices, generally reducing them, on the basis of market design flaws. The price correction authority was controversial among market participants and was substantially limited by the commission. A fundamental problem in the ancillary services markets is the method of clearing the markets. Each eligible generator submits a bid into one or more of these markets, which have a daily bidding deadline, the prior day. In the reserves markets, each bid is then subject to an often complex bid evaluation, which determines a ranking on the basis of expected costs, including opportunity costs and, in the case of automatic generation control, penalties for shifts upward along the bid curve during dispatch. The daily markets are then cleared during real-time dispatch in the following order: energy, automatic generation control, ten-minute spinning reserve, ten-minute non-spinning reserve, and thirty-minute operating reserve. This order reflects both technical requirements and the expectation that higher value uses of generation capacity will clear at higher prices.

9.2.3.2.4 Pennsylvania NewJersey Maryland (PJM) markets

From April 1997 to April 1999, PJM operated its energy market with cost-based bids and financial settlement at real-time prices. A uniform system-wide energy price was calculated until April 1998, when locational marginal prices were established. On May 1999 market-based bidding began and on June 2000, the energy market was divided into a day-ahead and real-time market. PJM has been authorised by the commission to operate a bid-based market for energy, energy imbalances, and regulation and frequency response. PJM has also been authorised to provide an operating reserve service (including spinning and supplemental reserves), reactive supply and voltage control. New York and New England have bid-based markets for operating reserves, where PJM is still planning its own such markets and PJM has bid-based markets for installed capacity.

9.2.4 Research into Reserve Procurement and Pricing

Reserve procurement in electricity markets is an almost new problem, which emerged after restructuring of the electricity industry. Most publications related to this area of research are from the last decade. In fact, the ideas behind this research are to facilitate reliable and secure power systems. Reliability improvement can be attained through procuring electrical energy associated with different types of reserve. Spinning reserve is a major part of the operating reserve procured by most generating units. Utilities use deterministic techniques to assess the operating reserve requirements of their systems through unit commitment approaches. Usually, deterministic approaches did not consider important factors such as machine characteristic, size, unavailability and response rate of the units. Therefore, inconsistent reserve allocation between pool members was addressed [8].

Incorporating probabilistic assessment of operating reserve, Chowdhury *et al.* [9] showed the effectiveness of this approach by applying it to both isolated and interconnected configurations. Purchasing reserves requirements through export/import was another solution to provide sufficient reserve to ensure the reliability of interconnected generation systems [10]. Reserve requirement assessment via an energy approach, proposed by Chowdhury [11], introduced the " *two risk concepts* ". It is associated with a probabilistic technique regarding possible assistance from neighbours for reserve provision based on capacity assistance. Considering risk parameters involving the reserve provision issue, a Composite System Operating States Risk (CSOSR) method, which is influenced by the risk of transmission outages and generation outages was carried out by Khan [12]. In the CSOSR technique, the effectiveness of calculation of risk value by applying probabilistic unit commitment incorporating transmission facility limitations is shown.

In power systems, considering reserve procurement, the economic reallocation of generating units aims to minimise generation costs, while maintaining a margin for responsive reserve. Bobo *et al.* [13] proposed a technique maintaining sufficient responsive reserve (based on the definition of Houston Lighting & Power Co., HL&P) across all generating units. They showed that HL&P generation criterion is more capable of responding to frequency deviations. Reserve allocation applying a Response Health approach was suggested by Fotuhi-Firoozabad *et al.* [14]. In this approach, the committed units are loaded to satisfy the generating system response risk associated with an acceptable response health probability.

The provision of a very short-term reserve to prevent frequency deficit, following unforecasted on-line unit outage, could be an important issue in small utilities. Frequency restoration, through generator response and load shedding, incorporates minimising operating costs while considering security of supply as well as economic performance. O'Solivan *et al.* [15] showed that the minimum frequency reached following a contingency would be a more appropriate measure of system security. It can be imposed as a frequency-based reserve constraint on the economic dispatch problem [15]. In [16], Marangon-Lima *et al.* proposed a new approach for spinning reserve requirements evaluation for multi-area systems introducing several risk indices for the whole system and for each control area. The effectiveness of their proposed method in comparison with deterministic approaches is shown.

Another stochastic approach was implemented for reserve assessment through a probabilistic simulation network by Halilcevic [17]. Risk indices for different levels of system load were defined, taking into account the values of spinning reserve and ready reserve, for reliability improvement in power systems. In continuation, Halilcevic *et al.* [18] introduced a stochastic-deterministic approach to calculate on-line ready reserve for rapid start-up generating units in power systems. They applied an artificial neural network technique to assess the appropriate amount of ready reserve, aiming to minimise the potential reduction in quality of electric power to be delivered to the customers.

As far as the new competitive electricity markets are concerned, the balance between costs and security is a dominant factor efficient operation and planning policies in electric utilities. To procure reserve in the new electricity market environment the proposed methodology by O'Sullivan *et al.* [19] enables the cost of providing sufficient reserves to be calculated and assessed against the risk and cost of load shedding. In fact, their technique is closely associated with the proposed economic dispatch for small isolated power systems. The dispatch costs are calculated and compared with the expected value of load shedding, following the loss of any generating unit.

Based on the new policy in competitive electricity markets, procurement of ancillary services is a duty of Independent System Operators (ISOs). To facilitate the required quality of energy via reserve procurement, reserve markets must support the energy markets. A competitive ancillary service mechanism based on the California energy and AS markets with the aim of day-ahead and hour-ahead markets was implemented by Singh *et al.* [20]. They introduced a reserve auction mechanism, including capacity reservation, associated with the estimated probability that reserve will be utilised in real time. For ancillary service settlement the interaction between various markets is also considered.

One of the major requirements in deregulated electricity markets is contingency reserves provision through ancillary service (AS) markets. In this regard Liu *et al.* [21] proposed an algorithm incorporating the rational buyer for AS to procure reserve requirements in restructured electric power markets. Previous ancillary service provision techniques were based on a sequential procedure from the highest quality service market to the lowest one, while the proposed method performs an exhaustive search to find the minimum cost procurement plan by keeping the total requirements for all services unchanged. This approach shows a significant saving of overall cost. To provide circumstances for generating unit participation in ancillary service markets for power system reliability and security operation, utilities must provide appropriate payment mechanisms for generating units as well as independent power producers. In fact, all utility participants including generators, loads, Independent System Operators may have the ability to procure reserves by entering the reserve markets. In order to maximise the expected value of profit for a participant, Allen *et al.* [22] introduced a formulation that determines how much energy can be sold in the spot market and how much capacity should be held in reserve. In this regard, they considered two different payments, one payment for energy delivered and another payment for reserve allocated. They showed that the price of reserve at equilibrium would balance the supply and demand for reserve.

9.3 Joint Dispatch for Reserve Provision and Pricing

In this section different dispatch methodologies including merit order dispatch, sequential dispatch and simultaneously dispatch are compared. A flexible joint energy and reserve dispatch algorithm is used for dispatching and pricing contingency reserves to take advantage of the flexibility of generating units for producing real generation as well as reserves [23].

9.3.1 Application of JEROD to Deal with Reserve Provision and Pricing

The bid-clearing price model for dispatching and scheduling is formulated as a constrained optimisation problem. The objective is to minimise the sum of costs to energy and ancillary service offers. To minimise the expected cost of reserves as an ancillary, the ISO cannot consider only capacity reservation, it must estimate the probability that reserve will be utilised in real time. Such probability factor may be considered in the spinning reserve cost function by a parameter ρ_i^u [20] . In fact there is a risk in dispatching, that there might be load shedding following the first contingency [24, 25]. An estimate of the amount of unserved load may be calculated by considering the shortfall in MW between the lost generator and the amount of spinning reserve from the on-line units and demand behaviour. Furthermore, providing more spinning reserve will decrease the probability of utilising peaking unit, resulting lower costs. The amount of saving depends on the percentage of spinning reserve. The expected cost of unserved energy may be calculated in terms of load curtailment because of the loss of generation unit, the penalty price of lost load and the probability of reserve utilisation (ρ_i^u). This probability may be strongly related to the reliability of generation units and transmission contingencies [26]. It is assumed that there are **n** competing generation companies (GenCos) with individually bidding blocks both for energy and reserve:

[E_i^j , eb_i^j] J = $1,\dots, m_e$ the *ith* GenCo energy quantity-price bids

[R_i^k , rb_i^k] $k = 1,\dots, m_r$ the *ith* GenCo reserve quantity-price bids

Where:

E_i^j : energy quantity offered for *jth* band by unit *I;*

eb_i^j : energy price offered for *jth* band by unit *I;*
R_i^k : eapacity Reservation quantity offered for *kth* band by unit *I;*
rb_i^k : eapacity Reservation price offered for *kth* band by unit *I;*
m_e : number of energy bid bands;
m_r : number of reserve bid bands;

The generation cost of energy for *ith* unit $\boldsymbol{GC_i(P_i)}$ can be written as:

$$GC_i(P_i) = \sum_{j=1}^{b} E_i^j eb_i^j + (P_i^g - \hat{E}_i^b) eb_i^{b+1} \tag{9.1}$$

$$\text{for } \hat{E}_i^b \le P_i \le \hat{E}_i^{b+1} \text{ and } \hat{E}_i^b = \sum_{j=1}^{b} E_i^j \text{ and } b \le m_e - 1$$

where P_i^g is real power generation by unit *i.*
The reserve cost function for *ith* unit is defined as:

$$\begin{aligned} RC_i(SR_i, \rho) = \sum_{k=1}^{b} R_i^k (rb_i^k + \rho rb_i^{m_e}) + \\ \sum_{k=1}^{b} (SR_i - \hat{R}_i^b)(rb_i^{b+1} + \rho eb_i^{m_e}) \end{aligned} \tag{9.2}$$

$$\text{for } \hat{R}_i^b \le SR_i \le \hat{R}_i^{b+1} \quad \hat{R}_i^b = \sum_{k=1}^{b} R_i \text{ and } b \le m_r - 1$$

where SR_i is spinning reserve capacity provision by unit *i.*
The objective of joint energy and reserve dispatch is described by

$$Min \left(\begin{aligned} & \sum_{i=1}^{nd} \{ \sum_{j=1}^{b} \left[E_i^j eb_i^j + (P_i^g - \sum_{j=1}^{b} E_i^j) eb_i^{b+1} \right] + \\ & \sum_{k=1}^{b} \left[\begin{aligned} & R_i^k (rb_i^k + \rho eb_i^{m_e}) + \\ & (SR_i - \sum_{k=1}^{b} R_i^k)(rb_i^{b+1} + \rho eb_i^{m_e} \end{aligned} \right] \} \end{aligned} \right) \tag{9.3}$$

The associated constraints can be physical as well as operational security. The physical constraints ensure that unit minimum and maximum set points are not exceeded. Also, the total power supplied by dispatchable and non-dispatchable units must equal to the total demand, $\sum_{l=1}^{m} D_l$ plus the total system losses, P_{loss}. Ramp rate limitation is considered a major factor when providing SR for ten minutes. The operational security constraint as System Reserve Requirement SRR is taken into account as well as individual spinning reserve contributions for committed units [27].

9.3.1.1 Physical Constraints

Maximum and minimum generation capabilities associated with spinning reserve capacity are considered to be inequality physical limitations. Supply, demand plus system losses are equality physical constraints

$$P_i \le P_i^{\max} \tag{9.4}$$

$$P_i + SR_i \ge P_i^{\min} \tag{9.5}$$

$$\sum_{i=1}^{N} P_i = \sum_{j=1}^{m} D_j + P_{losses} \tag{9.6}$$

9.3.1.2 Operational Security Constraints

In this model unit Ramp Rate (RR) capability is considered as well as Spinning Reserve Requirement (SRR). A generating unit has limits on its ability to move from one level of MW generation to another within a specified time period. Generation ramping up capability can be defined as $P_{i2} - P_{i1} = \Delta P_i = \tau RR_i$, where ΔP_i is the difference between two levels of output during the dispatch period. τ denotes the reserve provision time constant, which is ten minutes for TSR. On the other hand spinning reserve availability can be defined as the room equivalent to the extra generation capacity for each generating unit, $P_i^{\max} - P_i$. Finally, reserve unit contribution commitment can be considered as a percentage of maximum generation capacity [28], which may be mandated by reliability regulation mechanisms, such as $\alpha_i P_i^{\max}$. Applying all the above limits for reserve provision, it can be written as

$$SR_i = \min\left[\left(P_i^{\max} - P_i\right), \left(\alpha_i P_i^{\max}\right), \left(10 * RR_i\right)\right] \tag{9.7}$$

System reserve requirement is defined as:

$$SRR_{need}^{\min} \le \sum_{i=1}^{Nd} SR_i \le SRR_{need}^{\max} \tag{9.8}$$

9.3.2 Numerical Case Study

9.3.2.1 Six-unit Test System

In this case, the system demand is assumed to be 1040 MW and the system reserve requirement 100MW. The energy bidding blocks, ancillary service bidding blocks associated with energy offer bands and ancillary service offer bands are shown in Table 9.5. Unit ramp rate constraint and MW generation capacity are displayed in Table 9.5 as well.

Four different scenarios are implemented on this system: Merit Order Dispatch [29] (MOD), Sequential Dispatch (SQD)[30], Joint Dispatch with fixed reserve percentage (JD_fx), and Flexible Joint Dispatch (FJD) [24]. The dispatching results of four cases are provided in Tables 9.6 and 9.7, for both energy and spinning reserve respectively.

Table 9.8 shows the energy output associated with spinning reserve of each unit for all cases. In the MOD method a multi-product power market may be viewed as having independently energy and reserve dispatch only by considering individual band limits of unit capacities. Bidding blocks may be ranked in a bid-price order to apply the MOD method, while the energy is dispatched in the energy market [30]. The same process is used to dispatch the spinning reserve in the AS market. Under the assumption of no coupling between energy market and ancillary service market, then each product may be dispatched separately. In such a case, no one can guarantee the feasibility as well as optimality of the dispatching problem. For example, as Table 9.8 shows, unit 3 and unit 4 provide 108 MW and 570 MW energy and reserve, while their maximum capacities are 100 MW and 520 respectively, which is infeasible.

One of the methods by which the issue of coupling between energy and ancillary can be addressed is the SQD approach. SQD, as an extension of the merit order based dispatch approach, may consider the same generating capacity to produce energy and ancillary services, while it cannot handle energy and reserve interdependency [31]. Compared to the merit order dispatch results, the sequential dispatch approach respected the generating unit limits, but produced correspondingly higher Energy Market Cost (EMC) as well as Energy Market Clearing Price (EMCP). These results are recorded in Table 9.9.

Table 9.5. Six-unit test system data

Unit	Energy Offer						Ancillary Service Offer		Ramp Rate MW/min	MW Limit
	Band 1		Band 2		Band 3		Band 1			
	MW	Price	MW	Price	MW	Price	MW	Price		
G1	5	13	7	23	5	27	5	1.5	5	17
G2	80	14	60	26	60	28	20	3	5	200
G3	70	11	15	22	15	25	10	2.5	7	100
G4	400	12	60	21	60	24	50	2.75	10	520
G5	200	15	40	17	40	23	50	3.5	10	280
G6	50	12	30	27	30	29	20	1.25	5	110

Table 9. 6. Energy dispatch results

Case	G1 (MW)	G2 (MW)	G3 (MW)	G4 (MW)	G5 (MW)	G6 (MW)
MOD	12	80	98	520	280	50
SQD	12	158	90	470	230	80
JD_fx	12	133	90	475	280	50
FJD	12	80	98	520	280	50

Table 9.7. Spinning reserve dispatch results

Case	SR1 (MW)	SR2 (MW)	SR3 (MW)	SR4 (MW)	SR5 (MW)	SR6 (MW)
MOD	5	15	10	50	0	20
SQD	5	15	10	50	0	20
JD_fx	5	20	10	45	0	20
FJD	5	43	2	0	0	50

Table 9.8. Comparison of different dispatch approach

Unit	GEN_Limit	MOD	SQD	JD_fx	FJD
G1 (MW)	17	17	17	17	17
G2 (MW)	200	95	173	153	123
G3 (MW)	100	108	100	100	100
G4 (MW)	520	570	520	520	520
G5 (MW)	280	280	230	280	280
G6 (MW)	110	70	100	70	100

Fundamentally, in the case of multiple products, the challenge is in handling appropriate interactions among the separate products. Joint energy and reserve dispatch can handle inter-dependencies among the coupled products. In this method, the problem is formulated in the context of constrained optimisation considering various physical and operational constraints. This approach may provide coordinated dispatch among energy and ancillary services to achieve the most economic solution, while the security and reliability are considered. Using this approach two different alternatives related to the contribution of each generating unit may be included for reserve provision. First it is assumed that each unit may contribute a fixed fraction of its capacity, via Joint Dispatching, called JD_fx. Second it can be said that the only restriction on reserve provision is the Ramp Rate (RR) limitation, which should be applied as a physical constraint for spinning reserve capacity. The model, which considers the flexibility of reserve contribution with regard to Flexible Joint Dispatching, is called FJD. Regarding the results illustrated in Table 9.9, it can be seen that in the JD_fxd model, the Energy Market Costs (EMC) and Reserve Market Costs (RMC) are greater than similar data in the case of FJD. Energy Market Clearing Price (EMCP) and Reserve Market Clearing Price (RMCP), reported in Table 9.9, shows that the EMCP and RMCP for the JD_fxd model are higher than for FJD model.

Table 9.9. Energy and reserve costs and prices

Case	Tot_Cost	EMC	RMC	EMCP	RMCP
MOD	£15711	£15471	£240	£25/MW	£3/MW
SQD	£16095	£15855	£240	£28/MW	£3/MW
JD_fxd	£15810.5	£15569	£241.25	£26/MW	£4.75/MW
FJD	£15675	£15471	£204	£25.5/MW	£3/MW

9.3.2.2 Contingency Reserve Settlements

In this section, the effects of probability of reserve utilisation are investigated. On the California ancillary services markets all bidders submit a separate capacity bid and energy bid. Successful bidders are paid for capacity reservation and they are also paid for energy, if they are called on, to provide it in real time [21 32]. Based on the proposed method, all units submit their reserve bids associated with the ramp rate capabilities and their energy bids. The ISO may consider a reserve utilisation factor upon the first and/or second contingency and derives the probability of calling the units to provide reserve in real time. The successful bidders are paid the capacity price adjusted by the reserve utilisation facto. Two different reserve utilisation factors are assumed and applied to the models. Table 9.10 shows the result of two cases considering different ρ .

As Table 9.10 illustrates, successful bidders may receive different prices for capacity payment regarding reserve utilising factor (ρ) as well as the last band price for energy. The EMCP corresponding to the first contingency is also greater than that for the second contingency.

Table 9.10. Spinning reserve and capacity payment for successful bidders

Unit	Spinning Reserve Price for Successful Bidders			
	ρ = 0.3		ρ = 0.1	
	MW	ASCP	MW	ASCP
G1	5	£ 3.3/MW	5	£ 3.1/MW
G2	30	£ 3/MW	43	£ 3/MW
G3	15	£ 3.9/MW	2	£ 3.3/MW
G4	0	N/A	0	N/A
G5	0	N/A	0	N/A
G6	50	£ 2.9/MW	50	£ 2.7/MW
EMCP	£26/MW		£25.8/MW	

9.4 Development of Option Pricing Mechanism for Reserve Markets

9.4.1 Derivative Securities and Financial Contracts

Financial contracts, which are very common in commodity markets, can be applied in energy and especially electric power markets to protect market participants against financial risks. In this section the definition of financial derivatives and their attributes are discussed, while the major types of these contracts are explained.

9.4.1.1 What is a Derivative?

A derivative is commonly defined as a financial instrument designed in such a way that its price relates to the value of a particular underlying asset. In fact, a derivative is a contract, established by more than two parties, where payments are based on some agreed-price upon a benchmark. Individual parties can create a derivative product by means of mutual agreements. The prices of derivatives are not arbitrary and as long as the price of the underlying asset changes, they also will change. Derivatives are also risk-shifting devices to hedge financial price risks. Investment managers who are willing to enhance their returns may be interested in using these instruments for risk hedging. Initially, derivatives were used to reduce exposure to changes in foreign exchange rates, interest rates, or stock indices [33 34]. Recently, derivatives have become increasingly more important in the world of finance. There is a great deal of interest in applying these instruments to physical commodities, such as energy products [35]. Futures, forward, option, and swap contracts are examples of financial securities mainly used for risk hedging [33 36]. As has already been mentioned, a derivative is a contract whose value is a function of a spot price. In fact, forward, futures, and options are derivatives that can be settled based on the spot price. For example, a forward price is a derivative product in the sense that is a function of the spot price behaviour at some future point in time. An option on either the spot price or the forward price is a derivative contract, which can be based on the spot or forward price as well [37].

9.4.1.2 Forward Contracts

A forward contract is an agreement between two parties to buy or sell a specified quantity of an asset at a preset price with delivery at a specified time and place. The preset price will be referred to as the delivery price. The delivery price yields a fair price for future delivery of the commodity. The party who agrees to buy the underlying asset is said to have a long position, while the party who agrees to sell the underlying asset is said to have a short position. Forward contracts are settled at the delivery date or the so-called maturity date. The holder of the short position delivers the specified quantity of a commodity at a specified place and in return, receives from the holder of the long position a cash payment at an agreed price [33,34,36,37].

9.4.1.3 Futures Contracts

A futures contract is an agreement to buy or sell a specified quantity of an asset at a preset price at a specified time and place. This part of the definition of a futures contract is the same as that of a forward contract. Unlike forward contracts, futures contracts are usually traded on an exchange through a clearing-house that acts as a middleman to each transaction. In this sense, it can be said that futures contracts are regulated, while forward contracts are unregulated [33,34,36,37].

9.4.1.4 Option Contracts

An option is a contractual agreement between two parties to exchange a commodity at a prespecified price on a future date. However unlike forward and futures contracts, despite the option to fix a price for later use to limit losses and hedges, the fixed price is necessary, if it is better to use prices currently available in the spot market [38]. There are two basic types of option contracts: Call Options (CO) and Put Options (PO). A call option gives the holder the right to buy a commodity at a stated price, the so-called exercise price or strike price, on or before a stated date, called the maturity or expiration date. Call Options give the holder the right to buy, or to call the commodity away from someone else. Conversely, a Put Option gives the holder the right to sell a commodity at a stated price on or before a stated date. Put Options give the holder the option to sell, or to put the commodity to someone else [33,34,36].

In fact, an option gives the holder the right to do something, but the holder does not have to exercise this right. This is the main difference between option contracts and forward/futures contracts, in which the holder is obliged to buy or sell the underlying commodity [37,38]. Note that there is no cost to enter into a forward/futures contract, while there is a premium associated with the cost of acquiring an option contract. Since each option can be viewed as a type of insurance contract in financial markets to hedge against risks, this premium is the price paid for risk hedging [39]. Generally, Call Options and Put options are defined in one of two different ways:

- European option: which can only be exercised at the maturity date of the option.
- American option: which can be exercised at any time up to and including the maturity date.

Based on experience obtained in several countries, it can be inferred that electricity spot prices are very variable and difficult to forecast accurately. The electricity price behaviour reflects high and low demand hours and shows which generating units are at the margin for each hour. The demand side does not want to be without supply because of the damage it would cause to consumer activities. Thus the necessity for applying financial contracts to provide electric power transactions stabilised against uncertainties [40,41,35].

9.4.1.5 Why Option Contracts are Needed for Electricity and Ancillary Services?

There has been a tremendous increase in the awareness and activities of financial derivatives in commodity markets in recent years. Option markets are the most popular with financial institutions. Modelling and predicting option prices are very important for market efficiency. The contingent claim valuation of physical assets and financial derivatives depends very much on the characteristics of the stochastic process, which may make the price trend very volatile [42]. Energy commodity markets grew rapidly as deregulation and restructuring of the electricity industry took place in the UK, USA and all around the world. It can be said that the global trend of electricity restructuring may expose the portfolios of utilities' assets through risk management and asset valuation [43].

In fact, accurate modelling of electrical energy spot prices is needed for financial risk management. Furthermore, to prevent the electrical energy supply from collapsing, electricity has to be balanced instantaneously. Since there is no inventory for electricity to smooth the supply demand imbalance, the electricity spot price is very volatile. On the other hand, among the various types of derivative security contracts, option contracts consider price volatility as a major issue[37]. Electricity derivatives have payoffs dependent on the spot price of electricity, which is volatile, the electrical energy and ancillary services may require exotic option contracts rather than standard ones. In fact, exotic options are derivative securities with more complicated payoffs than the standard ones, either European or American call and/or put options. Exotic options, which are traded in every market, should be designed to meet particular needs with regards to market efficiency [44]. Competition may imply electricity spot price spikes, in the case of supply shortages and large demand variations. To facilitate perfect competition, exotic option contracts in electricity markets will consequently induce an optimal level of reliability in power systems [42]. Therefore, new non-standard or exotic options may present a basic challenge to quantitative risk management in the electricity industry.

9.4.2 Option Structure and Option Evaluation

The holder of an option will pay the option premium for the right the option gives him (her). In the case of either exercising the right or not exercising the right, the option will expire at the expiration date. The relative value of the underlying price at the time of the option exercise in relation to the option's strike price determines whether the option will be exercised or not.

If this relative value is greater than one, the option is referred to as being in-the-money, and the option will be exercised. If the relative value is less than one, the option is referred to as being out-of-the-money, and the option will not be exercised. There is an indifference case when the relative value is exactly one, and the option is referred to as being at-the-money. Figure 9.3 shows the profit of a call and a put option, while generally, it can be said that:

The Call Option will exercise only if " SP > Xc ".
The Put Option will exercise only if" SP < Xp ".

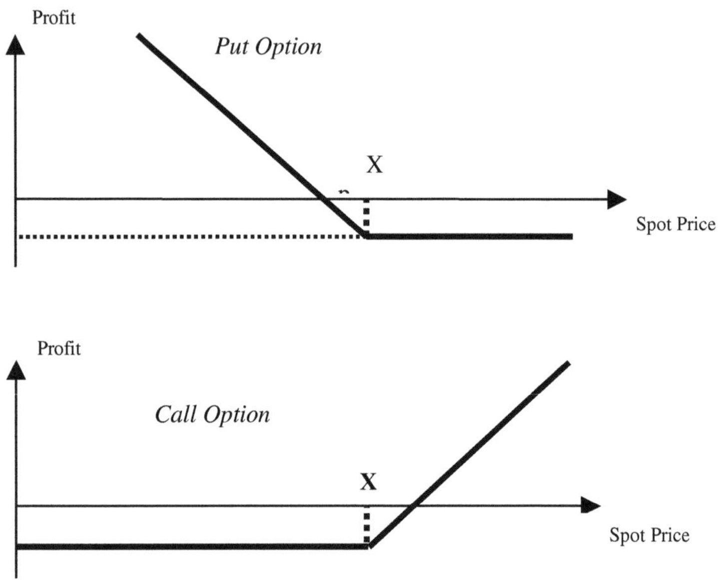

Figure 9.3. Profit for put and call options

An important characteristic of reserves, as distinct from pure financial options, is that they are procured not based on any dynamic hedging strategy. It can be said that they are based on preset standards defined by regulatory bodies such as North American Electric Reliability Council (NERC) in USA or Office of Gas and Electricity Markets (OFGEM) in the UK that determine reserve requirements. Based on the most famous option pricing methodology, modified Black Scholes closed-form equations are adopted:

$$c = SN(d1) - Xe^{-r(T-t)}N(d2) \tag{9.9}$$

where $N(d_i)$ is the commutative probability of a normal distribution.

$$p = Xe^{-r(T-t)}N(-d2) - SN(-d1) \tag{9.10}$$

$$d1 = [\ln(S/X) + (r + \sigma^2/2)(T-t)]/\sigma\sqrt{(T-t)} \tag{9.11}$$

$$d2 = [\ln(S/X) + (r - \sigma^2/2)(T-t)]/\sigma\sqrt{(T-t)} \tag{9.12}$$

where c is call option price, p is put option price, X is strike price, S is spot price, $(T-t)$ is time difference, r is rate of return, and σ is price volatility.

9.4.3 Application of Standard Options for Reserve Procurement and Pricing

In this section a three-phase algorithm associated with a standard form of option is proposed to solve the problem of finding the call option and/or put option price for reserves [45]. Figure 9.4 shows the procedure for calculating a call option (and put option) of spot price for spinning reserve. In phase one, solving an Optimal Power Flow (OPF) or Joint Dispatch (JD) problem goes the spot price of spinning reserve as well as the real power price for different reserve requirements. In phase two there is a modification of the spinning reserve spot price that leads to statistical calculations to derive the average and standard deviation of the modified spinning reserve spot price. Phase three includes computation of call option and put option prices of spinning reserve using modified Black Scholes closed form equations assuming a European option pricing mechanism.

In Figure 9.4, the following symbols are used:

RR_i : System reserve requirement.

$\lambda_{p|RR_i}$: Real power spot price considering reserve.

λ_{RR_i} : Spinning reserves spot price.

S : Average spot price of spinning reserve, $\overline{\lambda}_{RR_i}$.

X : Strike price of spinning reserve.

σ : Price volatility of spinning reserve.

$(T-t)$: Time difference, from now until exercise time.

srf_i : Spinning reserve factor for each GenCo.

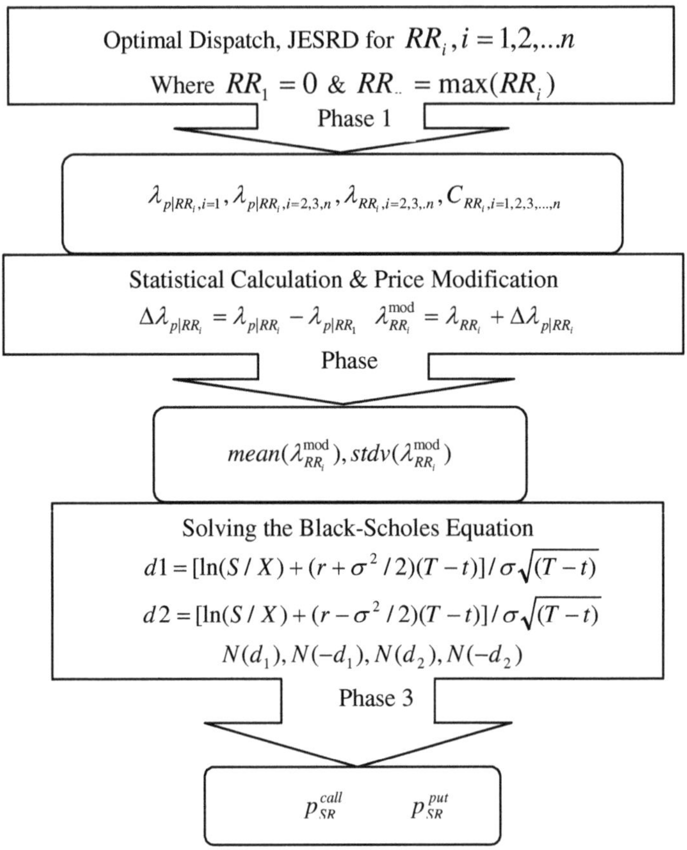

Figure 9.4. Computational flow diagram

9.4.4 Case Study and Results Analysis

In order to demonstrate the effectiveness of the proposed model, a modified 30-bus IEEE test system with six GenCos. is considered. An optimal dispatch solution is performed from $RR_1 = 0$ to $RR_{13} = 130$ increasing by 10 MW. The results ,which are shown in Table 9.11, include the difference in real power spot prices between the base case ($RR_1 = 0$) and the price of real power corresponding to each amount of $RR_i \neq 0$, and spinning reserve spot price.

From the results shown in Table 9.11, the volatility of spinning reserve price can be calculated. Based on the Black Scholes equations, the call option ($call\_opt.$) and the put option ($put\_opt.$) price of spinning reserve are calculated, where $call\_opt. = 0.3$ and $put\_opt. = 0.075$. The final price of spinning reserve considering the call option can be determined from the following relation:

$$p_{SRi} = D\lambda_{P|RR_i} + \lambda_{SR_i|RR_i} + [call\_opt. - \min(0, put\_opt.)] \qquad (9.13)$$

where $[call\_opt. - \min(0, put\_opt.)]$ is the net call option price of reserve. Figure 9.5 shows the variations of spinning reserve price considering the call option.

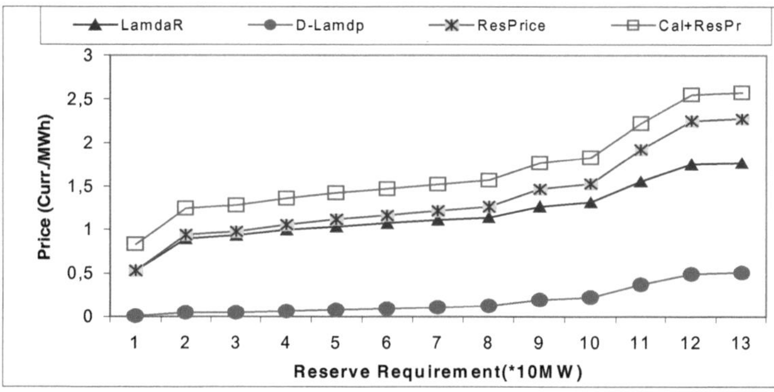

Figure 9.5. Spinning reserve price variations

Table 9.11. Spinning reserve requirements and price for IEEE 30-bus system

RR_i (MW)	$D\lambda_{P\|RR_i}$ (£/MW)	$\lambda_{SR\|RR_i}$ (£/MW)
10	0.0048	0.5289
20	0.0436	0.8973
30	0.048	0.9309
40	0.0635	0.9921
50	0.0801	1.0357
60	0.094	1.0721
70	0.1078	1.1085
80	0.1217	1.1449
90	0.1965	1.2675
100	0.2203	1.3099
110	0.3692	1.5503
120	0.4958	1.7539
130	0.5056	1.7696

As the model shows, spinning reserve as a security and reliability constraint has a considerable impact on the real power price. Therefore it can be said that spinning reserve price should be influenced by the price difference between cases with spinning reserve constraint and without spinning reserve constraint. Thus the price of spinning reserve includes three terms: marginal cost of reserve, differential price of real power and net call option price for capacity reservation.

9.5 References

[1] K. W. Cheung, P.Shamsollahi, and S. Asteriadis, "Functional Requirements of Energy and Ancillary Service Dispatch for the Interim ISO New England Electricity Market", IEEE Power Engineering Society Winter Meeting, Vol.1, 1999, pp.269-73, Piscataway, NJ, USA
[2] E. Hirst and B. Kirby, "Unbundling Generation and Transmission Services for Competitive Electricity Markets", National Regulatory Research Institute, Ohio University, USA, Jan. 1998
[3] L. L. Lai, "Power System Restructuring and Deregulation Trading Performance and information technology", John Wiley & Sons, 2001
[4] E. Hirst and B. Kirby, "Creating Competitive Markets for Ancillary Services", Oak Ridge National Laboratory, Energy Division, USA, Oct. 1997
[5] D. M. Logan, "Are Generation Capacity Markets Working?", IEEE Winter Power Meeting, Panel Session, Singapore, Jan. 2000
[6] New Electricity Trading Arrangements for England & Wales, http://www.neta.org.uk
[7] M. Rashidi-Nejad, Y. H. Song, M. H. Javidi-Dasht-Bayaz, "Contingency Reserve Pricing via a Joint Energy and Reserve Dispatching approach", Energy Conversion & Management, Elsevier, Accepted 27 Jan. 2001
[8] N. Nicholson and M. J. H. Sterling, "Optimization Dispatch of Active and Reactive Generation by Quadratic Programming", IEEE Trans. on PAS, Vol. 92, pp.644-654, 1973
[9] N. Chowdhury and R. Billinton, "A reliability Test System for Educational Purposes-Spinning Reserve Studies in Isolated and Interconnected Systems", IEEE Trans. on Power Systems, Vol. 6, No. 4, pp.1578-1583, Nov. 1991
[10] N. Chowdhury and R. Billinton, "Export/ Import of Spinning Reserve in Interconnected Generation Systems", IEEE Trans. on Power Systems, Vol. 6, No. 1, pp.43-49, Feb. 1991
[11] N. Chowdhury, "Energy Method of Spinning Reserve Assessment in Interconnected Generation Systems", IEEE Trans. on Power Systems, Vol. 8, No. 3, pp.865-871, Aug. 1993
[12] M. E. Khan, "Composite System Spinning Reserve Assessment in Interconnected Systems", IEE Proc.-Gener. Transm. Distrib. Vol. 142, No. 3, May 1995
[13] D. R. Bobo and D. M. Mauzy, "Economic Generation Dispatch with Spinning Reserve Constraints", IEEE Trans. on Power Systems, Vol. 9, No. 1, pp.555-559, Feb. 1994
[14] M. Fotuhi-Firuzabad, R. Billinton and S. Aboreshaid, "Spinning Reserve Allocation Using Response Health Analysis", IEE Proc.-Gener. Transm. Distrib., Vol. 143, No. 4, July 1996, pp. 337-343
[15] J. W. O'Sullivan and M. J. O'Malley, "Economic Dispatch of a Small Utility With a Frequency Based Reserve Policy", IEEE Trans. on Power Systems, Vol. 11, No. 3, pp.1648-1653, Aug. 1996
[16] J. W. Marangon-Lima and A. M. Leite-da-Silva, "Spinning Reserve Requirements for Multi-Area Systems", 12th Power Systems Computation Conference, Dresden, Aug. 1996

[17] S. S. Halilcevic, "Assessment of the Ready-Reserve Power in Power Systems", IEEE Power Engineering Review, Jan. 1999
[18] S. S. Halilcevic and F. Gubina, "An On-Line Determination of the Ready Reserve Power", IEEE Trans. on Power Systems, Vol. 14, No. 4, pp.1514-1519, Nov. 1999
[19] J. W. O'Sullivan and M. J. O'Malley, "A new Methodology for the Provision of Reserve in an Isolated Power System", IEEE Trans. on Power Systems, Vol. 14, No. 2, pp.519-522, May 1999
[20] H. Singh and A. Papalexopoulos, "Competitive Procurement of Ancillary Services by an Independent System Operator", IEEE Trans. on Power Systems, Vol. 14, No. 2, pp. 498-504, May 1999
[21]Y. Liu, Z. Alaywan, S. Liu, and M. Assadian, "A Rational Buyer's Algorithm Used for Ancillary Service Procurement", IEEE Power Engineering Society Winter Meeting, Conference Proceedings Vol.2, pp.855-60, 2000
[22] E. H. Allen and M. D. Ilic, "Reserve Markets for Power System Reliability", IEEE Trans. on Power Systems, Vol. 15, No. 1, pp.228-233, Feb. 2000
[23] M. Rashidi-Nejad, "Procurement and Pricing of Reserves via Joint Dispatch and Financial Derivatives " , Chapter 5, Ph.D Thesis, Electronic and Computer Engineering Department, Brunel University , London, UK, 2001
[24] M. Rashidi-Nejad, Y. H. Song, M. H. Javidi-Dasht-Bayaz, "Joint Energy and Spinning Reserve Dispatching and Pricing", IFAC Symposium on Power Plants & Power Systems Control, pp.477-482, Brussels, Belgium, Apr. 2000
[25] C. Concordia, L. H. Fink, and G.Poullikkas, "Load Shedding an Isolated System", IEEE Trans. on Power Systems, Vol. 10, No. 3, Aug. 1995
[26] H. Singh, "Auction for Ancillary Services", Decision Support Systems, Vol. 24 p183-191, 1999
[27] M.E. Flynn, M.J. O'Malley, P.Brown and M.Power "Balance Between Generating Plant Costs and System Load Following Requirements", CIGRE, 21, rue d'Artois F-75008 Paris, p. 400-03 London Symposium 1999
[28] B. Eua-Arporn and B. J. Cory, "The Application of Deterministic and Probabilistic Methodologies to Spinning Reserve Policy in Thermal Generation Scheduling", IFAC Control of Plants and Power System, Munich, Germany, 1992
[29] G. B. Sheble, "Real-Time Economic Dispatch and Reserve Allocation Using Merit Order Loading and Linear Programming Rules", IEEE Trans. on Power Systems, Vol. 4, No. 4, Oct. 1989
[30] W. Ongsakul and J. Tippayachai, "Constrained Economic Dispatch by Micro Genetic Algorithm Based on Migration and Merit Order Load Solution", International Conference on Electricity Utility Deregulation and Restructuring and Power Technologies 2000, p. 510-17, 4-7 April 2000, City University, London
[31] K. W. Cheung, P. Shamsollahi, and D. Sun, "Energy and Ancillary Service Dispatch for the Interim ISO New England Electricity Market", IEEE (0-7803-5478-8/99)
[32] http://www.calpx.com/publications
[33] R. Jarrow and S. Turnbull, "Derivatives Securities", International Thomson Publishing, 1996
[34] A. Konishi and R. Dattatreya, "The Handbook of Derivative Instruments", McGraw-Hill Book Company, 1991
[35] D. Pilipovic, "Energy Risk, Valuing and Managing Energy Derivatives", McGraw –Hill Publisher, 1998
[36] D. Winstone, "Financial Derivatives, Hedging with Futures, Forwards, Options and Swaps", Chapman & Hall, 1995
[37] J.C. Hull, "Options, Futures, & Other Derivatives", Prentice-Hall International, Inc., 2000

[38] P. Wilmott, J. Dewynne, and S. Howison, "Option Pricing, Mathematical Models and Computation", Oxford Financial Press, 1993
[39] J. C. Cox and M. Rubinstein, "Options Markets", Prentice-Hall International, Inc., 1985
[40] M. V. F. Pereira, M. F. McCoy, and H. M. Merrill, "Managing Risk in the New Power Business", IEEE Computer Applications in Power, Vol. 13, No. 2, pp.18-24, Apr. 2000
[41]]Task Force 38.05.09, "Electra", Cigre, No. 194, Feb. 2001
[42] S. Soft, "Power System Economics, Designing Markets for Electricity", IEEE-Wiley Press, Nov. 2001, http://www.stoft.com/x/book/index.shtml
[43] C. Celik, "Electricity Market Models Around the Globe", Department of Economics, Istanbul Bilgi University, Istanbul 2001
[44] S-J Deng, B. Johnson, and A. Sogomonian, "Exotic electricity options and the valuation of electricity generation and transmission assets", Decision Support Systems, Vol. 30, pp. 383–392, 2001
[45] M. Rashidi-Nejad, Y. H. Song, and M. H. Javidi, "Option Pricing of Spinning Reserve in a Deregulated Electricity Market", IEEE Power Engineering Review Journal, July 2000

10. Ancillary Services (II): Voltage Security and Reactive Power Management

G.A. Taylor, S. Phichaisawat, M.R. Irving and Y.H. Song

The focus of this chapter is on voltage security as a function of ancillary services that can be described in the context of reactive power markets and associated pricing mechanisms. First, an introduction to voltage security and reactive power management is presented and the consequences of recent electricity industry restructuring in the UK are highlighted. Second, an overview of recently developed reactive power markets is presented and several examples are discussed in detail. Finally, several recently developed algorithmic techniques are presented in detail. The techniques will introduce the concepts of transition-optimisation and congestion management in the context of a complete framework for the monitoring, assessment and control of voltage security.

10.1 Introduction

10.1.1 Reactive Power and Voltage Control

Reactive power and voltage control plays an important role in supporting the real power transfer across a large-scale transmission system [1]. In an open access system, the importance of this support is even greater as the power transfer is increased and the associated voltages then become a bottleneck in preventing additional power transfer [1]. In simple terms, the most important aim of reactive power dispatch is to determine the sufficient amount and correct location of reactive support in order to maintain a secure voltage profile [2 4]. Reactive support is generally provided by the switching of shunt reactors, the positioning of transformer taps and the reactive power outputs of generators [1]. Focusing on the mathematical problem of optimal reactive dispatch, original objective functions were oriented economically and involved the minimisation of real power transmission losses, an important case in point was the formulation of Peschon *et al.* in

1968 [5]. More recently, a number of optimisation techniques that simultaneously improved voltage profiles and minimised losses have been developed [2 4].

10.1.2 Monitoring and Assessment of Voltage Security

Over the last two decades voltage security has emerged as a major concern with regard to the secure operation of power transmission systems world-wide [6]. Voltage security is now considered on a par with thermal and stability constraints with regard to the limiting of real power transfers [6]. This is partly a consequence of the newly deregulated market environments that have recently been introduced to electricity industries over the same time period. Power systems are now operating under very different conditions, particularly with regard to the distribution and location of generation. A case in point is the large-scale UK power transmission system of England and Wales that is owned and operated by National Grid (NG). Over the last decade the majority of new generation has been located in the North whereas the majority of load remains in the South. For these reasons, an increasing amount of reactive compensation equipment has been installed by NG in order to maintain the required voltage profiles throughout the transmission system. Consequently, the secure and economic control of reactive compensation equipment combined with generator reactive power outputs is an important consideration, particularly when considering the development of reactive power markets [7].

Over the last decade a number of algorithmic methods have been proposed for on-line voltage security analysis and control. It is important to note that several of the proposed methods have been implemented, such as the voltage security monitoring system in TEPCO (Japan) [8] and the on-line voltage stability assessment package developed by Powertech Labs Inc. of Canada [9]. However, many areas need to be further addressed including accuracy, speed and practicality. Furthermore, it is important to note that all of the methods proposed so far place a strong emphasis on analysis and assessment, but not control. The following sections describe potential components of a framework that considers analysis, assessment and control of voltage security at the same time.

10.1.3 Transition-optimised Reactive Power and Voltage Control

Voltage control and reactive power optimisation has been researched extensively as a static snap-shot problem, but relatively little research has addressed the time-domain aspects of the problem [10]. In the following sections of this chapter, techniques that have been applied to the time-domain aspects of the problem are discussed. The discussion establishes the standard voltage control problem in a transition-optimised sense, which considers practical constraints such as the number of reactive control actions allowable within a time-domain or the time interval required between actions performed. A simple theoretical example of transition-optimisation is also presented [11].

10.1.4 Voltage Security and Congestion Management

Alleviating transmission congestion is another complex task in both power system planning and operation. This task has also become more important as the restructuring of the electricity industries has progressed. In order to manage such a complex problem, many factors must be considered. Voltage security is one significant aspect that might worsen after congestion is relieved. Severe congestion without effective management may lead at best to uneconomical operation or at worst to a blackout of the whole power system. It is the responsibility of the System Operator (SO) to alleviate transmission congestion while keeping the system secure. In the following sections of this chapter, techniques that have been applied to congestion management and voltage security are presented in detail. In previous research by Yu and Song [12,13], voltage security and environmental constraints are considered simultaneously using a linear programming (LP) method for short-term generation scheduling. The objective is to minimise both the MW output and penalty costs subject to a number of constraints. A Continuation Power Flow (CPF) method is employed to evaluate the voltage security margin. It was also highlighted that the line flow limit is an important factor when considering voltage security and the LP method was employed to investigate dynamic load dispatch using an objective function that minimises the MW output cost only.

10.2 Reactive Power Markets and Pricing Mechanisms

As a consequence of the global restructuring of electricity industries over the last decade, a wide range of markets and pricing mechanisms have been proposed and adopted with regard to reactive power. The proposal and adoption of such markets also requires improved techniques for efficient reactive power management and voltage security. An overview of such markets and pricing mechanisms is now presented.

10.2.1 Examples of Reactive Power Markets

In this section examples of reactive power markets that currently exist in several key countries will be presented. There are in fact a larger number of examples that could be presented [14], but only the following three examples are presented here. This can be justified because the following examples highlight current international trends in reactive power market development.

It is important to note that there are two sides to the development of reactive power markets [15]:

i. Payment for reactive power service providers.

ii. Allocation of reactive service costs to transmission system users.

Most discussions concerning reactive power market development focus on I, but it is important to note that ii should be addressed also.

10.2.1.1 England and Wales (UK)

Restructuring of the electricity industry in the UK began in 1990, when ownership and operation of the transmission system in England and Wales was vested to National Grid. Northern Ireland and Scotland own and operate their transmission systems separately for historical and geographical reasons. At that time it was decided that all generators of 50 MW capacity or more were required to provide mandatory reactive power support service. This was a stipulation of the Grid Code, which governs connection to the National Grid transmission system, as agreed with the regulator of the UK electricity industry.

Gradual restructuring of the electricity industry has continued in the UK over the last decade. It is possible to focus on the developments of reactive power markets in England and Wales over that period of time. Initially, an interim arrangement was introduced in 1990 with regard to compensating generators for supplying reactive power [16]. Compensation was based on a fixed payment reflecting the amount of reactive power produced in the year 1989 1990, prior to the onset of restructuring and privatisation. This arrangement was considered temporary and was to be reviewed after the completion of a new metering system that was capable of measuring reactive power flows [17]. When this was achieved an alternative scheme was proposed during the mid-1990 that was based on both MVAr utilisation and capability components. An essential feature of the proposed scheme was its potential for development to a full reactive power market mechanism [16].

The scheme was introduced in phases from 1997 2000. Again, all generators of 50 MW capacity or more were required to provide a mandatory reactive power support service. In order to receive payment for this service, generators must enter into either a default payment mechanism (DPM) or a tendering system. The tendering system consists of either an obligatory reactive power service (ORPS) or an enhanced reactive power service (ERPS). As an incentive for generators to enter the tendering system, the DPM ratio of capability to utilisation was introduced as 80:20 in 1997 and was them modified incrementally to 0:100 up to the year 2000 [17,18]. Alternatively, the generators can tender specific prices for capability and utilisation with regard to ORPS or ERPS [17,18]. The rounds for tenders have been held every six months since 1997 and the ninth tender round was for contracts effective from 1 April 2002. The tenders are for supplying a reactive service for a period of twelve months or more. If a tender is accepted the generator or provider of the reactive power service has a reasonably guaranteed level of income. The income a generation unit could receive from a reactive power service is dependent upon both the number of service providers and the demand within a particular area. On average approximately 50% of tenders have been accepted over the last five years and therefore approximately 50% of eligible generation units have reached bilateral market agreements with regard to reactive power services over that time. If a tender is rejected and the generation unit is eligible it enters the DPM. Tenders for ERPS can only be submitted by generation units with a reactive power capability in excess of that stipulated by the Grid Code. Up until now only a small number of generation units have tendered an ERPS and the majority of generation units have tendered an ORPS. A detailed account of submitted and accepted tenders for all tender rounds from 1997 2002 is available from National Grid [19]. It should be noted that the ORPS does not include reactive power services from synchronous compensators or SVCs. However, ERPS can include services from such compensators and indeed any other plant that can generate or absorb reactive power.

Finally, the cost of providing reactive power services is currently recovered by National Grid via use of system charges in the balancing mechanism [20]. The costs of reactive power contracts are recovered by NG via the daily Incentivised Balancing Cost (IBC). These are the costs that NG is incentivised to manage and are the basis upon which incentive payment to (or from) NG is calculated. The reactive power contract costs are included in the daily Balancing Services Contract Costs Allocation (BSCCA) that is a component of the IBC. Finally, it should also be noted that black start capability, frequency response and warming contracts are also included in the BSCCA.

10.2.1.2 New York (US)

New York independent system operator (NYISO) was formed, as a non-profit making organisation, in 1998 as part of the restructuring of the electricity industry in New York State. In the US the electricity industries and ISOs of all states are regulated by the Federal Energy Regulatory Commission (FERC). NYISO is responsible for operating the transmission system in New York State. The transmission owners are customers of the NYISO and purchase ancillary services as required. NYISO administrates and offers six ancillary services including reactive power support. In operating the transmission system NYISO is required to procure reactive power services from generators. Reactive power services are specified as ancillary services and therefore qualify for payment for the provision of such services. In order to qualify for payments, suppliers of voltage support services must provide a resource that has an Automatic Voltage Regulator (AVR) and has passed reactive power capability testing in accordance with the NYISO procedures and standards. The NYISO directs the suppliers to operate within their tested reactive capability limits. The scheduling of voltage support services is the responsibility of the NYISO and the transmission owners. NYISO also coordinates the power system voltages throughout the NY state control area. The transmission owners are responsible for the local control of the reactive power resources that are connected to their network [21].

Dependent upon the nature of the ancillary service, NYISO provides the service at either embedded cost-based or market-based prices [21]. Reactive power services are provided at embedded cost-based prices, while other services such as frequency response are provided at market-based prices [21]. Suppliers of reactive power services, namely generators and synchronous condensers, are paid by the NYISO as a consequence of the following costs where applicable;

- Total annual embedded costs.
- Attributable Lost Opportunity Cost (LOC).
- Prior year adjustment costs.

The total embedded costs are based upon the individual capital investment, operating and maintenance expenses of a reactive power service provider. Attributable LOC is based upon the amount of real power reduction that the NYISO dispatches in order to enable a generator to produce or absorb more reactive power. Prior year adjustment costs are the total prior year payments made to reactive power service providers less the payments received from transmission customers and owners [21].

The cost is recovered by NYISO on a monthly billing basis via an hourly charge for transmission usage. The hourly charge is calculated by dividing a sum of project payments (based upon the above costs) by an annual forecast of transmission usage [21].

10.2.1.3 Australia

Historically, Australia's electricity industry has been predominately governed at the state level, with a relatively modest amount of regulation at the national level. This structure is different from the UK, but has some similarities with the US. The main reason for such a structure in Australia is due to the historically weak transmission interconnection between states. For these reasons dual restructuring has also occurred at both a state and national level in Australia. The restructuring is in accordance with the previously mentioned international trend of restructuring electricity industries to place a greater reliance on market forces and less dependence on government with regard to investment [22].

The restructuring of Australia's electricity industry has moved toward privatisation in an incremental fashion since 1991. In October 1994 all eight states and territories endorsed a report concerning the restructuring of Australian electricity industries [22]. Subsequently, the National Electricity Code (the Code), which establishes the rules and procedures for operating in the competitive national electricity market, was approved in November 1996. The Australian national electricity market (NEM) has been introduced in stages, since May 1997, in order to develop fully competitive electricity markets for generation and retail supply. Although referred to as a national market, NEM includes only the following five geographically connected states; Victoria, New South Wales, South Australia, Queensland and the Australian Capital Territory. Tasmania will be included following the completion of its grid interconnection with mainland Australia. Western Australia and the Northern Territory will not participate in the market due to geographical and cost factors. As stated earlier, Australia has undergone dual restructuring and each state has restructured independently at its own pace. A national electricity generation pool was introduced when the two leading states Victoria and New South Wales began trading wholesale electricity in 1998. The Australian Capital Territory participated in the linked market via New South Wales [22].

NEMMCO was established in 1998 as the company responsible for the day-to-day operation and administration of both the national transmission system and the wholesale spot market. NEMMCO operates on a self-funding, non-profit making basis and the five interconnected states are shareholders in the company [23]. NEMMCO is responsible for ensuring that the power system is operated in a safe, secure and reliable manner. In order to fulfil this obligation, NEMMCO controls the key technical characteristics of the system. These include six ancillary services and NEMMCO is required to purchase these services under ancillary service agreements [23]. Therefore, NEMMCO can also be regarded as an ISO, but at a national level, and as such is responsible for reactive power provision as an ancillary service in a similar fashion to the NYISO. With regard to reactive power provision scheduled generators are required to provide obligatory amounts of reactive power. NEMMCO is able to utilise this obligatory reactive power provision and the generators receive no payment. NEMMCO can only employ the obligatory reactive power service from generators that are already committed. Under normal operating circumstances NEMMCO cannot commit a generator to provide additional reactive power services [24].

The provision of a network transmission service requires reactive power support, therefore the Transmission Network Service Providers (TNSP) must provide a significant amount of reactive power support in order to ensure such a provision. The reactive power support provided by the TNSP is available to NEMMCO to utilise free of charge. As NEMMCO is responsible for ensuring that the power system is operated in a secure manner it will utilise all obligatory generator reactive power support and all TNSP reactive power support in order to maintain the security of the power system. Where NEMMCO requires additional reactive power support it is procured via a contract tender process [24]. Suppliers of such additional reactive power services are paid by NEMMCO as follows;

- Generators – availability plus compensation fee.
- Synchronous compensators – availability plus enabling fee.

The availability fee is related to a suppliers readiness in providing the service, the compensation fee is related to opportunity cost to a supplier and the enabling fee is related to the start-up of the service by a supplier. The costs associated with the procurement of such contracts are recovered from the transmission network customers, in proportion to their energy usage [14].

It is important to note that a boundary of responsibility is required between the TNSP and NEMMCO with regard to how much reactive power support should be provided by the TNSP and how much should be contracted by NEMMCO. Furthermore, the TNSP does not pay for the additional reactive power services that are contracted by NEMMCO. For these reasons the lack of exposure of TNSP to a reactive power market is a current issue in the NEM [24].

10.2.2 Analysis of Reactive Power Markets

In recent years, many studies have been performed in order to determine effective reactive power management in newly developed reactive power markets. A number of key references are briefly described in this section. Reference [25] address the problem of reactive power procurement from a SO perspective, such that the reactive capability curves of generators and the opportunity costs are taken into account. Reference [26] proposes reactive power pricing mechanisms based on capability and contributions to system performance. In addition, mathematical indices are developed to indicate new reactive source allocation. Reference [16] describes reactive power pricing based on the NG system in England and Wales by focusing on modelling requirements and associated OPF development.

10.3 Transition-optimised Reactive Power Control

10.3.1 Introduction

The objective of the research presented here is to develop transition-optimised voltage and reactive power control within a framework that can also consider voltage security issues directly. In order to achieve such objectives the framework is developed from existing algorithms and software. The research described in this section addresses these issues and seeks to provide a 'transition-optimised' voltage and reactive power control method. Essentially this approach would consider voltage and reactive power control as a time-based scheduling problem, with the intention of avoiding unnecessary changes in status and output of reactive control plant. The time-transitions of reactive output and voltage targets would follow an optimised trajectory, in an analogous manner to the schedule of generator MW outputs produced by unit commitment programs.

A recent review of reactive power and voltage control techniques illustrates that voltage control and reactive power optimisation has been researched extensively as a static snap-shot problem, but relatively little research has addressed the time-domain aspects of the problem [10]. However, in recent years the problem has been formulated in a transition-optimised sense that is consistent with the approach that is described in this chapter. In one case the objective function is the minimisation of control actions over a short-term operational period (1 day) [27]. In this approach, the entire problem is decomposed into master and slave levels, which are based upon the cardinal operating points over a day. The master level deals with the minimisation of the depreciation cost of compensators and transformer taps while satisfying operating constraints over all operating points. The slave level considers the minimisation of active power losses while satisfying system security constraints for each operating point independently [27]. In another transition-optimised approach the minimisation of energy loss over intervals of time has been described [28, 29]. However, in this formulation the transition of discrete variables is governed by the selection of time intervals. In this manner a near optimal transition of discrete variables can be achieved by ensuring that the discrete variables remain constant over a time interval. Furthermore, each time interval consists of a number of sub-intervals and the continuous variables are determined to remain constant for each sub-interval. The sizes of the time intervals and sub-intervals are determined in a heuristic fashion from the associated load forecasts [28] and real-time constraints with regard to the frequency of running optimal reactive power flow analyses. It is important to note that in the above cases the underlying emphasis is on loss minimisation and not voltage security. The objective of this research is to develop transition-optimised voltage and reactive power control within a framework that can also consider voltage security issues directly. The following sub-sections describe recent algorithmic developments towards achieving this goal.

10.3.2 Algorithmic Procedure

The multi-objective function that has been employed is a weighted combination of active generation cost minimisation and additional objectives, such as the maximisation of the voltage collapse proximity indicator for the whole system. It is also possible to apply a variety of standard and non-standard constraints to the problem variables. A complete description of both the multi-objective approach and the derivations of non-standard constraints that have been adopted in this research has been presented by Chebbo and Irving [30, 31].

10.3.2.1 Objective Function

The active generation cost minimisation term of the multi-objective function is defined as follows:

$$F_P = \sum_{i=1}^{N} \sum_{j=1}^{G} c_j \left(P_j^i\right) \Delta t^i \tag{10.1}$$

where N is the number of time intervals Δt^i, G is the number of generators with active power control, c_j is the cost function and P_j^i is the generator active power output. It is possible to apply a successive linear programming approach to loss minimisation by linearising (1) as follows:

$$\Delta F_P = \sum_{i=1}^{N} C_{\text{Slack}} \, \Delta P_{\text{Slack}}^i \, \Delta t^i \tag{10.2}$$

where C_{Slack} is the linear cost coefficient of the slack generator. $\Delta P_{\text{Slack}}^i$ is the incremental update of the active power output of the slack generator for time interval Δt^i.

It is important to note that the maximisation of the voltage collapse proximity indicator for the whole system can be included within the linearised multi-objective function as follows:

$$\Delta F_V = \sum_{i=1}^{N} \sum_{j=1}^{C} \Delta \lambda_j^i \tag{10.3}$$

where C is the number of contingencies associated with a time interval and λ_j^i is a scalar update of the nodal voltage collapse proximity indicator for a contingency at a time interval [32].

10.3.2.2 Transition Constraints

The Combined Active and Reactive Dispatch (CARD) algorithm of Chebbo and Irving [30] applied transition-optimisation to active power control via ramping constraints on the active power outputs generators. However, transition-optimisation was not considered with regard to reactive power and voltage control. As a first step towards implementing a transition-optimised capability with regard to reactive power and voltage control, transition constraints can be included with regard to the discrete independent variables such as transformer tap positions and switchable shunt capacitors or reactors. It is assumed that continuous independent variables such as generator voltages (or reactive power outputs) do not require transition optimisation as they can be adjusted more frequently.The incremental updating of the transformer tap positions can be constrained as follows:

$$\sum_{i=2}^{N}\sum_{j=1}^{M} \quad 0 \le T_j^i - T_j^{i-1} \le 0 \quad \text{and}$$
$$\sum_{j=1}^{M} \quad T_j^L - T_j^U \le T_j^1 \le T_j^U - T_j^L \tag{10.4}$$

where M is the number of transformers, T_j^1 are the initial transformer tap positions and T_j^i are subsequent transformer tap positions. T_j^U and T_j^L are the upper and lower tap position limits for the j transformer. The above constraint equations are required to determine the optimal tap setting for each transformer over N time intervals. Similarly constraint equations can be developed for other independent variables that may require transition-optimisation.

It is important to note that the intervals could also be regarded as sub-intervals and that the transition-optimisation of transformer tap positions or switchable shunts could then be performed for a number of periods that relate to different load conditions. Furthermore, the generator voltages (or reactive power outputs) are determined for every interval, whether it is a sub-interval or not. It is also important to note that the equality constraints of (10.4) could be modified to inequality constraints that represent narrower operating limits at different load levels. Obviously, this would then permit limited transition of reactive compensation equipment over the time intervals.

10.3.2.3 Solution Algorithm

The main steps of the solution algorithm can be summarised as follows:

i. Perform a Newton Raphson load-flow [33] for the base and defined contingency cases at each time step.
ii. For every contingency case at each time step include an additional set of variables and constraints that relate to those of the base case of that time step.
iii. Linearise the objective function and the constraints around the operating

point [30].

iv. Execute the sparse Linear Programming (LP) algorithm [34] to optimise the linearised system.

v. Iterate between the LP and load-flow until the solution converges.

The algorithm converges if the following criteria are simultaneously satisfied:

- No violation of constraint limits.
- Iterative changes of control variables are within specified tolerances.
- Iterative change of the objective function is within specified tolerances.

10.3.3 Case Studies

In order to demonstrate the algorithmic procedures the Ward and Hale 6-bus model is analysed. This model has been studied by a number of researchers with regard to the development of algorithmic procedures for transition-optimised voltage and reactive power control [2,29]. For the purposes of this study the shunt capacitors at buses 4 and 6 are fixed and are not included as control variables in the optimisation procedure. The case studies presented in this paper are concerned with active power loss minimisation as an overall objective function. Active power loss minimisation can be achieved by employing the active generation cost minimisation term of the multi-objective function, as described by Equation (2). In order to minimise losses by reactive power and voltage control exclusively, only the slack generator has active power control and the active power outputs of the remaining generators are fixed. In this way it is possible to minimise the active power losses by voltage and reactive power control when linearisations of the active and reactive power flow equality constraints are included in the formulation of the LP system [30]. Hence, the active output of the slack generator is sensitive to independent variables such as generator voltages and transformer tap positions. It should be noted that the loss minimisation is independent of the cost coefficient associated with the slack generator. Finally, it should also be noted that it is also possible to minimise the losses using a branch based approach [2].

Models of the Ward and Hale power system are illustrated in Figure 10.1 with regard to the full load situation without loss minimisation and in Figure 10.2 with regard to the full load situation with loss minimisation. The data employed in the studies presented here is consistent with that of Sharif *et al.* [29] and Mamandur and Chenoweth [2]. The overall objective in the original case study was to minimise active power losses, while constraining the voltages from rising as the load decreases. The objective can also be viewed as an energy loss minimisation when active power losses are integrated over the time intervals [28,29]. Consistent objective functions and voltage constraints have been adopted in this research in order to test the implementation of algorithmic procedures. Constraint, initial and final variable values are presented in Tables 10.1 and 10.2 for the two case studies. Initial variable values that violate constraints are highlighted via a preceding asterisk. In both case studies the load is gradually decreased over all four intervals. There is a 10% decrease in the second and third intervals with regard to the preceding intervals. For the final interval the decrease is only 3%. As expected the voltages at the load buses rise as the load decreases, however it is important to note that this effect is masked in the final solution when the voltage constraints are enforced with regard to the load buses. It should also be noted that no ramping constraints are applied with regard to the active power output of the slack generator.

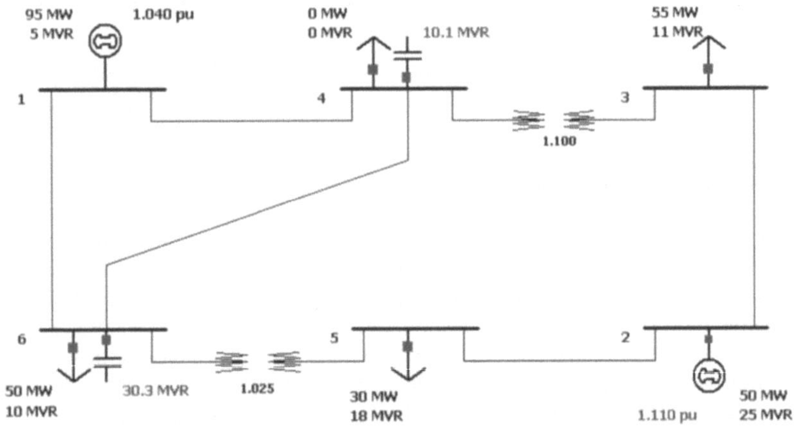

Figure 10.1. Full load without loss minimisation: 9.72 MW losses.

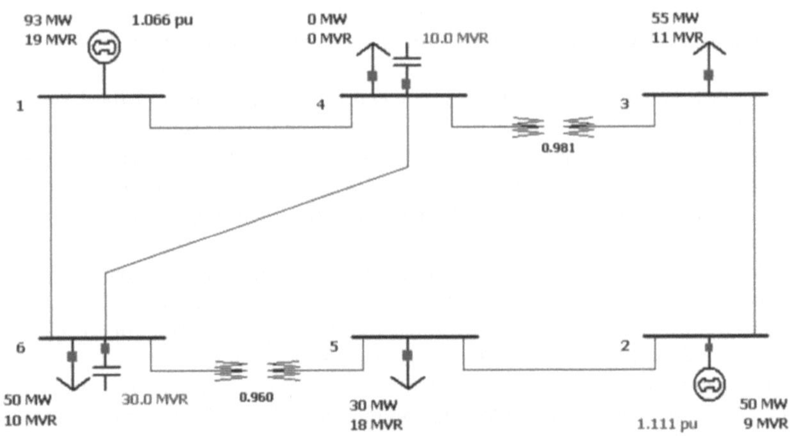

Figure 10.2. Full load with transition-optimised losses: 8.48 MW losses.

10.3.3.1 Case Study I

In the first case study no transition constraints are applied with regard to the two transformer tap positions T_{4-3} and T_{6-5} for the four load intervals that are under consideration. Hence, the series of transformer tap values and generator voltages that minimise system MW losses are given in Table I. As indicated by Table 10.1 the losses are minimised for all four load intervals and the violations of the voltage constraints at each load interval have been removed.

Table 10.1. Variable limits and results for case study I

Variables		Limits		100%Load		90%Load		81%Load		78.6%Load	
		Lower	Upper	Initial	Final	Initial	Final	Initial	Final	Initial	Final
Independent	T4-3	0.9	1.1	1.1	**0.981**	1.1	**0.978**	1.1	**0.975**	1.1	**0.974**
	T6-5	0.9	1.1	1.025	**0.954**	1.025	**0.96**	1.025	**0.962**	1.025	**0.963**
	V1	1	1.1	1.04	1.066	1.04	1.043	1.04	1.026	1.04	1.02
	V2	1	1.15	1.11	1.111	1.11	1.111	1.11	1.106	1.11	1.106
Dependent	Q1	-20	100	5.32	19.94	-2.47	10.54	-8.34	4.56	-9.76	2.66
	Q2	-20	100	24.67	8.34	20.19	8.26	16.53	6.71	15.6	6.7
	V3	0.9	1	0.901	1	0.916	1	0.928	1	0.931	1
	V4	0.9	1	*1.003	1	*1.019	0.995	*1.032	0.991	*1.035	0.989
	V5	0.9	1	0.954	1	0.973	1	0.988	1	0.991	1
	V6	0.9	1	*1.005	0.999	*1.026	1	*1.042	1	*1.046	1
System MW losses				9.72	8.47	8.16	7.38	7.07	6.57	6.82	6.38

Table 10.2. Variable limits and results for case study II

Variables		Limits		100%Load		90%Load		81%Load		78.6%Load	
		Lower	Upper	Initial	Final	Initial	Final	Initial	Final	Initial	Final
Independent	T4-3	0.9	1.1	1.1	**0.981**	1.1	**0.981**	1.1	**0.981**	1.1	**0.981**
	T6-5	0.9	1.1	1.025	**0.96**	1.025	**0.96**	1.025	**0.96**	1.025	**0.96**
	V1	1	1.1	1.04	1.066	1.04	1.043	1.04	1.029	1.04	1.022
	V2	1	1.15	1.11	1.111	1.11	1.11	1.11	1.1	1.11	1.1
Dependent	Q1	-20	100	5.32	19.3	-2.47	10.54	-8.34	5.57	-9.76	3.56
	Q2	-20	100	24.67	8.87	20.19	8.27	16.53	5.73	15.6	5.97
	V3	0.9	1	0.901	1	0.916	0.997	0.928	0.994	0.931	0.992
	V4	0.9	1	*1.003	1	*1.019	0.995	*1.032	0.992	*1.035	0.989
	V5	0.9	1	0.954	1	0.973	1	0.988	1	0.991	1
	V6	0.9	1	*1.005	0.996	*1.026	1	*1.042	1	*1.046	0.998
System MW losses				9.72	8.48	8.16	7.39	7.07	6.59	6.82	6.4

10.3.3.2 Case Study II

In the second case study the transition constraints of Equation (10.4) are applied with regard to the transformer tap positions for the four load intervals that are under consideration. For the purpose of this study it is assumed that no constraints are required on the transition of generator voltages over the load intervals studied. Furthermore, in order to simplify the analysis equal time-lengths are also assumed with regard to each load interval. The algorithmically determined optimum tap positions for each transformer over all the load transitions are in bold text in Table 10.2. As indicated by Table 10.2 the losses are again minimised for all four load intervals and the violations of the voltage constraints at each load interval have been removed. However, it is important to note that the additional transition constraints on the transformer tap positions T_{4-3} and T_{6-5} do not impair the minimisation of the losses too severely. In fact the total losses were reduced by 9.4% for the unconstrained case and by 9.2% for the constrained case. Indeed such an impairment could be acceptable when the switching of transformer tap positions is avoided or simplified. The consequential benefits are that the switching sequence is simplified and excessive depreciation of plant is avoided.

The iterative behaviour of the transition-optimised algorithm is described in Table 10.3. The algorithm converges in three LP iterations and the minimisation of the losses at all load intervals is clearly illustrated.

Table 10.3. Convergence of LP iterations for loss minimisation

Iteration	System Losses MW			
	100% Load	90% Load	81% Load	78.6% Load
0	9.72	8.16	7.07	6.82
1	8.61	7.49	6.67	6.49
2	8.52	7.37	6.56	6.37
3	8.48	7.39	6.59	6.4

10.3.4 Concluding Remarks

The initial algorithmic developments that have been investigated and reported in this chapter indicate the suitability of developing existing algorithms and software [30] to achieve the objectives of this research project. However, as described below further research and a number of additional algorithmic developments are required to satisfy the overall objectives of this research project.

It is important to note that the solution algorithm employed in this research solves the constrained optimisation problem without using decomposition. However, it is possible to solve voltage and reactive power transition-optimisation problems using decomposition strategies with regard to time intervals and discrete/continuous independent variables [27,28]. Therefore further research is planned to develop the solution algorithms of Chebbo and Irving [30] to include decomposition techniques. It will then be possible to perform comparative studies with and without using decomposition strategies to determine the relative robustness and efficiency of such algorithms.

In this chapter transformer tap positions are considered as pseudo-discrete variables and have been included within the LP as continuous variables. Such an approach is not viable for fully discrete variables such as the switching states of shunt capacitors [11]. For this reason further research and development is required to include suitable heuristic and/or algorithmic techniques to solve such integer programming problems. Further research and development is also required with regard to multi-objective analyses that consider loss minimisation and/or voltage stability issues by combining Equations (10.2) and (10.3), load patterns that exhibit a wider range of variation and include sub-intervals, and contingency constrained analyses. Large scale models of the NG UK transmission system that include additional reactive compensation equipment such as static VAr compensators and mechanically switched capacitors will also be studied.

In some cases it may be more appropriate to alleviate voltage problems by applying a combined active and reactive transition-optimised approach [30]. Such an approach has not been considered at this stage, but may be considered in the future. However, it should be noted that any additional rescheduling of active controls significantly increases the computational effort required.

10.4 Congestion Management and Voltage Security

10.4.1 Introduction

In recent years a number of congestion management studies have been performed. Yu and Song [12] have introduced a combination of inter-zonal/intra-zonal schemes and fixed transmission rights to minimise the MW adjustments of generators and loads. Wang and Song [35] have applied Lagrangian relaxation (LR) to the congestion management problem, such that both MW and MVAr adjustments are minimised. In [36], the LR method is also used to deal with the congestion problem across interconnected regions, in this case the minimisation of MW generator adjustment is the main objective. However, it should be highlighted that although [35] and [36] introduced sophisticated concepts to address both congestion management and minimization of related adjustments, voltage security has not been considered.

This section presents an approach that minimises congestion charges and reactive power costs by coordinating congestion and reactive power management, while also considering the voltage security margin. CPF [37] is applied and coordinated with ac OPF. Both MW and MVAr terms are included in the objective function. With the proposed method, the SO can relieve the congestion while maintaining the voltage security.

10.4.2 Nomenclature

The symbols and abbreviations used in this section are shown below:

k_1, k_2 and k_3 : Constant values

σ : Step size for the CPF iterations

V_{mn} : Minimum voltage (p.u.) in the system

$\Delta P_{G_i}^{+}$: Incremental change of active generation at bus i

$\Delta P_{G_i}^{-}$: Decremental change of active generation at bus i

$\Delta P_{D_i}^{-}$: Decremental change of active load at bus i
Q_{Ginj_i} : Reactive generation injected at bus i
Q_{Gabs_i} : Reactive generation absorbed at bus i
$P_{G_i}^{0}$: Initial value of active generation injected at bus i
$P_{D_i}^{0}$: Initial value of active load at bus i
S_{ij} : Apparent power flow in line $i-j$
PF_i : Power factor of load at bus i
V_i : Voltage magnitude at bus i
θ : Phase angle
Q_{sh_i} : Reactive shunt at bus i
NG : Number of generator participants
ND : Number of load participants
$C_{PG_i}^{+}(\Delta P_{G_i}^{+})$: Cost function of $\Delta P_{G_i}^{+}$
$C_{PG_i}^{-}(\Delta P_{G_i}^{-})$: Cost function of $\Delta P_{G_i}^{-}$
$C_{PD_i}^{-}(\Delta P_{D_i}^{-})$: Cost function of $\Delta P_{D_i}^{-}$
$C_{inj_i}(Q_{Ginj_i})$: Cost function of Q_{Ginj_i}
$C_{abs_i}(Q_{Gabs_i})$: Cost function of Q_{Gabs_i}
Cost/h : Cost per hour

10.4.3 Algorithmic Procedure

Transmission congestion influences the system voltage profile and voltage security. In order to relieve congestion, loads might be curtailed and some generators might reduce their power output though this action is not an ideal solution. While the transmission system is congested, power might flow in a more problematic direction in order to keep heavily load lines within their limits. Some generators might increase their power output and it is therefore difficult during transmission congestion to ensure a secure system voltage, especially under heavy load or in the event of a contingency condition. Reactive power compensation has a close relationship with voltage security, affecting the power flows in transmission lines and also playing a significant role in power market arrangements. Consequently, there is a need to solve the congestion problem while considering voltage security within a competitive market environment.

10.4.3.1 Mathematical model

The algorithmic approach presented in this section coordinates ac OPF with CPF. The objective function as stated in Equation (10.1) minimizes the reactive power cost and the incremental/decremental changes related to congestion charges by indicating a willingness to change load and generation levels.

$$\begin{aligned} Min\ F = & \sum_{i \in NG} \left(C^{+}_{PG_i}(\Delta P^{+}_{G_i}) + C^{-}_{PG_i}(\Delta P^{-}_{G_i}) \right) \\ & + \sum_{i \in ND} \left(C^{-}_{PD_i}(\Delta P^{-}_{D_i}) \right) \\ & + \sum_{i \in NG} \left(C_{inj_i}(Q_{Ginj_i}) \right) + \sum_{i \in NG} \left(C_{abs_i}(Q_{Gabs_i}) \right) \end{aligned} \tag{10.5}$$

Subject to the following constraints:
Power flow constraints at each node:

$$P^{0}_{G_i} + \Delta P^{+}_{G_i} - \Delta P^{-}_{G_i} - (P^{0}_{D_i} - P^{-}_{D_i}) - P_i(V, \theta) = 0 \tag{10.6}$$

$$Q_{Ginj_i} - Q_{Gabs_i} - Q_{D_i} + Q_{sh_i} - Q_i(V, \theta) = 0 \tag{10.7}$$

Voltage limit constraints:

$$V_i^{\min} \le V_i \le V_i^{\max} \tag{10.8}$$

Limit constraints for generators:

$$0 \le \Delta P^{+}_{G_i} \le P^{\max}_{G_i} - P^{0}_{G_i} \tag{10.9}$$

$$0 \le \Delta P^{-}_{G_i} \le P^{0}_{G_i} - P^{\min}_{G_i} \tag{10.10}$$

$$0 \le Q_{Ginj_i} \le Q^{\max}_{G_i} \tag{10.11}$$

$$0 \le Q_{Gabs_i} \le Q^{\min}_{G_i} \tag{10.12}$$

Limit constraints for consumers:

$$0 \le \Delta P^{-}_{D_i} \le P^{0}_{D_i} - P^{\min}_{D_i} \tag{10.13}$$

$$Q_{D_i} = Q^{0}_{D_i} - \Delta P^{-}_{D_i} * \tan \cos^{-1} PF_i \tag{10.14}$$

Line flow limit constraints:

$$S_{ij}^{\min} \le S_{ij} \le S_{ij}^{\max} \tag{10.15}$$

Shunt capacitor/reactor constraints:

$$Q_{sh_i}^{\min} \le Q_{sh_i} \le Q_{sh_i}^{\max} \tag{10.16}$$

From (10.1), it can be seen that the incremental/decremental changes only affect MW generation and MW consumption. This is because the active power is more expensive and more difficult to adjust than reactive power. Furthermore, by neglecting incremental/decremental changes of reactive power, the burden on the SO and the computational analysis can be reduced. However, it should also be noted that all cost functions are piecewise linear functions that can be unsymmetrical as shown in Figure 10.1.

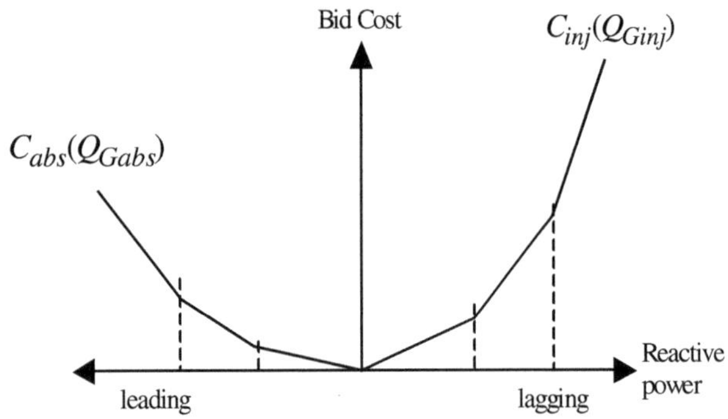

Figure 10.1. Reactive bid curve

Figure 10.2 shows a typical generator capability curve. The shaded area represents the approximated available energy. Lagging reactive power indicates that the generators are injecting reactive power into the system. Leading reactive power means that the generators are absorbing reactive power. Both reactive power injection and absorption are regarded as positive quantities. The reactive power limits are assumed to remain constant.

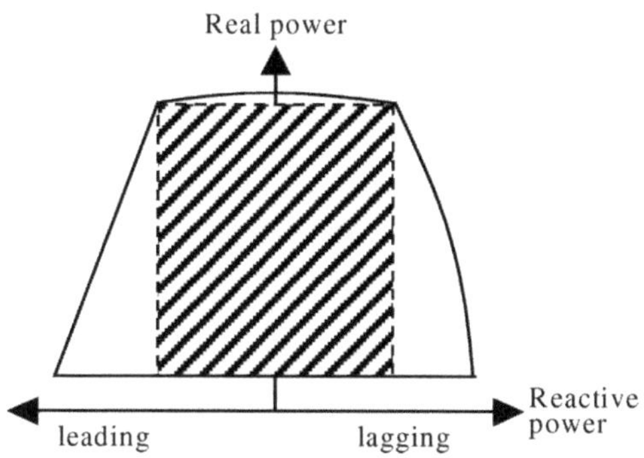

Figure 10.2. Capability curve of a generator

10.4.3.2 Computational Procedures

In order to alleviate network congestion while performing reactive power management, the proposed algorithm is applied and coordinated via the procedure shown in Figure 10.3.

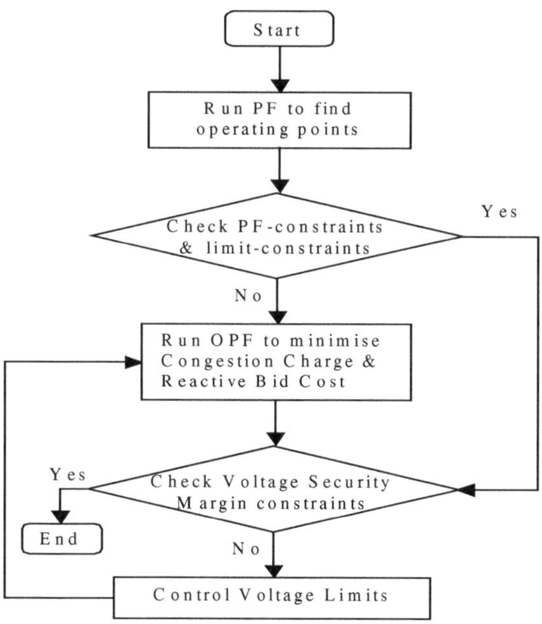

Figure 10.3. Flow chart diagram

As illustrated in Figure 10.3, a typical ac power flow is performed in order to evaluate the operating point. Once the operating point is obtained, it is checked for constraint violations, especially voltage constraints and thermal limit constraints on transmission lines. The voltage security margin constraints are not checked unless the standard constraints are accepted. If the standard constraints are not satisfied, the OPF is performed with regard to the objective function as described in (10.1). The network congestion is alleviated at this step. Once the standard constraints are accepted the voltage security margin is analysed using the CPF approach. Using the CPF method that is based on the predictor and corrector concept, the operation point can be traced over the P-V curve as shown in Figure 10.4. It is important to note that tracing the operating point by CPF independently from OPF simplifies the algorithm. Following the CPF analysis; if the voltage security margin is not within the acceptable level, the voltages at some buses are adjusted.

The voltage at the most insecure bus that produces the largest voltage change [37] will be the first to be increased, even if there are other buses with unacceptable security margins. In this chapter, the size of the voltage increase is determined by extrapolation and interpolation methods with the initial step-size to reach the required security margin. In practice, the initial step-size selected can be obtained from experience or historical operation. The next step is to return to the OPF to minimise the objective function and to make sure that all standard constraints are within their limits. The algorithm is converged when all costs are minimised and all constraints, including voltage security margin constraints, are satisfied.

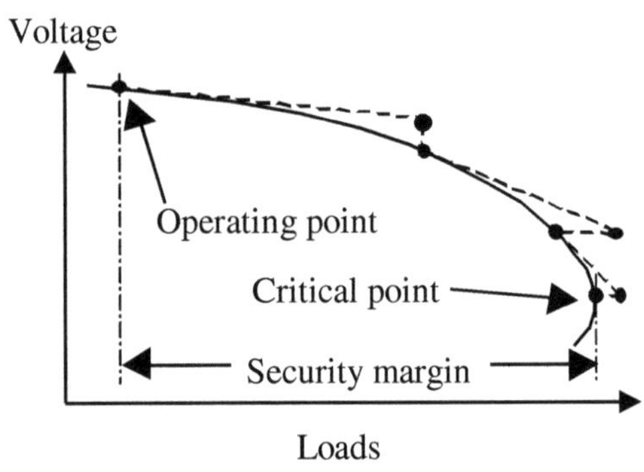

Figure 10.4. Load voltage curve by using the CPF

One step that affects the speed and accuracy of the CPF algorithm is the choice of step size used to update the parameter values. The approach presented here improves the step-size evaluation by proposing the following step-size function:

$$\sigma = k_1 e^{-k_2(1-|V_{mn}|)} + k_3 \tag{10.17}$$

Equation (10.17) is a function of the voltage at the bus that is most insecure. The step-sizes obtained for the first several iterations are large and are controlled by the specified constants k_1 and k_2. However, when approaching the critical point, the step-sizes become smaller and are dominated by the constant k_3. In practice, the constants can be obtained by testing at heavy load conditions, before applying the proposed algorithm. The constants can then be used for a wide range of load levels. The characteristic curve of the step-size function is shown in Figure 10.5.

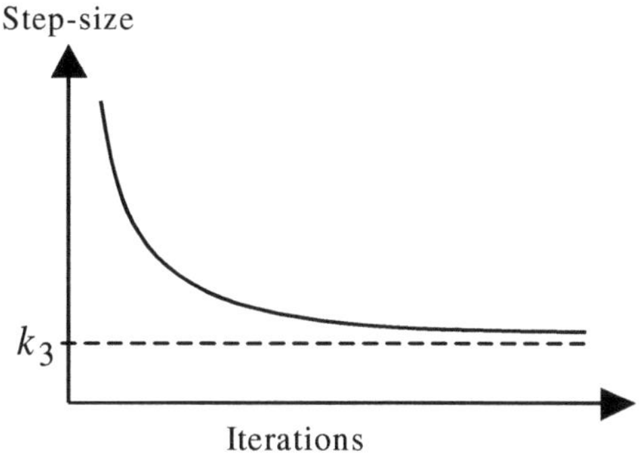

Figure 10.5. Characteristic of the step-size function

10.4.4 Computational Case Studies

In this section, modified IEEE 30-bus and 57-bus systems are analysed computationally. The percentage MW security margin target is 150% of the operating point, with 2% and 0 % as allowable upper and lower errors, respectively. To calculate the step-size, constant k_1 is set to 5.0 for the 30-bus system and 1.5 for the 57-bus system. Constants k_1, k_2 and k_3 are set to 5.0, 15.0 and 0.01, respectively for both test systems. For the voltage control, the initial voltage change at the most insecure bus is 0.01 p.u. From the results, bus 26 in the 30-bus system and bus 31 in the 57-bus system are the most insecure buses. Table 10.4 shows the development of the security margin, the total cost and the voltage at the most insecure buses for the 30-bus system.

Table 10.4. Security margin at each iteration for the 30-bus system

Iterations	Security Margin (%)	Cost (Cost/h)	Voltage (p.u.)
1	117.52	34.06	0.95
2	131.70	37.85	0.96
3	151.12	43.08	0.97

From Table 10.4, notice that the security margin improves from 117.52% to over 150%. This improvement is worth 9.02 Cost/h. In other words, it costs around 0.27 Cost/h for each per cent of security margin increase, by assuming that the results are linear within the small change. The step-size of the voltage change is calculated by extrapolation. Furthermore, the voltage at iteration 3 becomes 0.971 p.u. when extrapolated to meet the target security margin of 150%. For the 57-bus system, the results are similar. However, the total costs are not increased as much due to the smaller changes in reactive power generation required to improve the security margin. Furthermore, there is no reversion of reactive power injection or absorption. Figures. 10.6 and 10.7 show the P-V curves at the most insecure bus of the 30-bus and the 57-bus systems, respectively. At iteration 3 for the 30-bus system and iteration 5 for the 57-bus system, the security margins have been developed to meet the 150% target.

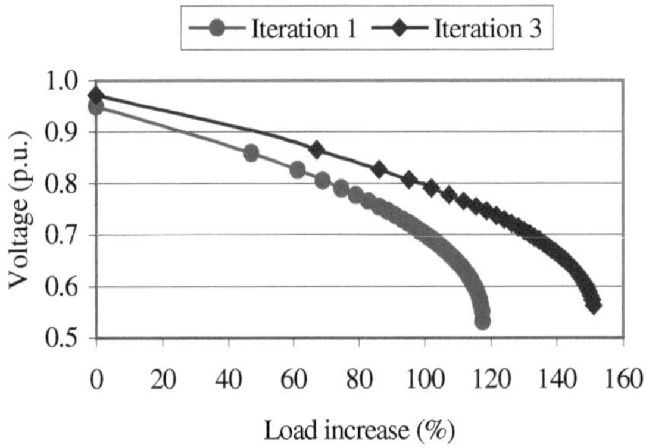

Figure 10.6. Comparison of P-V curves for the 30-bus system

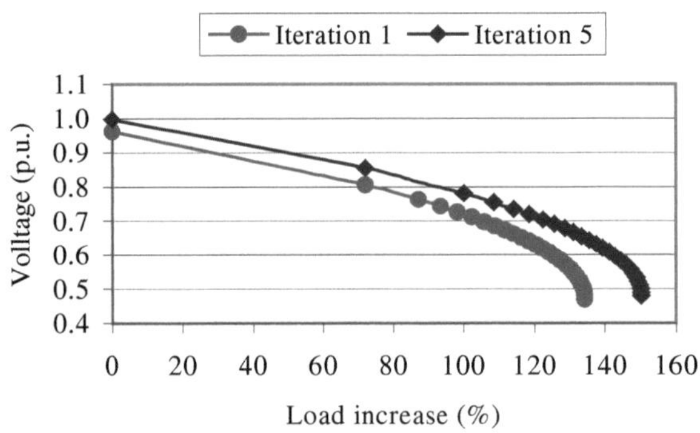

Figure 10.7. Comparison of P-V curves for the 57-bus system

For the 30-bus system, the generator at bus 13 is the only generator that undergoes a reversal in its Mvar generation status. The generator reverts from injection to absorption of reactive power as shown in Figure 10.8. All other generators continue to inject reactive power into the system, even though in some cases reactive power production is reduced. These changes lead to an improvement in transmission congestion and security margin. Before initiating the congestion management procedure, the voltages at some generation buses might be high. In order to reduce these high voltages, reactive power generation is reduced. To illustrate this point in the analysis, consider the voltage at bus 13, which falls from 1.079 p.u. at the first iteration to 0.963 p.u. after the third iteration. Therefore, the reactive power at some generators could be reduced during the procedure. For the 57-bus system, reactive power generation is shown in Figure 10.9.

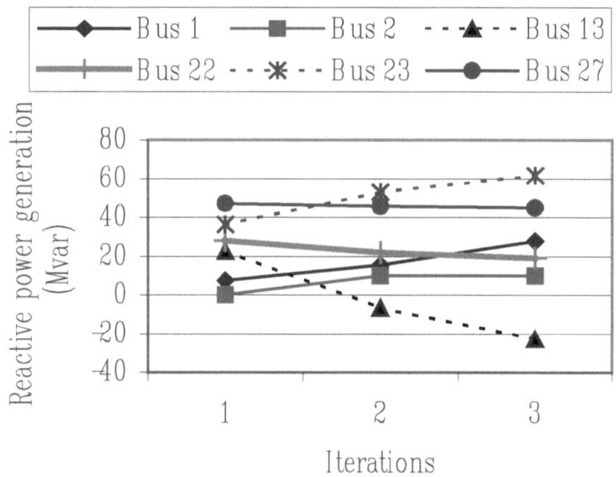

Figure 10.8. Reactive power generation at each iteration for the 30-bus system

Figure 10.10 shows the comparison of power flows in lines between iterations 1 and 3. At the initial condition before performing iteration 1, lines 21-22, 25-26 and 25-27 are congested. After the first iteration, all congested lines are relieved with lines 21-22 and 25-27 at their limits, but the security margin is unacceptable. After iteration 3, only line 21-22 is operated at its limit and the voltage margin is well over 150%. In Figure 10.11, notice that the algorithm does not reduce power flows in all lines. Some lines carry more power while line congestion is relieved and reactive power management is performed. That is because when power generators and loads are rescheduled to optimise the costs and to keep the system secure, some generators may increase generation and then power flows change. Otherwise, loads must be reduced, which would be much more expensive. For example, power flow in line 15-23 increases from 24 MVA to 43 MVA. That is because the generator at bus 13 reverts from injection to absorption of reactive power, while reactive power flow on line 15-23 does not change direction and actually increases.

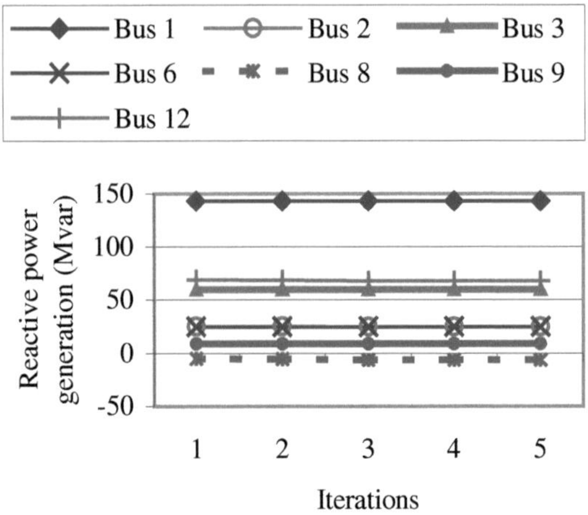

Figure 10.9. Reactive power generation at each iteration for the 57-bus system

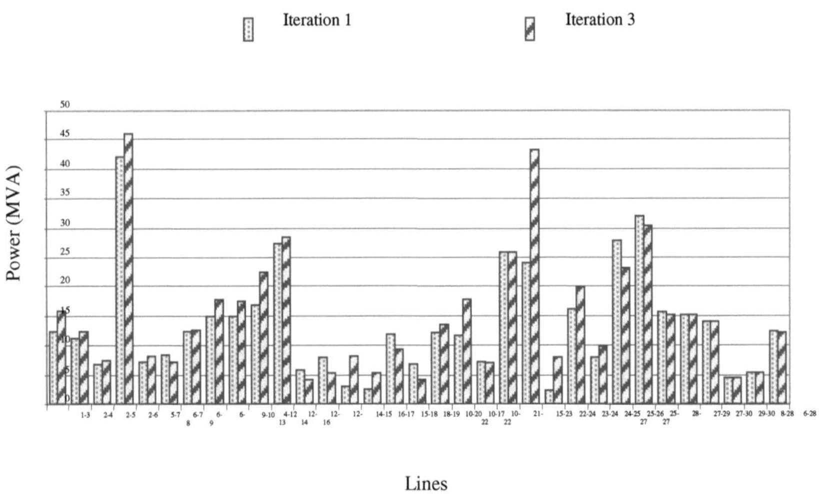

Figure 10.10. Power flow in lines of the 30-bus system

10.4.5 Concluding Remarks

This section describes the application of an algorithm that coordinates OPF with CPF to minimise congestion and reactive power charges, while also improving/maintaining the voltage security margin. The step-size function of the CPF method has been enhanced to improve the accuracy of the results and the speed of the calculation. The proposed algorithm can alleviate network congestion while maintaining the voltage security margin by keeping the security margin at an acceptable level. The algorithm can be employed by the SO in order to operate the transmission system reliably and economically under severe technical conditions in a competitive market environment. Whatever the severe technical condition, heavy load or line outage or even both, the SO can ensure system security at all times in percentage terms of the security margin. Finally, the modified IEEE 30-bus and 57-bus systems have been analysed computationally. The results are satisfactory, however, the studies do not consider contingencies or the impact of FACTS controllers. Research is currently underway to develop algorithms that can consider such cases.

Acknowledgment

The authors of this chapter gratefully acknowledge financial support from National Grid, UK and Chulalongkorn University, Thailand.

10.5 References

[1] T.J.E Miller, Reactive Power Control in Electric Systems, John Wiley & Sons, 1982.

[2] K.R.C. Mamandur and R.D. Chenoweth, "Optimal control of reactive power flow for improvement in voltage profiles and for real power loss minimization," IEEE Trans. on Power Apparatus and Systems, vol. PAS-100, no. 7, pp. 3185-3193, 1981.

[3] R. Mota-Palomino and V.H. Quintana , "Sparse reactive power scheduling by a penalty function – linear programming technique", IEEE Trans. on Power Systems, vol. 1, No. 3, pp. 31-39, 1986.

[4] J. Qui and S.M. Shahidehpour, "A new approach for minimizing power losses and improving voltage profile", IEEE Trans. on Power Systems, vol. 2, No. 2, pp. 287-295, 1987.

[5] J. Peschon, D.S. Piercy, W.F. Tinney, O.J. Tveit and M. Cuénod, "Optimum control of reactive power flow", IEEE Trans. on Power Apparatus and Systems, vol. PAS-87, No. 1, pp. 40-48, 1968.

[6] IEEE Special Publication, Voltage Stability of Power Systems: Concepts, Analytical Tools, and Industry Experience, 90TH0358-2-PWR, IEEE Press, 1990.

[7] H.B Wan, M.E. Bradley, A.O. Ekwue and A.M. Chebbo, "Method for alleviating voltage limit violations using combined DC optimisation and AC power flow technique", IEE Proc.-Gener. Transm. Disturb., vol.147, No. 2, pp. 99-104, 2000.

[8] M. Suzuki, S. Wada, M. Sato, T. Asano, and Y. Kudo, "Newly developed voltage security monitoring system," IEEE Trans. on Power Systems, vol. 7, no. 3, pp. 965-973, 1992.
[9] B. Gao, G.K. Morison, and P. Kundur, "Towards the development of a systematic approach for voltage stability assessment of large-scale power systems," IEEE Trans. on Power Systems, vol. 11, no. 3, pp. 1314-1324, 1996.
[10] G.A. Taylor, Y.H. Song, M.R. Irving, M.E. Bradley and T.G. Williams, "A review of algorithmic and heuristic based methods for voltage/VAr control", in IPEC'01, Singapore, vol. 1, pp. 117-122, 2001.
[11] G.A. Taylor, M. Rashidinejad, Y.-H. Song, M.R. Irving, M.E. Bradley and T.G. Williams, "Algorithmic techniques for transition-optimised voltage and reactive power control", in PowerCon 2002: IEEE-PES/CSEE, Kunming, China, vol. 3, pp. 1660-1664, 2002.
[12] I.K. Yu and Y.H. Song, "Short-term generation scheduling of thermal units with voltage security and environmental constraints," IEE Proc. Gener., Transm. and Distrib., vol. 144, no. 5, pp. 469-476, 1997.
[13] Y.H. Song and I.K. Yu , "Dynamic load dispatch with voltage security and environmental constraints," Electric Power Systems Research, vol. 43, pp. 53-60, 1997.
[14] J. Zhong, 'Design of Ancillary Service Markets: Reactive Power and Frequency Regulation,' Tech. Rep. 392L, School of Elec. And Comp. Engg., Chalmers University of Technology, 2001.
[15] E.L. da Silva, J.J. Hedgecock, J.C. Mello and J.C. Luz, 'Practical cost-based approach for the voltage ancillary service,' IEEE PE-041 PRS, July, 2001.
[16] N.H. Dandachi, M.J. Rawlins, O. Alsac, M. Prais and B. Stott, "OPF for reactive pricing studies on the NGC System," IEEE Trans. Power Systems, vol. 11, no. 1, pp. 226-232, 1996.
[17] National Grid, 'Obligatory and Enhanced Reactive Power Services: Methodology Document for the Aggregation of Reactive Power Metering,' June, 2000.
[18] S. Ahmed and G. Strbac, 'A method for simulation and analysis of reactive power market,' In Power Industry Computer Applications, pp. 337-341, 1999.
[19] National Grid, 'National Grid Reactive Market Report: Ninth Tender Round for Obligatory and Enhanced Reactive Power Services,' May, 2002.
[20] National Grid, 'A Guide to Balancing Services Use of System (BSUoS),' March, 2001.
[21] NYISO, 'Ancillary Services Manual', July, 1999.
[22] US Dept. of Energy, 'Electricity Reform Abroad and US Investment', September, 1997.
[23] NEMMCO, 'An Introduction to Australia's National Electricity Market,' 2002.
[24] R. Pritchard, NEMMCO Information Centre, Personal Communication, 2002.
[25] K. Bhattacharya and J. Zhong, "Reactive power as an ancillary service," IEEE Trans. Power Systems, vol. 16, no. 2, pp. 294-300, 2001.
[26] J.A. Momoh and J. Zhu, "A new approach to VAr pricing and control in the competitive environment," in System Sciences, the Thirty-First Hawaii International, vol. 3, pp. 104-111, 1998.
[27] Y.Y. Hong and C.M. Liao, "Short-term scheduling of reactive power controllers," IEEE Trans. on Power Systems, vol. 10, no. 2, pp. 860-868, 1995.
[28] S.S. Sharif, J.H. Taylor, and E.F. Hill, "Dynamic online energy loss minimisation" IEE Proc.-Gener. Transm. Distrib., vol. 148, no. 2, pp. 172-176, 2001.
[29] S. S. Sharif, J. H. Taylor, and E. F. Hill, "On-line optimal reactive power flow by energy loss minimization," in IEEE Decision and Control, Kobe, Japan, pp. 3851-3856, 1996.
[30] A.M. Chebbo and M.R. Irving, "Combined active and reactive dispatch. Part 1: Problem formulation and solution algorithm," IEE Proc.-Gener. Transm. Distrib., vol. 142, no. 4, pp. 393-400, 1995.

[31] A.M. Chebbo, M.R. Irving, and N.H. Dandachi, "Combined active reactive dispatch. Part 2: Test results," IEE Proc.-Gener. Transm. Distrib., vol. 142, no. 4, pp. 401-405, 1995.

[32] A.M. Chebbo, M.R. Irving, and M.J.H. Sterling, "Voltage collapse proximity indicator: Behavior and implications," IEE Proc.-C, vol. 139, no. 3, pp. 241-252, 1992.

[33] M.R. Irving and M.J.H. Sterling, "Efficient Newton-Raphson algorithm for load flow calculation in transmission and distribution networks," IEE Proc. C, vol. 134, no. 5, pp. 325-328, 1987.

[34] R.M. Dunnett, "A guide to the linear programming subroutines SDRS2 and SDRS4 (Version 1.5)," Tech. Rep. NGC TSD/PSBM/206/95, 1995.

[35] X. Wang and Y.H. Song, "Apply Lagrangian relaxation to multi-zone congestion management," in Proc. IEEE Power Engineering Society Winter Meeting, pp. 399-404, 2001.

[36] X. Wang, Y.H. Song and Q. Lu, "Lagrangian decomposition approach to active power congestion management across interconnected regions," IEE Gener., Transm. and Distrib., vol. 48, pp. 497-503, 2001.

[37] V. Ajjarapu and C. Christy, "The continuation power flow: A tool for steady state voltage stability analysis," IEEE Trans. Power Systems, vol. 7, pp. 416-423, Feb. 1992.

11. Load and Price Forecasting via Wavelet Transform and Neural Networks

I.K. Yu and Y.H. Song

Short-term load forecasting (STLF) plays an important role in the operational planning and the security functions of an energy management system. The STLF is aimed at predicting electric loads for a period of minutes, hours, days or weeks for the purpose of providing fundamental load profiles to the system. Over the years, considerable research effort has been devoted to STLF and various forecasting techniques have been proposed and applied to power systems. Conventional methods based on time series analysis exploit the inherent relationship between the present hour load, weather variables and the past hour load. Autoregressive (AR), moving average (MA) and mixed autoregressive and moving average (ARMA) models are prominent in the time series approach. The main disadvantage is that these models require complex modelling techniques and heavy computational effort to produce reasonably accurate results [1].

The emergence of artificial intelligence (AI) techniques in recent years, effective utilisation of AI in the context of ill-defined processes have led to their application in STLF as expert-system type models. These models are discrete and logical in nature, and use the knowledge of an expert to develop forecasting rules. Transforming an experts knowledge to a set of mathematical rules is, however, often very difficult. Artificial neural network (ANN) based models have been found to be the most popular for load forecasting applications. The advantage of ANN over statistical models lies in its ability to model a multivariate problem without making complex dependency assumptions among input variables. The performance of ANN based techniques, however, depends strongly on the relationship between the patterns used in training the networks and the expected forecast patterns. If the diversity or the inconsistency between training patterns and expected forecast patterns is strong, the forecast errors of the ANN technique may be relatively high [2].

On the other hand, the electricity supply industry is undergoing unprecedented restructuring world-wide and there is a growing interest in the prediction of system marginal price (SMP) under the competitive market structure of deregulated power systems. The prediction of SMP improves the financial performance of an independent power producer bidding in the day-ahead market. All generators participating in the day-ahead bidding of the power pool face the challenge of how to set the prices of their generating units. The SMP is a major portion of the payment that generators obtain from the power pool for their scheduled generating units. Consequently, there is an increasing demand for SMP prediction in the deregulated power pool, however, few papers have been published on the topic due to the complexity and newness of the problem. Recently, a neural network based technique for the prediction of SMP was proposed for the UK power pool [3].

This chapter presents conventional forecasting techniques in brief followed by novel load and price forecasting techniques via the wavelet transform and neural networks with case studies to help comprehension. The wavelet transform is a recently developed mathematical tool for signal analysis. It has been applied successfully in astronomy, data compression, signal and image processing, earthquake prediction, and so on. The basic concept in wavelet analysis is to select a proper wavelet (mother wavelet) and then perform an analysis using its translated and dilated versions. There are many kinds of wavelets that can be used as the mother wavelet, such as the Haar wavelet, Meyer wavelet, Coiflet wavelet, Daubechies wavelet, Morlet wavelet, etc. Similar to the Fourier transform, there are different wavelet transforms; the called continuous wavelet transform (CWT), also known as integral wavelet transform (IWT), discrete wavelet transform (DWT), and fast discrete wavelet transform (FWT). The Fast wavelet transform is known as Multi-resolution Analysis (MRA) or the Mallat pyramidal algorithm [4].

11.1 Load Forecasting and Conventional Techniques

There are two major types of forecasting models: time-series and regression models. In the first type, prediction of the future is based on past values of a variable and/or past errors. The objective of such time-series forecasting methods is to discover the pattern in the historical data series and extrapolate that pattern into the future. The most commonly used models for electric load forecasting are the linear regression models. In some implementations weather parameters have also been accounted for. Other approaches included a dynamic model for on-line load forecasting and a state space model.

11.1.1 Time-series Models

The idea of the time-series approach is based on the understanding that a load pattern is nothing more than a time-series signal with known seasonal, weekly, and daily periodicities. These periodicities give a rough prediction of the load at the given season, day of the week, and time of the day. The difference between the prediction and the actual load can be considered as a stochastic process. By analysis of this random signal, more accurate prediction results may be obtained. The techniques used for the analysis of this random signal include Kalman filtering, the Box Jenkins method, the ARMA model, and the spectral expansion technique. The Kalman filter approach requires estimation of a covariance matrix. The possible high non-stationarity of the load pattern, however, typically may not allow an accurate estimate to be made. The Box Jenkins method requires the autocorrelation function to properly identify the ARMA model. This can be accomplished by using pattern recognition techniques. A major obstacle here is its slow performance [5].

11.1.1.1 Auto-regressive (AR)

Smoothing methods base their forecasts on the principle of averaging (smoothing) past errors by adding a percentage of the error to a percentage of the previous forecast.

Mathematically, single smoothing methods are of the form

$$F_{t+1} = F_t + \alpha(X_t - F_t) \tag{11.1}$$

Equation (11.1) can be expanded by substituting

$$F_t = F_{t-1} + \alpha(X_{t-1} - F_{t-1})$$

Thus,

$$F_{t+1} = F_{t-1} + \alpha(X_{t-1} - F_{t-1}) + \alpha(X_t - F_t) \tag{11.2}$$

Substituting for F_{t-1} in the first term of (11.2) gives

$$F_{t+1} = F_{t-2} + \alpha(X_{t-2} - F_{t-2}) + \alpha(X_{t-1} - F_{t-1}) + \alpha(X_t - F_t) \tag{11.3}$$

The results of further expanding this substitution should be clear. Given some initial forecast, call it F_{t-2}, new forecasts can be obtained by adding a percentage of the errors between the actual and forecast values (e.g. $X_{t-2} - F_{t-2}$) to this initial forecast. Since some of the errors will be negative and some positive, the final forecast, F_{t+1}, will be close to the actual pattern of data, on average.

In multiple regression, the causal or explanatory model is of the form

$$Y = a + b_1 X_1 + b_2 X_2 + b_k X_k + u \tag{11.4}$$

In Equation (11.4), $X_1, X_2, \cdots, X_k$ can represent any factors. Suppose, however, that these variables are defined as

$$X_1 = Y_{t-1},\ X_2 = Y_{t-2},\ X_3 = Y_{t-3}, \cdots,\ X_k = Y_{t-k}$$

Equation (11.4) then becomes

$$Y_t = a + b_1 Y_{t-1} + b_2 Y_{t-2} + \cdots + b_k Y_{t-k} + u_t \tag{11.5}$$

Equation (11.5) is still a regression equation, but differs from Equation (11.4) in that the right-hand side variables of Equation (11.4) are different independent factors, while those of Equation (11.5) are previous values of the dependent variable Y_t. These are simply time lagged values of the dependent variable, and therefore the name AR is used to describe equations or schemes of the form of Equation (11.5).

11.1.1.2 Moving Averages (MA)

The time-series technique of moving averages consist of taking a set of observed values, finding the average of those values, then using that average as the forecast for the next period. The actual number of past observations included in the average must be specified at the outset. The term moving average is used because as each new observation becomes available a new average can be computed by dropping the oldest observation from the average and including the newest one. The new average is then used as the forecast for the next period. Thus the number of data points from the series used in the average is always constant and includes the most recent observations.

With a small n, the forecasts will trail the pattern, lagging behind it by a few periods. Algebraically, the techniques of forecasting with moving averages can be represented as follows:

$$F_{t+1} = (X_t + X_{t-1} + \cdots + X_{t-n+1})/n = \frac{1}{n}\sum_{i=t-n+1}^{t} X_i, \qquad (11.6)$$

where t is the most recent value, and $t+1$ is the next period, for which a forecast is desired. It can be seen from this equation that in order to compute the moving average one must have the values of the past n observations. A somewhat shorter and easier way to state this equation of calculating the moving average is

$$F_{t+1} = \frac{X_t}{n} - \frac{X_{t-n+1}}{n} + F_t \qquad (11.7)$$

In order to apply MA, only a limited amount of historical data, n, is needed. Then either Equation (11.6) or (11.7) can be used to compute the forecast for future periods. Thus, both the data requirements and the computational requirements for applying MA to a single time series are minimal. However, the accuracy of the forecast obtained through the MA method is usually low. In practice the MA technique is not used extensively because the method of exponential smoothing has all the advantages of MA plus a few additional ones that make it more attractive in those situations where MA would be suitable.

11.1.1.3 Mixed Auto-regressive and Moving Average (ARMA)

ARMA model have been studied extensively by George Box and Gwilym Jenkins (1970), and their names frequently have been used synonymously with general ARMA processes applied to time-series analysis, forecasting, and control. AR models were first introduced by Yule (1926) and later generalised by Walker (1931), while MA models were first used by Slutzky (1937). It was the work of Wold (1938), however, that provided the theoretical foundations of combined ARMA processes, and thus should be considered as the most valuable in the field.

The ARMA model is used to describe the stochastic behaviour of the hourly load pattern on a power system. The model assumes the load at the hour can be estimated by a linear combination of the previous few hours. Generally, the larger the date set, the better is the result in terms of accuracy. A longer computational time for parameter identification, however, is required.

To achieve greater flexibility in fitting the actual time series, it is sometimes advantageous to include both auto-regressive and moving average terms in the model. This leads to the mixed auto-regressive moving average model

$$X_t = \phi_1 X_{t-1} + \phi_2 X_{t-2} + \cdots + \phi_p X_{t-p} + e_t - \theta_1 e_{t-1} - \theta_2 e_{t-2} - \cdots - \theta_q e_{t-q} \quad (11.8)$$

$$\text{or} \quad \phi(B) X_t \text{-} \theta(B) e_t$$

which employs $p + q + 2$ unknown parameters μ ; $\phi_1, \cdots, \phi_p$; $\theta_1, \cdots, \theta_q$; σ_a^2, that are estimated from the data. This model may also be written in the form of the linear filter. In practice, it is frequently true that adequate representation of actually occurring stationary time series can be obtained with autoregressive, moving average, or mixed models, in which p and q are not greater than 2 and often less than 2.

11.1.2 Regression Model

Regression analysis is the most popular approach that has been applied and is still being applied to the forecasting of electric power demand. Regression analysis is concerned with modelling the relationships among variables. It quantifies how a response (or dependent) variable is related to a set of explanatory (independent, predictor) variables. The general procedure for the regression approach is as follows:

(1) Select the proper and/or available weather variables.

(2) Assume basic functional elements.

(3) Find proper coefficient for the linear combination of the assumed basic functional elements. Since temperature is the most important information of all weather variables, it is used most commonly in the regression approach (possibly non-linear). However, if additional variables such as humidity, wind velocity, and cloud cover are used, better results should be obtained. Most regression approaches have simple linear or piecewise linear functions as the basic functional elements [6].

In the Multiple Linear Regression (MLR) method, the electric demand is found in terms of explanatory variables such as population and economic variables that influence the electric demand. The electric demand model using this method is expressed in the form

$$Y(t) = \beta_0 + \beta_1 X_1(t) + \cdots + \beta_n X_n(t) + a(t) \tag{11.9}$$

where, $Y(t)$ = electric power demand

$X_1(t), \cdots, X_n(t)$ = explanatory variables correlated with $Y(t)$,

$a(t)$ = random variable with zero mean and constant variance, and

$\beta_0, \beta_1, \cdots, \beta_n$ = regression coefficients.

The explanatory variables of this model are identified on the basis of correlation analysis on each of these (independent) variables with the load (dependent) variable. Experience about the load to be modelled helps initial identification of the suspected influential variables. Estimation of the regression coefficients is usually performed using the least squares estimation technique. Statistical tests are performed to determine the significance of these regression coefficients. The t-ratios resulting from these tests determine the significance of each of these coefficients, and correspondingly the significance of the associated variables with these coefficients. Several special cases of the general regression model are given below.

(1) $Y(t) = \beta_0 + a(t)$ (constant mean model)

(2) $Y(t) = \beta_0 + \beta_1 x(t) + a(t)$ (simple linear regression model)

(3) $Y(t) = \beta_0 \exp\lfloor \beta_1 X(t) + a(t) \rfloor$ (exponential growth model)

(4) $Y(t) = \beta_0 \beta_1 X(t) + \beta_2 X(t)^2 + a(t)$ (quadratic model)

(5) $Y(t) = \beta_0 \beta_1 \ln(X_2(t) + \beta_2 \ln(X_2(t)) + \ldots + \beta_n \ln(X_2(t)) + a(t)$
(logarithmic model)

(6) $Y(t) = \beta_0 X_1(t) + \beta_2 X_2(t) + a(t)$
(linear model with two independent variables)

(7) $Y(t) = \beta_0 X_1(t) + \beta_2 X_2(t) + \beta_{11} X_1(t)^2 + \beta_{22} X_2(t)^2 + \beta_{12} X_1(t) X_2(t) + a(t)$
(quadratic model with two independent variables)

11.2 Novel Methods for Short-term Load Forecasting

A wavelet transform based STLF approach is described in this section [7]. In general, over 50% of the consumption of the electrical energy is by industry, the other heat, which is more sensitive to weather conditions, is consumed by individuals.. The component of the load curve produced by industry has a rather regular profile and exhibits low-frequency changes. On the other hand, the consumption individual consumers may be highly irregular, leading to high-frequency components. Based on such load properties, the wavelet transform is selected as one alternative and used to establish a novel model of STLF. From the methodological point of view, wavelet transform techniques provide a multi-scale analysis of the signal as a sum of orthogonal signals corresponding to different time scales, allowing a kind of time-scale analysis.

11.2.1 Wavelet Transform Applications

11.2.1.1 Wavelet Transform Analysis

Wavelet transform analysis is applied to forecast high-frequency components of daily load. Wavelet theory provides a unified framework for a number of techniques, which have been developed independently for various signal-processing applications. In particular, the wavelet transform is of interest for the analysis of non-stationary signals, because it provides an alternative to the classical short time Fourier transform (STFT) or Gabor transform. The basic differences from the STFT, which uses a single analysis window, is that the wavelet transform uses short windows at high frequencies and long windows at low frequencies and is also related to time-frequency analysis. Thus the windowing of the wavelet transform is adjusted automatically for low or high frequencies and each frequency component is treated in the same manner without any reinterpretation of the results. Thus, the wavelet transform provides an alternative way of breaking a signal down into its constituent parts [8].

In the wavelet transform, the original domain is the time domain. The transformation process from time domain to time scale domain is a wavelet transform, technically known as signal decomposition because a given signal is decomposed into several other signals with different levels of resolution. From these decomposed signals, it is possible to recover the original time domain signal without losing any information. This reverse process is called the inverse wavelet transform or signal reconstruction. To illustrate the transform, let $s(t)$ be the time domain signal to be decomposed or analysed. The dyadic wavelet transform of $s(t)$ is defined as follows.

$$DWT_{\varphi} s(m,n) = 2^{\frac{-m}{2}} \int_{-\infty}^{\infty} s(t)\, \varphi^{*}(\frac{t-n2^{m}}{2^{m}})dt \tag{11.10}$$

where, the asterisk denotes a complex conjugate, m and n are scale and time-shift parameters, respectively, and $\varphi(t)$ is a given basis function (mother wavelet). The dyadic wavelet transform is implemented using a multi-resolution pyramidal decomposition technique. In principle a recorded digitised time signal $S(n)$ is decomposed into its detailed $cD_1(n)$ and smoothed $cA_1(n)$ signals using filters *HiF_D* and *LoF_D*, respectively, as in Figure 11.1 [9]. Filter *HiF_D* has a band-pass response. Thus, the filtered signal $cD_1(n)$ is a detailed version of $S(n)$ and contains higher frequency components than the smoothed signal $cA_1(n)$, because filter *LoF_D* has a low-pass frequency response.

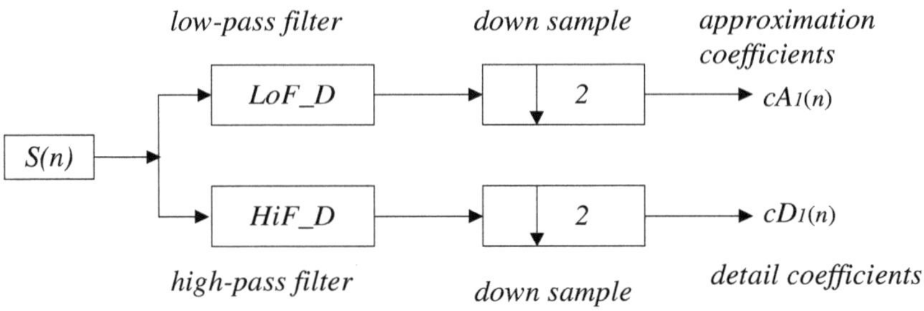

Figure 11.1. The first-scale signal decomposition of $S(n)$ in $cA_1(n)$ and $cD_1(n)$

The decomposition of $S(n)$ into $cA_1(n)$ and $cD_1(n)$ is a first-scale decomposition. It is possible to reconstruct the original signal from the coefficients of the approximations and details as in Figure 11.2 for the *NS* (Number of Samples) case for example.

It is also possible to reconstruct the approximations and details themselves from their coefficient vectors. The coefficient vector cA_1 passes through the same process used to reconstruct the original signal. However, instead of combining it with the level-one detail cD_1, a vector of zeros is fed in place of the details. This process yields a reconstructed approximation A_1, which has the same length as the original signal $S(n)$ and which is a real approximation of it. Similarly, the first-level detail D_1 can be reconstructed using an analogous process to that above. The reconstructed details and approximations are then true constituents of the original signal as

$$A_1 + D_1 = S \tag{11.11}$$

Note that the coefficient vectors cA_1 and cD_1, because they were produced by down sampling, contain aliasing distortion, and are only half the length of the original signal, cannot directly be combined to reproduce the signal. It is necessary to reconstruct the approximations and details before combining them. Extending this technique to the components of a multi-level analysis, it is possible to find that similar relationships hold for all the reconstructed signal constituents. That is, there are several ways to reassemble the original signal.

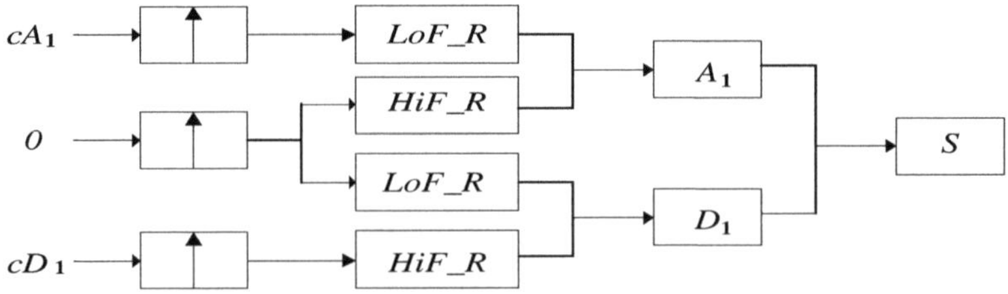

Figure 11. 2. Reconstruction of the approximations and the details

11.2.1.2 Load Forecasting Process by the Wavelet Transform

In this application, the three-scaled reconstruction technique as in Figure 11.3 is applied to predict load, and the second-order regression polynomial is used to describe the relationship between temperature and load.

$$L_F = a_0 + a_1 T^1 + a_2 T^2 \quad (11.12)$$

where L_F represents forecasted load, a_0, a_1 and a_2 are coefficients and T^1, T^2 are mean weighted temperatures. For convenience, the coefficients a_0, a_1, a_2 are included in A. Matrix algebra, for the load forecasting, can be used as follows:

$$L_F = TA + e \quad (11.13)$$

where, L_F is an $n \times 1$, T an $n \times k$, A a $k \times 1$ and e a $k \times 1$ matrix, respectively. In order to obtain the values of A, the sum of squared deviations must be minimized.

$$\sum e_i^2 = e'e = (L_F - TA)'(L_F - TA) \quad (11.14)$$

where, $e', (L_F - TA)'$ is the transpose of e. Thus the coefficients are finally calculated as

$$A = (T'T)^{-1} T' L_F \quad (11.15)$$

where, $(T'T)^{-1}$ is the inverse of $(T'T)$.

The process of STLF using the wavelet transform is shown in Figure 11.4. Input data consist of three hours interval loads, and weighted temperature data in which the regional load consumption rates are considered for the same period. The wavelet transform is performed, and the approximation and details determined. The regression coefficient is calculated using d_1, d_2, d_3 and weighted temperature, and the temperature applied wavelet coefficients are predicted. Load forecasting is then finally implemented using the previous days low-frequency and the forecasted high-frequency components.

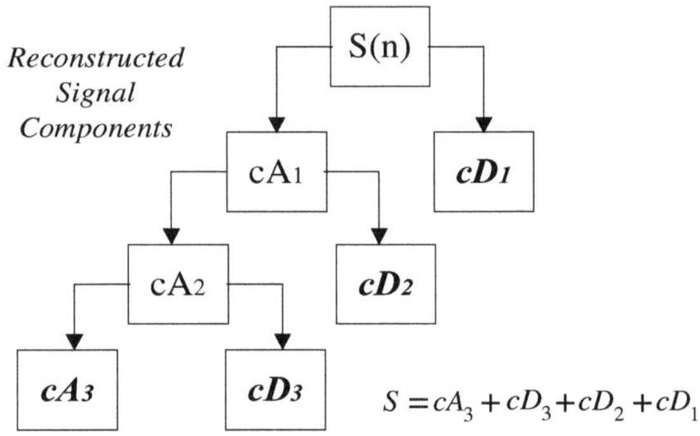

Figure 11.3. Reconstruction of the three scaled approximations and details

The accuracy of the forecasted results is estimated by the following percentage error calculation:

$$Percentage\ \ error = \frac{|actual\ \ load - forecasted\ \ load|}{actual\ \ load} \times 100(\%) \tag{11.16}$$

11.2.2 Kohonen-neural-network-based Approach

The approach described in the previous section decomposed a given signal into several scales at different levels of resolution. Then, the wavelet coefficients associated with high frequencies are adjusted or predicted using the conventional MR method in order to consider the weather conditions through a three-scale synthesis technique. The short-term (1-day ahead) load is then forecasted by reconstructing the predicted high frequencies and the previous day's low frequency from the wavelet-transformed signal. The results of wavelet transform based load forecasting has proven that the approach is relatively a good challenge, however, some problems remain unsolved because the historical data are classified depending only on calendar date. As a matter of fact, the characteristics of daily loads for each time interval, even for the same weekday, are represented differently due to vacations and special days, etc. The Kohonen neural network which is used for classifying the loads by self-organisation feature mapping (SOFM) [10] is introduced, and combined with the wavelet transform for the purpose of establishing a composite forecasting model. In addition, the 5-level decomposition of daily load curve is implemented in order to reflect weather sensitive components of the load more effectively. The Kohonen neural network model is based on the SOFM technique that transforms input patterns into neurons on the two-dimensional grid. There is no target output and learning the weights is unsupervised.

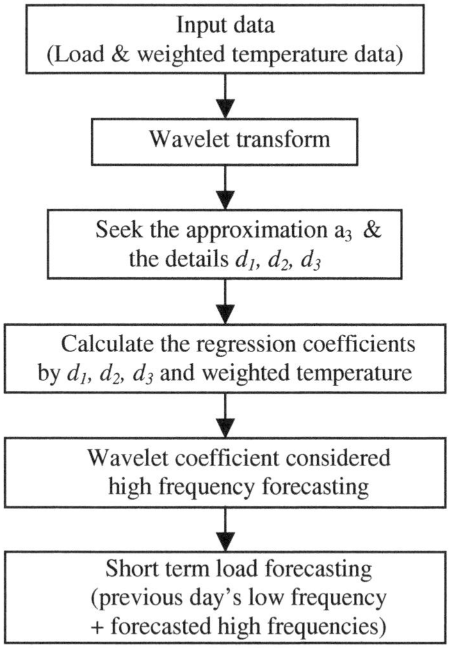

Figure 11.4. Flow of STLF using the wavelet transform

11.2.2.1 Architecture of the Kohonen Neural Network

The SOFM technique, which transforms input patterns into neurons on the two-dimensional grid, proposed by Teuvo Kohonen is a well-known unsupervised learning algorithm that produces a map composed of a fixed number of units. Self-organisation in networks is one of the most fascinating topics in the neural network field. Such networks can learn to detect regularities and correlations in their input and adapt their future responses to that input accordingly. The neurons of competitive networks learn to recognise groups of similar input vectors. The SOFM learns to recognise groups of similar input vectors in such a way that neurons physically close together in the neuron layer respond to similar input vectors. The advantages of the SOFM model are that the algorithm does not require the teacher's signals, that is, the model uses an unsupervised learning algorithm, the algorithm is not so complicated and the resulting mapping makes it visually easy to understand the input pattern [11].

The architecture for the Kohonen neural network is shown in Figure11.5. In the Figure, R, P, W_{ij}, b_i, S_i, n_i, a_i denote inputs, input vector, input weight matrix, biases, competitive layer neurons, net input and output, respectively. In this application, the weights are initialised to the centre of the input ranges.

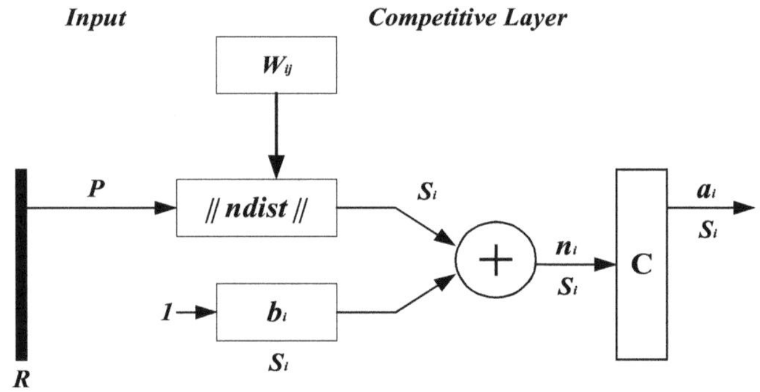

Figure 11.5. Kohonen neural network architecture

The dist box in the figure accepts P and W_{ij}, and produces a vector having S_i elements. The elements are the negative of the distances between the input vector and vectors W_{ij} formed from the rows of the input weight matrix. The net input n_i of a competitive layer is computed by finding the negative distance between P and the weight vectors and adding the biases b_i. If all biases are zero, the maximum net input a neuron can have is zero. This occurs when P equals that neuron's weight vector. The competitive transfer function accepts a net input vector for a layer and returns neuron outputs of zero for all neurons except for the winner, the neuron associated with the most positive element of net input n_i. The winner's output is 1. If all biases are zero then the neuron whose weight vector is closest to the input vector has the least negative net input, and therefore wins the competition to a_i [12].

11.2.2.2 Unsupervised Learning

A physical neighbourhood relation between the units is defined. Each unit is characterised by a parameter vector W_{ij} of the same dimension as the input space. After learning, each unit represents a group of individuals with similar features. The correspondence between the features and the units respects (more or less) the input space topology: similar features correspond to the same unit or to neighbour units. The final map is said to be a SOFM that preserves the topology of the input space. The learning algorithm takes the following form.

(1) Set learning step $s=0$ and at time 0, $W_i(0)$ is defined randomly for each unit i.

(2) Set $s=s+1$ and $t=t+1$ and then input signal $R(t)$, t is input pattern number.

(3) Calculate neuron $n^*(s)$ that have the closest distance, $D(s)$, to input vector $R(t)$. Each input is connected through a weight W_{ij} to all neurons of the competitive layer. An input vector R is presented to the map. The basic operation of the SOFM produces the Euclidean distance $D(s)$ which is caculated as follows:

$$D(s) = \min_i [D_i(s)] = \min_i \sum_j (R_j(t) - W_{ij}(s))^2 \tag{11.17}$$

(4) Find topological neighbourhood around $n^*(s)$.

(5) Update the weight for $n^*(s)$ and topological neighbourhood, using the following relations. Therefore, W_{ij} will then be modified in order to move the winning unit $n^*(s)$ and its physical neighbours toward (R).

$$W_{ij}(s+1) = W_{ij}(s) + \alpha(s)[R_j(t) - W_{ij}(s)], \text{ for } i \in n^*(s) \tag{11.18}$$

$$W_{ij}(s+1) = W_{ij}(s) \quad \text{otherwise} \tag{11.19}$$

where, $\alpha(s)$ is a small positive adaptation parameter (learning rate), progressively decreased during the learning. The neurons in a competitive layer distribute themselves to recognise frequently presented input vectors [13].

11.2.3 STLF by a Composite Model

The process of STLF using the Kohonen neural network and the wavelet transform is shown in Figure 11.6. Seasonal load data consisting of three hours interval for each season are used and then load patterns are classified into four classes (weekdays, Monday, Saturday and Sunday) by the SOFM.

The wavelet transform is then applied to the classified seasonal loads, and the approximation and details obtained. The regression coefficients are calculated using d_1, d_2, d_3, d_4, d_5, and the weighted temperature data in which the regional load consumption rates are considered for the same period. The next day's three-hour interval temperature considered wavelet coefficients are predicted. The load forecasting is then finally implemented using reconstruction of the previous day's low frequency and the forecasted high frequency components using the five-scale synthesis scheme.

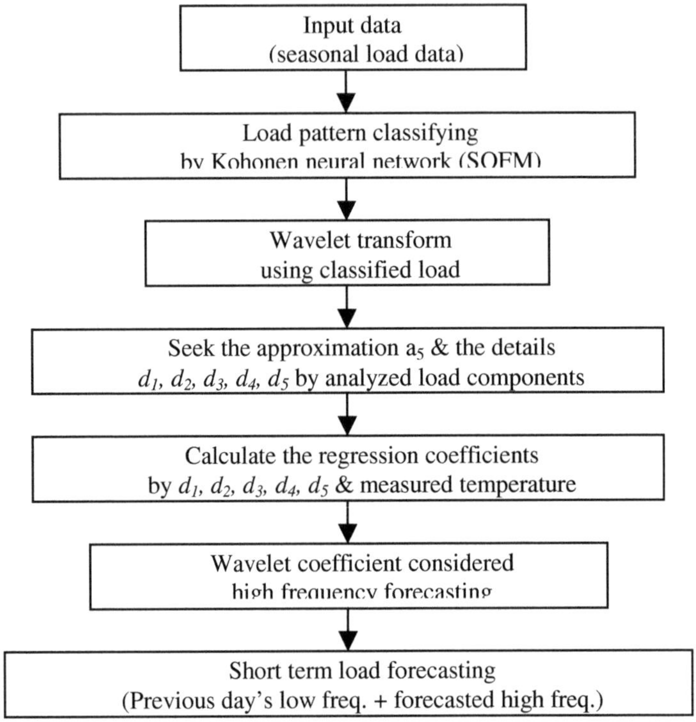

Figure 11.6. Flow of STLF using the composite model

11.2.4 Case Studies and Analysis

For the purpose of demonstrating the effectiveness of the wavelet transform based model and the composite model, STLF is performed using seasonal load and temperature data.

11.2.4.1 Case Study by Wavelet-transform-based Model

Figure 11.7 illustrates the three-level decomposition coefficients, one low and three high frequency components, of a daily load curve analysed using Daubechies DB4 wavelet transform.

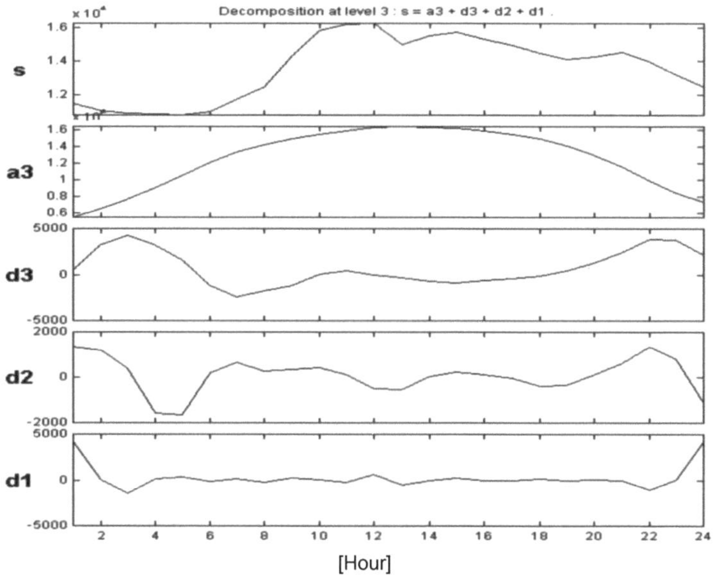

Figure 11.7. Three-level decomposition of a daily load curve

Table 11.1. Multiple regression coefficients of week days for the three-level decomposition

Hour	DETAILS OF HIGH FREQUENCIES								
	d1			d2			d3		
3:00	-15368	1717	-37	648	-22	-44	260	-31	1
6:00	4091	-497	10	-6660	782	-19	-2156	365	-6
9:00	1893	-209	4	1171	-74	62	-25577	2480	-50
12:00	-5391	492	-9	3104	-434	9	-36498	3153	-58
15:00	4067	-334	7	-5654	300	-4	-32063	2659	-48
18:00	5424	-507	9	-11752	1007	-18	-10754	976	-18
21:00	-555	131	-3	-12697	1269	-26	25234	-2200	47
24:00	-10288	1137	-24	-9312	973	-22	31532	-3014	66

The multiple regression coefficients for the decomposed three-level high frequencies are shown in Table 11.1: *d1, d2* and *d3* denote the detail coefficients of the three-level decomposition components.

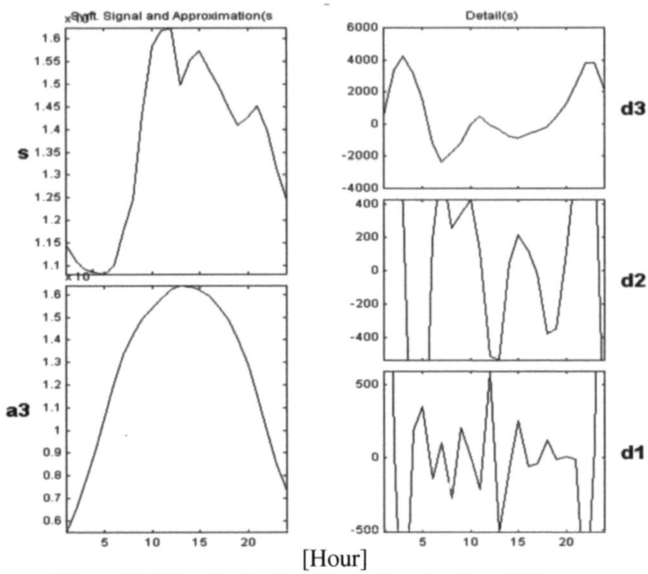

Figure 11.8. Synthesis of predicted three-level coefficients

Table 11.2. Mean percentage errors of weekdays in the summer

Hour	DB 2			DB 4			DB 10		
	lev 1	lev 2	lev 3	lev 1	lev 2	lev 3	lev 1	lev 2	lev 3
3:00	2.27	2.04	1.73	2.39	1.96	1.92	2.47	2.10	2.00
6:00	2.39	2.08	1.78	2.86	3.38	1.82	3.09	2.77	1.87
9:00	2.70	1.76	2.12	2.73	3.25	2.00	2.89	3.23	1.62
12:00	3.36	2.35	2.62	3.16	2.61	2.40	3.09	3.15	2.13
15:00	2.90	3.08	3.08	3.53	2.54	2.66	3.36	2.98	2.11
18:00	3.30	4.56	2.95	3.23	2.38	2.24	3.24	2.56	2.17
21:00	1.90	1.58	1.70	2.27	1.93	1.39	2.22	1.58	1.81
24:00	1.80	1.74	1.56	1.40	2.11	1.79	1.35	1.67	2.30
Mean(%)	2.58	2.40	2.19	2.70	2.52	2.03	2.71	2.51	2.00

Figure 11.8 shows the synthesis (reconstruction) of the predicted three-level coefficients by which the forecasted load is eventually obtained.

Table 11.2 shows the mean percentage error of weekdays for the predicted load. DB2, DB4 and DB10 denote Daubechies wavelet transforms, and lev1, lev2 and lev3 describe the level of high frequencies decomposed by the wavelet transform. From Table 11.2, it is clear that the mean percentage error of DB10 (lev3) is the lowest (2.0%).

The average percentage error for one month in the wintertime is shown in Table 11.3.

Table 11.3. Mean percentage errors of weekdays in the winter

Hour	DB 2			DB 4			DB 10		
	lev 1	lev 2	lev 3	lev 1	lev 2	lev 3	lev 1	lev 2	lev 3
3:00	2.89	2.82	2.88	0.59	0.57	0.55	0.57	0.55	0.55
6:00	2.31	2.07	2.03	0.91	1.00	0.73	0.98	0.89	0.72
9:00	1.07	1.15	1.00	1.06	1.09	0.82	1.09	1.04	0.87
12:00	1.58	1.58	1.30	0.83	0.73	0.65	0.86	0.78	0.65
15:00	1.50	1.31	1.24	1.01	0.81	0.73	0.93	0.87	0.73
18:00	1.52	1.34	1.35	1.24	1.27	1.11	1.19	1.29	1.21
21:00	1.84	1.87	2.16	1.08	1.01	0.83	1.03	0.95	0.92
24:00	2.88	3.10	3.28	0.79	0.85	0.75	0.85	0.85	0.75
Mean(%)	1.95	1.91	1.91	0.94	0.92	0.77	0.94	0.90	0.80

Table 11.4. Mean percentage errors of weekend days in the summer

DAY	SATURDAY			SUNDAY			MONDAY		
Hour	DB2 lev 3	DB4 lev 3	DB10 lev 3	DB2 lev 3	DB4 lev 3	DB10 lev 3	DB2 lev 3	DB4 lev 3	DB10 lev 3
3:00	1.08	1.08	1.12	1.3	1.23	1.24	2.77	2.3	2.25
6:00	1.21	1.21	1.58	1.15	1.01	1.26	2.25	2.26	2.41
9:00	1.00	1.00	1.59	0.88	1.02	1.52	1.49	1.56	2.01
12:00	0.93	0.93	1.65	1.12	1.27	1.78	1.44	1.65	2.2
15:00	1.03	1.03	1.87	1.05	1.29	1.85	1.14	1.41	1.97
18:00	1.18	1.18	2.07	1.82	2.14	2.51	1.27	1.69	2.2
21:00	1.25	1.25	1.98	1.38	1.62	1.81	1.59	1.92	2.34
24:00	1.53	1.53	2.06	1.8	1.88	1.83	1.97	2.11	2.3
Mean(%)	1.15	1.15	1.74	1.31	1.43	1.73	1.74	1.86	2.21

An interesting observation is shown: that the mean percentage error of DB4 (lev3) is the lowest in this case. That's why DB10 for the summer and DB4 for the winter are used as dominant wavelet transform coefficients in these cases. For weekend days, the mean percentage error is the lowest at DB2 (lev3). Weekend days (Saturday, Sunday and Monday) mean percentage error of 1.15%, 1.31% and 1.74%, respectively, are given as in Table 11.4.

Figure 11.9 compares results of the actual and forecasted load for one day, predicted by DB4 (lev3) wavelet transform, in the winter.

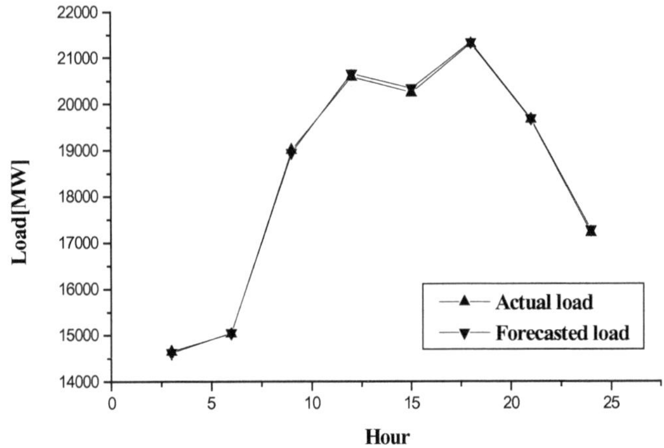

Figure 11.9. Comparison of actual and forecasted load for one day in the winter case study by the composite model

11.2.4.1.1 Classification of Daily Load Patterns

In order to classify the daily load patterns by the SOFM, the seasonal load data are normalised before neural network training, and the weights are initialised to the centre of the input range. The Kohonen learning rule is used to adjust the weights of the winning neuron. Therefore, the neuron whose weight vector is closest to the input vector is updated to be even closer. In this application, 10000 epochs training is implemented to classify the input vectors into four groups, effectively. During training each neuron in the layer closest to a group of input vectors adjusts its weight vector toward those input vectors. To the end, if there are enough neurons, every cluster of similar input vectors have a neuron which outputs 1 when a vector in the cluster is presented, while outputting a 0 at all other times. To determine the optimal size of the output layer for seasonal load forecasting, various network sizes are simulated. As a result of the simulation, the minimum network size that can classify the daily load pattern effectively is [7×7].

Figure 11.10 shows the distribution of the daily load pattern assigned to each output node of a [7×7] network for the summer time, as an example. As shown in Figure 11.10, the daily load curves are classified into four groups, *A*, *B*, *C* and *D*. Group *A*, which represents the weekdays load is distinguished into two groups *A-1* and *A-2*. It is noticed that the electric load of the group *A-2* is much lower than that of *A-1*, which is influenced by the summer vacation period and special days. The groups *B*, *C* and *D* represent the classified load for Saturday, Sunday and Monday, respectively. The two groups of Saturday and Sunday, influenced by the vacation period, are separated into two parts as well. Group *D* includes a few weekday's load besides that of Monday. Among the classified load patterns, *A-1* is adopted for weekdays load forecasting and *B-1*, *C-2* and *D* are used as input data for Saturday, Sunday and Monday's load forecasting, respectively.

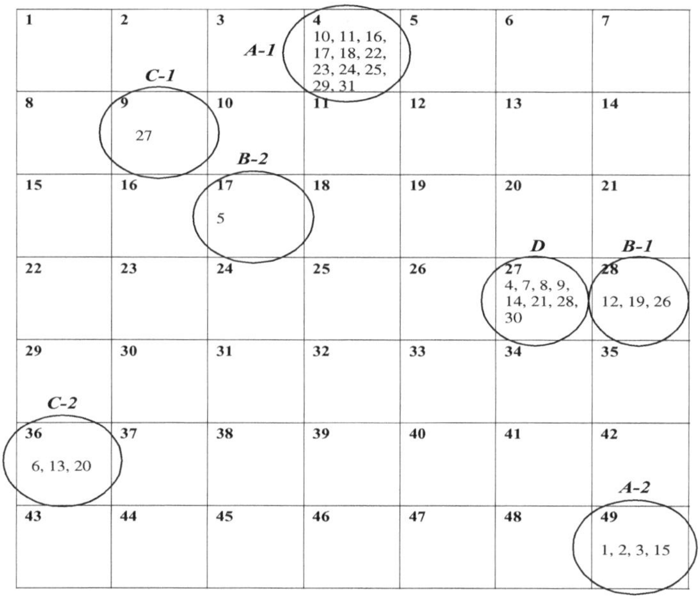

Figure 11.10. Mapping results by the SOFM

11.4.2.1.2 Numerical Results

The temperature based multiple regression coefficients for the decomposed five-level high frequencies are shown in Table 11.5. In the table, d1, d2, d3, d4 and d5 denote the detail coefficients of five-level decomposition components, respectively. One-day-ahead hourly load forecasting results obtained by the composite model are given in Table 11.6. From this table, it is clear that the composite (SOFM+WT) model guarantees better results than the wavelet transform (WT) alone. Mean percentage errors of 0.52 % and 0.49 % are given, respectively.

Figure 11.11 compares results of the actual and forecasted load for one-day, which are predicted by DB4 (lev5) wavelet transform, in the autumn. The forecasted load components are close to the actual load components.

The plot of mean percentage errors in January and February are shown in Figure 11.12. The mean percentage errors are higher on weekends than on weekdays. Especially during Sunday, the daily forecasted mean percentage error is 1.49%. It is possible to recognise, from the results of each study case, that the composite forecasting model of the Kohonen neural network and wavelet transform has contributed to reducing the forecasting error compared to the results of wavelet transform analysis alone.

The forecasted mean percentage errors for one week of all seasons are summarised in Table 11.7. The forecasting simulations for each season are performed using Daubechies DB2, DB4 and DB10, respectively. According to the simulation results, the DB2 for the winter, the DB4 for the Spring and Autumn, and the DB10 for the Summer are applied as dominant wavelet transforms in these study cases. The yearly mean percentage error of 1.08 % is represented in Table 11.7, the mean percentage error of 0.78 % for the wintertime is the smallest, and the summertime error of 1.53 % is the largest.

Table 11.5. Multiple regression coefficients of the five-level decomposition for a weekday

Ho	3:00	6:00	9:00	12:00	15:00	18:00	21:00	24:00
	526.5	1587.3	2815.3	4135.6	6868.3	3366.8	3201.1	102.7
d1	472.3	51.1	327.1	231.0	532.6	409.0	324.4	409.8
	10.7	1.0	7.1	4.4	10.3	8.1	6.3	9.4
	1673.5	2112.5	2915.1	5650.1	1371.2	4338.0	9069.6	2877.1
d2	232.8	50.2	226.0	165.9	286.4	168.1	432.5	61.9
	5.7	2.7	4.0	1.9	7.0	3.4	8.4	0.6
	874.9	1929.8	5723.4	8639.5	7286.5	2969.5	9304.2	1087.3
d3	90.2	162.5	69.5	85.3	80.9	7.2	737.7	215.5
	2.3	4.1	2.0	1.2	1.3	0.1	14.7	5.1
	227.0	2882.6	6090.1	11924.6	13403.5	11422.2	7152.4	4666.9
d4	267.8	289.5	220.3	84.2	101.0	64.1	1539.6	1208.0
	6.2	6.9	5.1	1.1	1.4	1.7	32.2	26.1
	189.7	953.8	1621.1	2508.2	2510.1	2189.6	1268.4	1152.2
d5	143.3	92.8	42.5	19.4	12.7	14.6	302.9	300.1
	3.3	2.2	1.0	0.2	0.1	0.4	6.3	6.5

Table 11.6. Mean percentage errors of daily load forecasting in the summer

Hour	Actual[MW]	WT[MW]	Error[%]	SOFM+WT [MW]	Error[%]
3:00	17925	17875	0.28	17850	0.42
6:00	18108	18130	0.12	18120	0.07
9:00	22112	22169	0.26	22150	0.17
12:00	23726	24063	1.42	24017	1.23
15:00	23913	24094	0.76	24170	1.07
18:00	24233	24128	0.43	24345	0.46
21:00	24249	24175	0.31	24319	0.29
24:00	20854	20729	0.60	20817	0.18
Mean percentage errors : WT - 0.52(%), SOFM+WT - 0.49(%)					

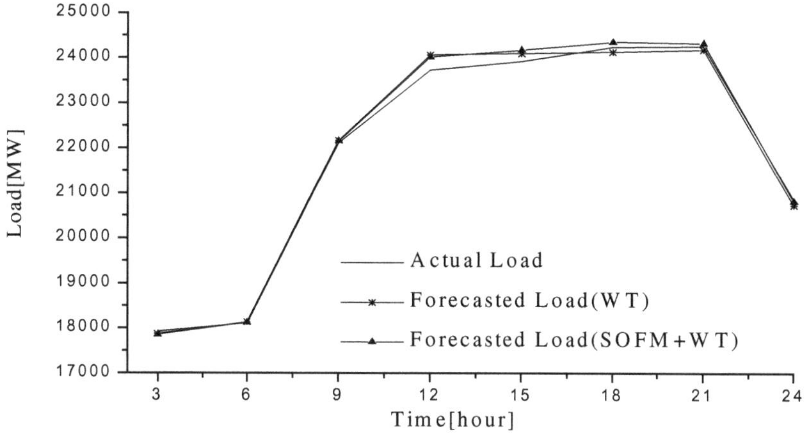

Figure 11.11. Comparison of actual and forecasted load for one day

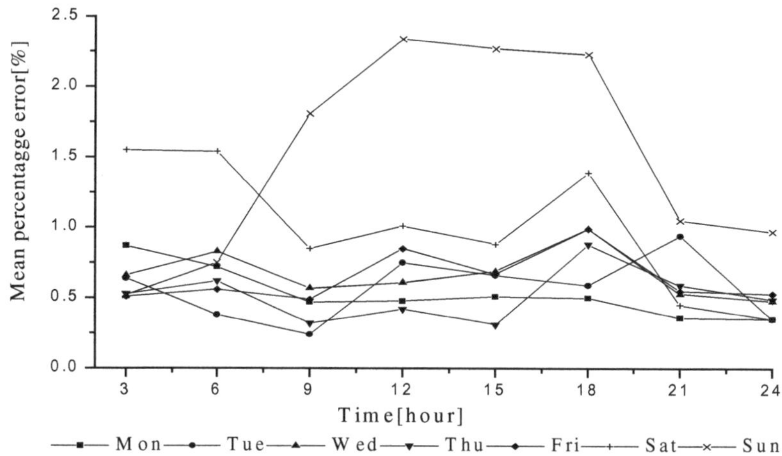

Figure 11.12. Mean percentage errors (%) in wintertime (Jan. & Feb.)

11.3 Electricity Price and Modelling

The deregulation of the electricity supply industry has introduced new opportunities for competition to reduce the cost and cut the price. It is a tremendous challenge for utilities to maintain an economical and reliable supply of electricity in such an environment, consequently, there are many technical issues coupling with economic aspects which need novel, sometimes even fundamentally new, theories. All electricity is sold to and bought through the power pool. The operation of such systems calls for consideration of a number of new issues: all generators participating in the day-ahead bidding of the power pool face the challenge of how to set the prices of their generating units. The SMP is a major portion of the payment that generators obtain from the power pool for their scheduled generating units. Essentially, it is possible for a high degree of competition to occur in the provision of generator sources. As a result, there is growing interest in the prediction of SMP and the development of proper models [14].

Table 11.7. Seasonal and yearly mean percentage errors

Date	Mon.	Tue.	Wed.	Thu.	Fri.	Sat.	Sun.	Season
Spring	0.97	0.62	0.69	0.96	0.78	1.36	1.83	1.03
Summer	2.19	1.71	1.43	1.03	1.32	1.22	1.81	1.53
Autumn	1.16	0.92	1.26	0.75	1.09	0.87	0.89	0.99
Winter	0.53	0.57	0.67	0.52	0.64	1.00	1.49	0.78
Year	1.21	0.96	1.01	0.82	0.96	1.11	1.51	1.08

11.3.1 Characteristics of the SMP

The SMP is influenced by numerous factors: historical SMP, electricity demand, generating availability, seasons, the time of day, the day of the week and overall market situation etc. Historical data showing daily and weekly cycle attributes of SMP curves are shown in Figure 11.13.

It is observed that the SMP curve follows the daily and weekly cycles. The trend in electricity demand also has influences on the trend of the SMP curve as depicted in Figure 11.14. The shape of the SMP curve follows the daily and weekly cycle such as the load behaviour for weekdays (Monday through Friday) with large random variations due to changes in the total bidding of generating units availability, industrial activities, community life style and so on. There are forty-eight SMP settlement periods in each scheduling day. It is noted that the demand is converted from MW to GW in Figure 11.14.

The SMP has different patterns on Saturdays, Sundays and public holidays [15].

- Saturday: The Saturday pattern is influenced by both the previous day, Friday, and last Saturday. The maximum SMP on Saturday is much less than on Friday. In other words the Saturday pattern follows the Friday pattern but not the value. Thus a filter is designed to shave the maximum SMP on Friday to the absolute maximum SMP on the previous Saturday in the whole training data set (see Figure 11.15).

- Sunday: There are normally two peaks on the summer curve. The SMP values between the two peaks are close to, or slightly greater than, two end off-peak values. It has been found that the Sunday pattern strongly follows the previous day, Saturday, and last Sunday.

- Public holidays: There are eight public holidays annually in England and Wales. They are New Year Day (1 January), Christmas and Boxing Day (25, 26 December), Good Friday and Easter Monday (in March or April), and three bank holidays in May and August. It is noted that the SMP patterns are different because these holidays fall in different seasons. Some are influenced by the SMP patterns of last weekend and the same day of last week. Some are influenced by the SMP patterns of last weekend and weekday.

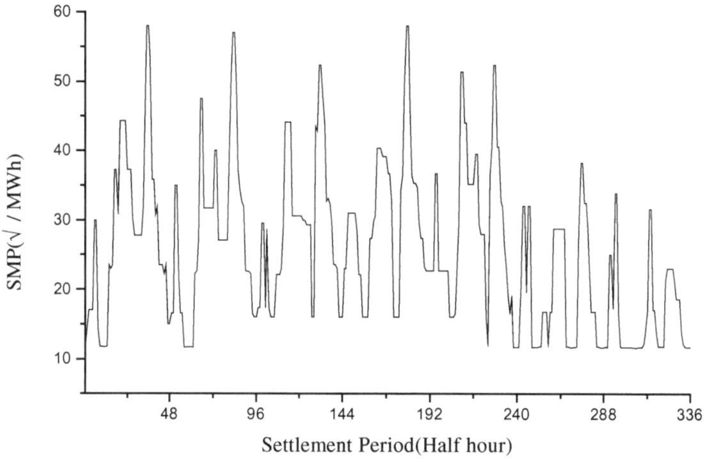

Figure 11.13. One-week SMP curve

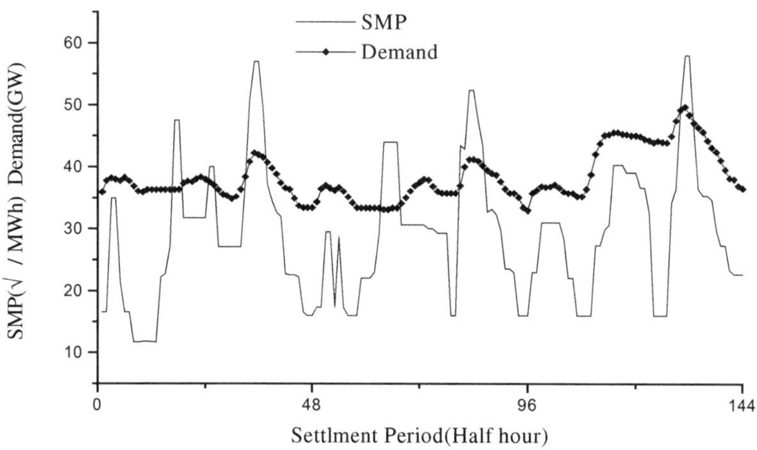

Figure 11.14. Trends in SMP and demand

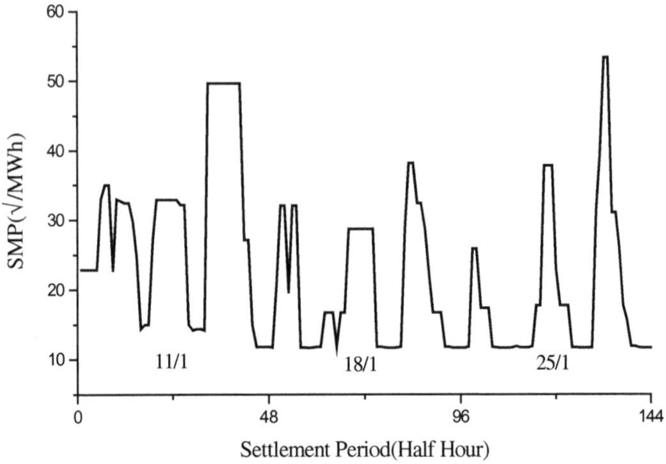

Figure 11.15. SMP features on Saturday

11.3.2 SMP Models

In order to understand the nature of power pool transactions, it is important first to understand day-ahead bidding. The definitions and notations for the day-ahead bidding settlement process in the pool are as follows.

- Scheduling Day : the period from 05:30 on one day to 05:00 on the next

- Settlement Period : half-hour

- Settlement Day : the period from 00:00 on one day to 00:00 on the next, i.e. calendar day

- genset : generating unit

- System Marginal Price : the highest price of qualified units scheduled in the unconstrained schedule in each Settlement Period

- Availability Declaration Period : the period from 21:00 on bidding day to 12:00 on the day after next.

Day-ahead bidding starts by 10:10am on each day. The grid operator produces forecast demand in MW for each settlement period in the availability declaration period. Each generator provides to the grid operator the offer data in respect of each centrally dispatched genset for the following scheduling day; this offer data covers:

- Offered availability;
- Price : start-up cost, no-load cost, incremental cost;
- Unit dynamic running characteristics: run-up rates, run-down rates, minimum and synchronizing generation;
- Declared inflexibility.

The GOAL (Generator Ordering And Loading) program schedules units from a merit order derived from the generator's offered prices and considers the factors; offered availability, declared inflexibility and dynamic running characteristics.

11.4 Forecasting the SMP

SMP forecasting, both on a long-term and a short-term basis, is becoming more and more important. In the short-term, knowledge of the estimated SMP helps market participants who bid into the spot price market or set up bilateral transactions to determine the bidding strategy of their generators. Accurate SMP forecasting will reduce the risk of under or over-estimating the revenue from the generators for the power companies and provide better risk management. In the long-term, accurate prediction will enable the power marketers and companies to make sound business decisions, however, there are few papers published on the topic due to the complexity and newness of the problem. Recently, an energy price forecasting method and a neural-network-based technique for the prediction of SMP proposed in [16-17], respectively. The wavelet-transform-based model is also presented by the authors [18].

11.4.1 Neural-network-based Model

The role of neural networks is to find the unknown mapping between input and output data. A multi-player feed-forward neural network with back-propagation algorithm is chosen for this. There are twelve neurons selected to represent the key features of the SMP in the input layer. However, selection of the optimal number of hidden layers and the optimal number of neurons in each layer is still an open issue. Generally speaking, there are two factors to consider: the complexity of the problem and the amount of training data available. Through a great deal of computation and modifications, it was found that the network given in Table 11.8 provides the best performance. This is a four-layer network, with twelve inputs, one output, and eight neurons in the first hidden layer and five neurons in the second hidden layer. The transfer function applied is the sigmoid function.

Table 11.8. Topology of neural network

Input neurons	D (i), T (i, t), L (i, t), A_B (i, t) smp (i-21, t), smp (i-14, t) smp (i-7, t-1), smp (i-7, t), smp (i-7, t+1) smp (i-1, t-1), smp (i-1, t), smp (i-1, t+1)
Output neuron	smp (i, t)

The symbols used in Table 11.8 have the following meaning.
i: the predicted day
t: the predict settlement period
D: Day of the week
T: Settlement period index
L: the estimated load
A_B: the estimated Table A/B.

There are a few factors which affect SMP prediction. These factors can be classified into two groups. One is related information and the other is historical SMP. In the related information group, there are four variables to be taken into account. They are day of the week, settlement period, the estimated load and Table A/B. Other factors like temperature have relatively indirect influence on the SMP and can be incorporated in the load variable. In the historical SMP group, there are two types of data used each time. One is three (i-21) and two (i-14) weeks ago at the same settlement period data which are smp (i-21, t) and smp (i-14, t). Another is one week ago and yesterday at the same and neighbouring settlement period data, which are smp (i-7, t-1), smp (i-7, t), smp (i-7, t+1), smp (i-1, t-1), smp (i-1, t) and smp (i-1,t+1). If i is a working Monday, smp (i-1, t-1), smp (i-1, t), and smp (i-1, t+1) are replaced by smp (i-3, t-1), smp (i-3, t) and smp (i-3, t+1). If i is a working day, but any of (i-1), (i-7), (i-14) or (i-21) is a public holiday, then it is replaced by the previous working day. Note that in the yesterday data vector, the last element smp $(i-1, t+1)|_{t=48}$ should be taken to be the smp (i, 1) as the rule. But this is not realistic because smp (i, 1) is the part of predicted data. Thus, smp (i-1, 48) is capped to smp (i-1, 47).

11.4.2 Wavelet-transform-based Model

The wavelet transform based SMP prediction model is shown in the Figure 11.16. The input data consist of settlement periods (forty-eight intervals) for four months (From January to April) of the UK and forecasted demands are considered for the same period. The wavelet transform is performed to seek the approximation and details. The regression coefficients are calculated using *d1, d2, d3, d4, d5*, and forecasted demand, and the demand applied wavelet coefficients are predicted. The SMP forecasting is then finally implemented using the previous day's low frequency and the forecasted high frequency components using the five-scale scheme.

11.4.3 Combined Model

In the processes of the approach, the historical SMP is first decomposed into approximate parts associated with low frequencies and several detail parts associated with high frequencies through wavelet transform. Then, the neural network trained by the low frequencies and the corresponding load data is used to predict the approximate parts of the future SMP. Finally, the short-term SMP is forecasted by summing the predicted approximate part and the weighted detail parts. Figure 11.17 illustrates this process, which is described in detail in the rest of this section [19].

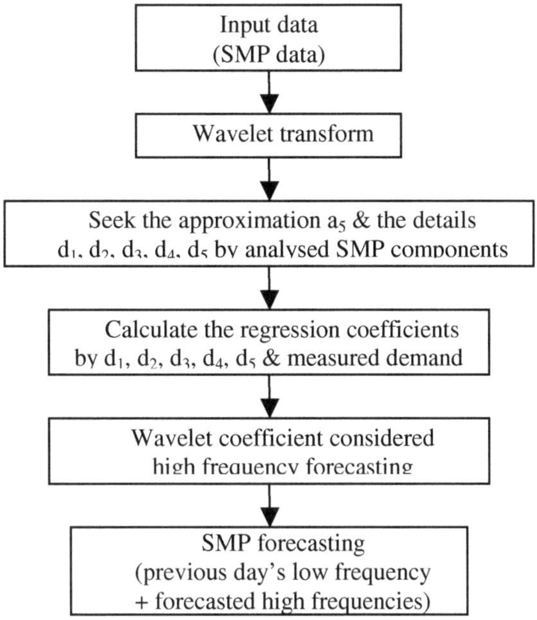

Figure 11.16. Flow of the proposed SMP predicting method by wavelet transform

11.4.3.1 Decomposing the SMP Data

Figure 11.18 illustrates the decomposition of the SMP of a given week. There are one low and four high-frequency components. From Figure 11.18, it is clear that the shape of the approximation of every day represents the trend of the normal SMP. There are usually several levels for each type of wavelet transform (DB1, DB2, DB4), such as level 1, level 2, level 3, and so on. The approximation shape of every level is usually different. The level whose approximation shape is the closest to the actual SMP profile should be selected.

11.4.3.2 Predicting the Approximation

1) Decide the type of the predicted day: The SMP-profile is different for each. The type of day is usually divided into weekday, weekend, and holiday. In this work, except holidays, the seven types corresponding to day of the week are included.

2) The approximate SMP of the predicted day is computed through seven ANN models. Every model corresponds to the approximation of every type of day. Because of the slow speed of convergence of the back-propagation network, the radial basis function (RBF) is applied. Table 11.9 lists the input nodes and output node of the model.

i : The i th time
K : The k th day
$D(k)$: Day index of a year
$W(k)$: Day index of a week
$L(i,k)$: Load the i th time of the k th day
$L(i-1,k)$: Load the $(i-1)$ th time of the k th day
$L(i+1,k)$: Load the $(i+1)$ th time of the k th day
$App_0(i-1,k-14)$: Approximation of SMP at the $(i-1)$ th time of the $(k-14)$ th day
$App_0(i,k-14)$: Approximation of SMP at the i th time of the $(k-14)$ th day
$App_0(i+1,k-14)$: Approximation of SMP at the $(i+1)$ th time of the $(k-14)$ th day
$App_1(i-1,k-7)$: Approximation of SMP at the $(i-1)$ th time of the $(k-7)$ th day
$App_1(i,k-7)$: Approximation of SMP at the i th time of the $(k-7)$ th day
$App_1(i+1,k-7)$: Approximation of SMP at the $(i+1)$ th time of the $(k-7)$ th day
$App_2(i-1,k)$: Approximation of SMP at the $(i-1)$ th time of the k th day
$App_2(i,k)$: Approximation of SMP at the i th time of the k th day

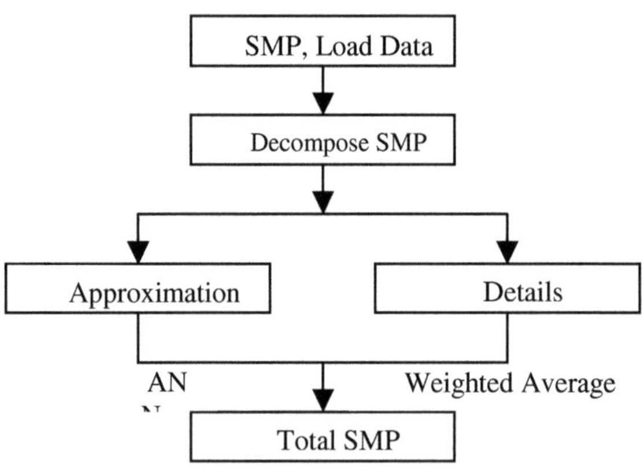

Figure 11.17. Flowchart of the proposed approach

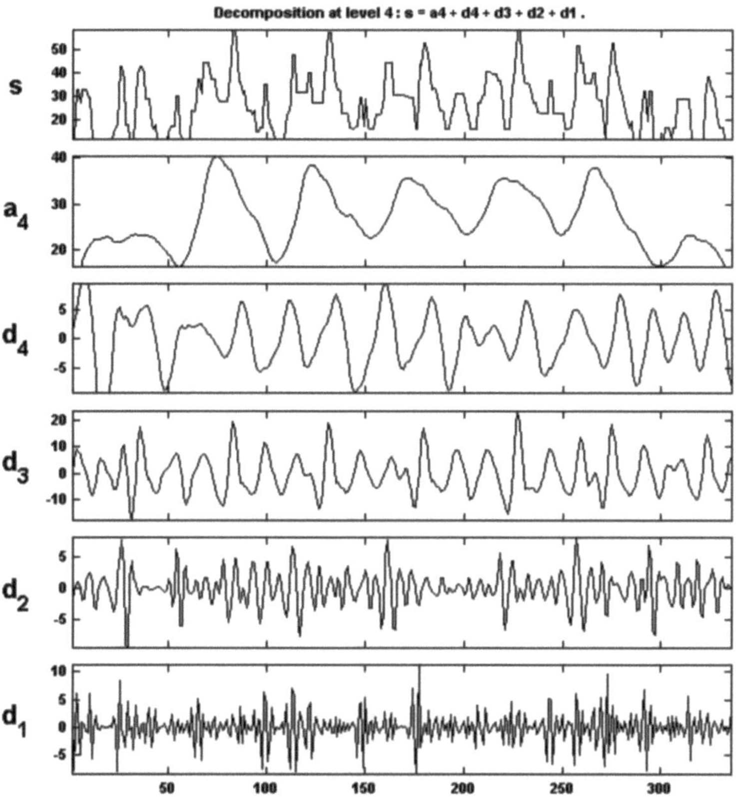

Figure 11.18. Decomposition of the SMP of one week

11.4.3.3 Estimating the Detail

Because the detail represents the high-frequency part of the SMP and it fluctuates frequently, it is difficult to forecast the detail by ANN. In this work, the weighted mean method is used to forecast the detail:

$$D(i,k)=\sum_{m=1}^{n} b_m \cdot \frac{D(i,m)}{n} \tag{11.20}$$

where, n : Last n day

$D(i,k)$: Detail at the i th time of the k th day

$D(i,m)$: Detail at the i th time of the m th day

b_m : Weight of the m th day. $\sum_{m=1}^{n} b_m = 1$

Table 11.9. Node list of the RBF network

Maximum model	
Input nodes	Output node
$D(k)$, $W(k)$, $L(i-1,k)$, $L(i,k)$, $L(i+1,k)$ $App_0(i-1,k-14)$, $App_0(i,k-14)$ $App_0(i+1,k-14)$, $App_1(i-1,k-7)$ $App_1(i,k-7)$, $App_2(i+1,k)$, $App_2(i-1,k)$	$App_2(i,k)$

11.4.3.4 Summing the Approximation and the Details

Having obtained the approximation and the details, the SMP can be predicted by the restructured levels of the wavelet transform.

$$SMP(i) = App(i) + D(i) \tag{11.21}$$

where $SMP(i)$ is the SMP at i th time of the predicted day.

11.4.4 Prediction Results and Analysis

11.4.4.1 Prediction Results by Neural- network- based Model

Table 11.10 reports some results for the different combinations of neural network node. Table 11.11 shows mean average percentage error (MAPE) on each individual weekday and the average MAPEs. The average MAPE of these five days, Average (5), is 12.86% and the average MAPE without Thursday Average(4) is 11.57%.

Table 11.10. Results of different combinations of neural network node

NN topology	12-13-1	12-10-1	12-9-1	12-8-1	12-7-1
MAPE (%)	16.29	16.01	16.86	14.84	17.97
NN topology	12-8-7-1	12-8-6-1	12-8-5-1		
MAPE (%)	15.53	14.97	12.87		
NN topology	12-7-7-1	12-7-3-1	12-6-3-1	12-5-3-1	
MAPE (%)	17.99	15.40	17.19	16.86	

11.4.4.2 Prediction Results by Wavelet-transform-based Model

An interesting observation is that the mean percentage error of weekdays is the lowest at DB2 (lev5). On the other hand, the mean percentage error of weekend days is lowest at DB1 (lev5). That's why DB2 on weekdays and DB1 (Haar) on weekend days are used as dominant wavelet transform coefficients in this work. Table 11.12 shows the mean percentage errors of weekdays and weekend days at lev5, which describes the level of high frequencies decomposed by the wavelet transform. It is clear that the mean percentage errors of DB2 (lev5) for weekdays and DB1 (lev5) for weekend days are the lowest.

Table 11.11. Daily and average MAPE

Day of the week	MAPE (5)
Monday	11.48
Tuesday	13.30
Wednesday	11.09
Thursday	18.06
Friday	10.39
Average (5)	12.86
Average (4)	11.57

Figure 11.19 compares results of actual and forecasted SMP for one day, which is predicted by the DB2 (lev5) wavelet transform. The mean percentage error is 6.21%. Figure 11.20 gives the mean percentage error for weekdays. The weekdays mean percentage error obtained is 6.24% at DB1 (lev5), 6.03% at DB2 (lev5) and 6.31% at DB4 (lev5). From settlement period 8 to 12, the percentage error is much higher than in other periods.

Table 11.12. Mean percentage errors of week days and weekend days at lev5

Day	Weekdays			Weekend days		
	DB1	DB2	DB4	Sat.	Sun.	Mon.
Mean percentage errors	6.24	6.03	6.31	11.11	2.49	9.72

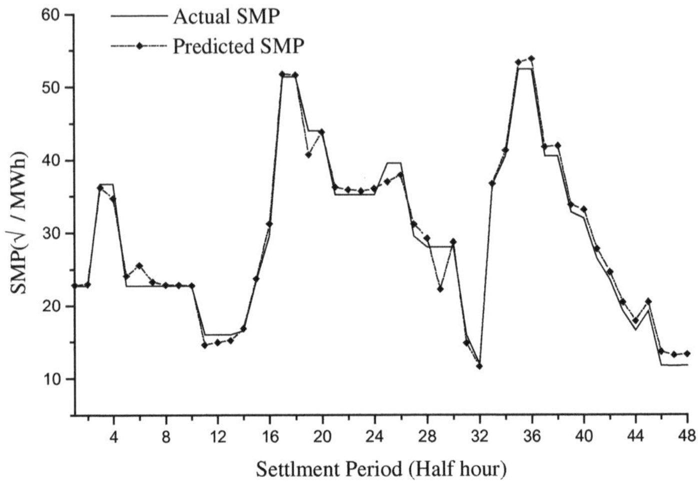

Figure 11.19. Comparison of actual and forecasted SMP for one-day

11.4.4.3 Prediction Results by Combined Model

The approach has been implemented and tested on practical SMP data. The interval of SMP data is half an hour. The holiday data are extracted. The data are divided into two parts: one part (about two months' data) is used to train the network; the rest is used to test the performance. The percentage error is defined as

$$error = \frac{|actual\ SMP - forecasted\ SMP|}{actual\ SMP} \times 100(\%) \tag{11.22}$$

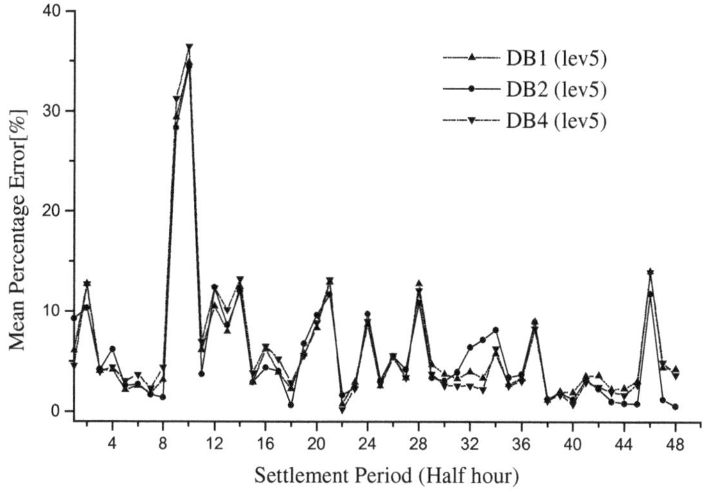

Figure 11.20. Mean percentage errors of weekdays

In this section, the DB4 and level 4 wavelet transform is used. The percentage error is summarised in Table 11.13.

Table 11.13. Summary of the errors (%)

	Jan. 27	Jan. 28	Jan. 29	Jan. 30	Jan. 31	Feb. 1	Feb. 1
Abs.Ave.error	6.09	4.12	3.59	3.46	4.48	6.31	7.44
STD error	5.15	3.21	5.94	3.43	4.06	7.02	7.36
RMS error	4.93	5.20	5.28	3.84	5.02	6.01	5.98

11.5 Summary

A novel composite technique for short-term load forecasting using the Kohonen neural network and wavelet transform is described in this chapter. The Kohonen neural network, which is used for classifying seasonal loads by the SOFM, is adopted so that the forecasting accuracy is improved. In addition, 5-level decomposition of the daily load curve is implemented in order to reflect the weather-sensitive component of loads more effectively. The forecasting simulations for each season are performed using Daubechies DB2, DB4 and DB10, respectively. According to the simulation results, DB2 for winter, DB4 for spring and autumn, and DB10 for the summer are applied as dominant wavelet transform. Yearly mean percentage errors of 1.03%, 1.53%, 0.99%, and 0.78% are obtained. The mean percentage error for weekdays (Tuesday through Friday) is 0.98%, and the mean percentage errors for Monday, Saturday and Sunday are 1.21%, 1.11% and 1.51%, respectively. The weekdays and weekend days load for the four seasons of 1995 have been forecast. The outcome of the study clearly indicates that the composite model comparing the Kohonen neural network and wavelet transform approach is an attractive and effective means of short-term load forecasting.

On the other hand, the SMP profile is a typical time series and, to some extent, similar to the load profile. Thus, many forecasting methods can be used to predict the short-term SMP. A novel method based on the wavelet transform and neural network is also presented. Results show that the method is encouraging. Judging from the prediction results given in the previous section, the combined model seems to the best choice so far. In general, the predicted SMP value is strongly coupled with the generator's bidding pattern, which in turn is dependent on available generation and transmission, the anticipated system power reserves (SPR), and system potential demand (SPD). So the SPR and SPD should be included to forecast the SMP. If the SPR is also included, the accuracy should be improved.

11.6 References

[1] I. Moghram and S. Rahman: "Analysis and evaluation of five short-term load forecasting techniques", IEEE Transactions on Power Systems, 4, 1989, 1484-1491.

[2] T. W. S. Chow and C.T. Leung: "Neural network based short-term load forecasting using weather compensation", IEEE Transactions on Power Systems, 11, 1996, 1736-1742.

[3] J. Bastian and J. Zhu: "Forecasting energy prices in a competitive market", IEEE Magazine – Computer Application in Power, 7, 1999, 41-45

[4] A. Graps: "An introduction to wavelets", IEEE Computational Science & Engineering, 2, 1995, 50-61.

[5] S. Makridakis and S. C. Wheelwright: "Forecasting: methods and applications", John Wiley & Sons, 1978.

[6] S. Vemuri, W. Huang and D. Nelson: "On-line algorithms for forecasting hourly loads of an electric utility", IEEE, Trans., on Power APP. and Syst., Vol. PAS-100, 1981, 3775-3784

[7] I. K. Yu, C. I. Kim, Y. H. Song: "A novel short-term load forecasting technique using wavelet transform analysis", Electric Machines and Power Systems, 28, 2000, 537-549.

[8] G. T. Galli, Heydt and P. F. Riberio: "Exploring the power of wavelet analysis", IEEE Computer Application in Power, 9, 1996, 37-41.

[9] M. Misiti, Y. Misiti, G. Oppenheim and J. M. Poggi: "MATLAB Wavelet Toolbox", The MathWorks, Inc., 1997, First Printing

[10] D. Niebur and A. Germond: "Power system static security assessment using the Kohonen neural network classifier", IEEE Transactions on Power Systems, 7, 1992, 865-872.

[11] T. Kohonen, O. A. E. Oja, O. Simula, A. Visa and J. Kangas: "Engineering application of the self-organizing map", Proceedings IEEE, 84, 1996, 1358-1384.

[12] H. Demuth and M. Beale: "Neural network toolbox for use with MATLAB", The MathWorks, Inc., 1998, Fifth Printing – Version 3.

[13] M. Hiroyuki, T. Yoshihito and T. Senji: "An artificial neural-net based technique for power system dynamic stability with the Kohonen model", IEEE Transactions on Power Systems, 7, 1992, 856-864.
[14] R. Sididqi and A. B. Kader: "POWERCOACH : An electricity trading advisor", IEEE Magazine – Computer Applications in Power, 7, 1994, 41-46.
[15] A. J. Wang and B. Ramsay: "A neural network based estimator for electricity spot-pricing with particular reference to weekend and public holidays", Neurocomputing 23, 1998, 47-57.
[16] J. Bastian and J. Zhu: "Forecasting energy prices in a competitive market", IEEE Magazine – Computer Applications in Power, 7, 1999, 41-45.
[17] A. Wang, B. Ramsay: "Prediction of system marginal price in the UK power pool using neural networks", Proc., Int. Conf. on Neural Networks, ISBN 0-7803-4122-8197, Houston, USA, 1997, 2116-2120.
[18] I. K. Yu, C. I. Kim and Y. H. Song: "Prediction of system marginal price of electricity using wavelet transform analysis", International Journal of Energy Conversion and Management, 43 (14) (2002) pp. 1839-1851
[19] S. J. Yao, Y. H. Song, L. Z. Zhang, and X. Y. Cheng: "Prediction of system marginal price by wavelet transform and neural network", Electric Machines and Power Systems, 28, 2000, 537-549

12. Analysis of Generating Companies' Strategic Behaviour

A. Mariorano, Y.H. Song and M. Trovato

In this chapter, a methodology to simulate the strategic behaviour of generating companies in an oligopolistic electricity market, by using strategic supply functions, is proposed. In particular, electricity producers are assumed to bid in a pool-based electricity market [13]. Every day, generators submit prices for each generating set for the following day and the transmission system operator calculates the operating schedules that will meet the forecast levels of demand at minimum cost, based upon the bid prices. Then, for each time interval, typically half-hour, all generating sets in the schedule are paid the market-clearing price, which varies with demand and is based on the bid of the most expensive set in normal operation during that time interval.

In this context, each firm sharing the market does not bid at marginal cost as it is in a perfect competitive market, but profit-maximising strategies are carried out [12,14,16,17,18,20,21,23]. In order to simulate the strategic competition among producers in the electricity market, the bidding process is expressed using linear supply functions. This means that each firm submits a linear supply schedule, relating the amount of power supplied to price. Nevertheless, marginal cost bids are evaluated and considered as desired supply function in order to determine the "market power" of producers, in terms of distance of the strategic bids from the desired marginal cost bids.

Another important aspect to be taken into account when analysing the electricity market, as already mentioned in the previous chapter, is the presence of private contracts between generating companies and customers. These private agreements may be carried out using financial derivatives like forwards contracts, futures and options contracts [7,8,9,12].

In the methodology developed, the general case of an asymmetric oligopoly is analysed and the strategic behaviour of producers is investigated, assuming linear supply functions. The results of the analysis are expressed in terms of market clearing price, profit-maximising value of the supply functions' slope, and hence real power output sold and profit made, of each producer sharing the market. Moreover, the proposed methodology allows the presence of private contracts, set up as Contracts for Difference (CfDs), to be taken into account, evaluating their effects on the market equilibrium conditions.

The effectiveness of the proposed model is tested on a wholesale electricity market composed by ten different generating companies.

12.1 The Electricity Marketplace

As shown in the previous chapter, there are three different basic approaches to carry out power trading in different nations and in different states of one country [1 6]. Power Exchange (PX) is an organisation that operates a marketplace for the power suppliers and customers and determines the price in an auction process. In the direct access approach, power suppliers and customers can make a contract to exchange power at certain prices. The Power Pool can be regarded as an agent that has the responsibilities of balancing the power supply and demand and also running the power system. Generally these three approaches are combined in different ways to form a composite system.

All these three approaches involve the presence of brokerage systems (auction markets) to trade energy. Such brokerage systems may not include all the financial and commodity brokerage aspects and attributes. Commodity brokering operations occur in a regulated environment where the procedures are set by commissions.

The main concept is that electric energy, as a commodity, can be analysed in a financial framework. Unlike any other commodities, electric energy cannot be stored in large amounts. Electricity has demand and supply that must be carefully balanced, creating the problem that instant demand requires instant generation. Additionally, the delivery path to follow cannot be chosen for energy when moving from one point to another without consideration of all other transactions. The origin and delivery points are fixed but the delivery path is variable based on physical laws and may even be impossible under particular operational constraints. The path followed by energy can cause problems when wheeling power across intermediate systems because of resulting operational problems, such as voltage dipping, reactive power flows, increased losses and reliability problems.

Nevertheless, when treated as a commodity, energy can be traded in a free market or exchange. The major difference between a free market and a centrally directed one is that the free market is a distributed optimal solution when appropriate rules are in place to provide a fair and level environment [19]. A free market is controlled by consumer demand and relative costs of production. The broad objectives of such exchanges are greater equality of opportunity, greater efficiency in markets and improvement in information flow. In the following, the main features of electricity auction markets are illustrated.

12.1.1 Auction Structures

In designing an electricity auction, the auctioneer must decide on such auction dimensions as: how to bundle demand into lots for auction, what the pricing rule will be and what the sequencing of the auction will be. United Kingdom, California and Australia have decided to bundle demand into vertical lots, which is explained in the following, to pay the same uniform price to every winning generator in a lot, and to have their generators submit their bids for all lots simultaneously. As is shown in [22], this auction structure does not guarantee efficiency in equilibrium.

In the following the different auctions dimensions and the possible alternatives within each dimension are identified.

12.1.1.1 Bundling of Demand into Lots

In the case of electricity, the basic object to be auctioned is 1 MWh of the forecast daily demand.

If it is assumed that the demand is deterministic and inelastic, the auctioneer can determine all the necessary information on the demanded amount of power by using the load curve, as shown in Figure 12.1, forecast for the next day. Alternatively, the load duration curve, Figure 12.2, can be used. This expresses the duration of a determined value of power requested as a function of the value itself and it can be easily calculated from the load curve.

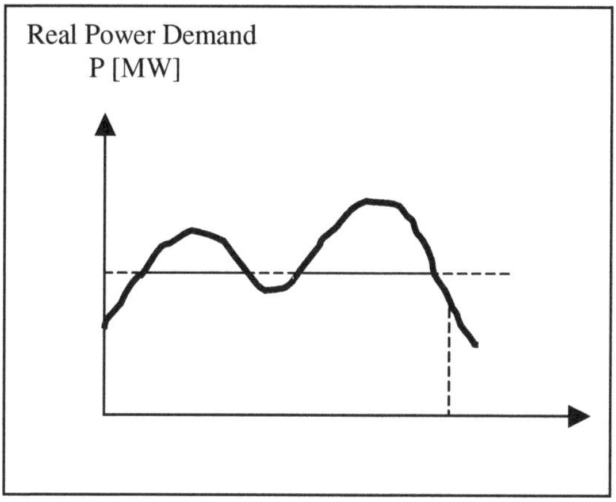

Figure 12.1. Daily load curve

When there are several objects to be auctioned, the auctioneer must decide how to bundle the objects in lots for auction. In particular, there are two possible ways to do this:

- the bidders submit one bid that applies equally to all objects;
- the object is divided into distinct lots for which bidders submit a separate bid for each lot.

In the context of an electricity auction, there exist two "natural" bundling forms: horizontal and vertical.

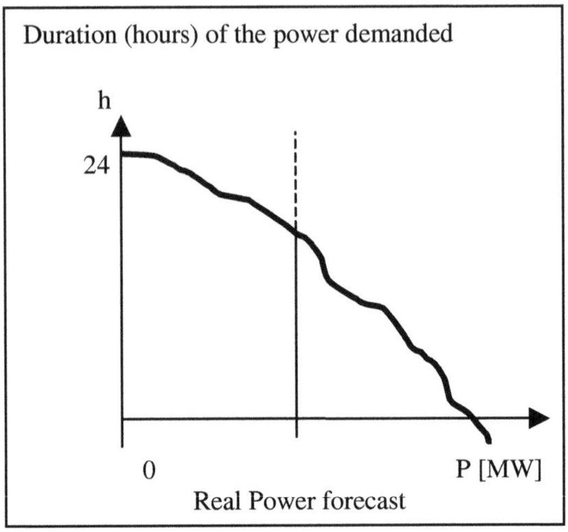

Figure 12.2. Load duration curve

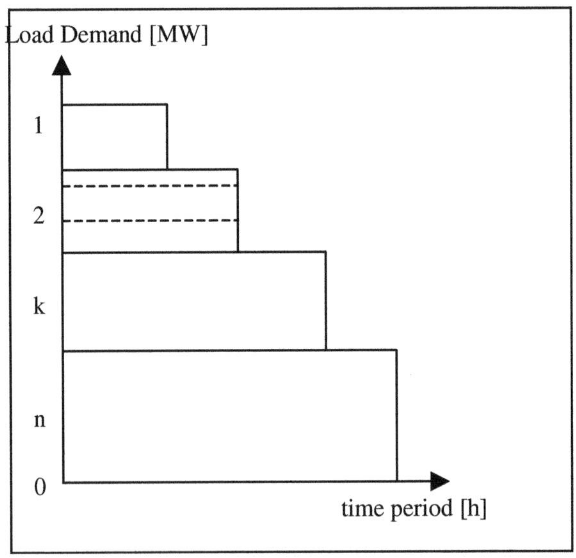

Figure 12.3. Horizontal auction demand lots

In a *horizontal auction*, see Figure 12.3, demand lots are formed by partitioning daily demand by its duration, that is considering a distinct lot for each duration t. Hence generators submit a supply curve for each lot, indicating the price at which they are willing to generate k megawatts for a duration of t hours, where k, t >0.

In a *vertical auction*, see Figure 12.4, daily demand is divided into t hourly demand lots, where each demand lot contains all the demand in hour t, $t = 1, \ldots, t$. For each hour t, generators bid the price at which they are willing to generate k megawatts during hour t, where $k, t > 0$.

While there exist countless ways to bundle demand, the two bundling forms identified here are the most practical and logical to examine in an electricity auction setting. As stated earlier, the electricity auctions in operation in the United Kingdom, Australia and California are vertical auctions, where generators must bid their generation into hourly markets. However, the decision to operate a plant is not made on an hourly basis. The physical characteristics of most generation plants requires planning its scheduling for a period of time. Therefore, allowing generators to bid for period of operating time seems to be a more natural way to design an electricity auction.

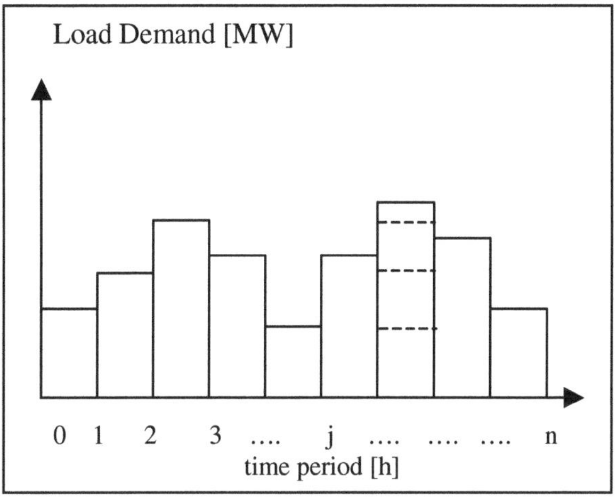

Figure 12.4. Vertical auction demand lots

12.1.1.2 Sequencing of Auctions

Where there is more than one demand lot to be auctioned, the auctioneer must decide how to sequence their sale. In a simultaneous auction, bids are submitted and allocation decisions for all demand lots are made simultaneously. Alternatively, in a sequential auction, demand lots are auctioned sequentially; before each new auction, the results of any previous auctions are made known.

12.1.1.3 Pricing Rule

Before the bidders submit their bids, they must know how the prices at which the transactions take place are to be determined. If there is more than one winner per lot, each with different bids, it is important to know if the winners all have to pay the same price or if different prices are available. The former pricing rule is called a uniform pricing rule and the latter a discriminatory pricing rule.

Under a uniform pricing rule, all winners in a lot are paid the highest accepted bid price. This is because an electricity auction is a procurement auction, in which the goal is to solicit bids from suppliers for a service.
Under a discriminatory pricing rule, each winner is paid its own bid price.

If a mandatory pool is present, in which the unconstrained dispatch of supply to meet the load demand is carried out, a discriminatory pricing rule is not usually used, since the main purpose of having a mandatory pool is to eliminate any price discrimination.

Figure 12.5 lists all the possible auction mechanisms given the dimensions and choices identified above.

The major trading objectives in the auction market are hedging, speculation and arbitrage. Hedging is a defence mechanism against loss and/or supply shortages. Speculation is assuming an investment risk with a chance for profit. Arbitrage is the crossing of sales and purchases between the markets.

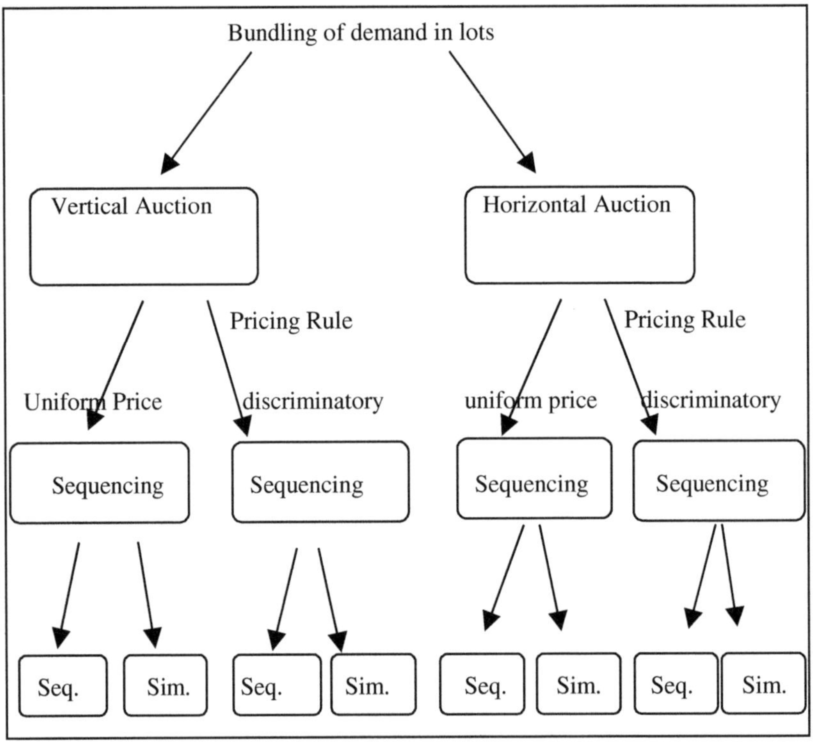

Figure 12.5. Possible auction mechanisms

12.2 Strategic Supply Functions

As is known, the strategy for a generating company is formally a function mapping price into a level of real power output independent of time T. Then, each firm submits its supply function to the grid dispatcher simultaneously, the day before. Finally, the dispatcher determines the spot price and each firm's supply by solving the price output pair that matches supply and demand at any moment. In this paragraph, the case of non-linear supply functions is analysed following [24].

Let us consider the simplest case of a symmetric duopoly. Let us assume that the load demand curve can be forecast with certainty for the twenty-four hours of the following day. The demand function can be expressed as follows:

$$D = D(p,t) \tag{12.1}$$

where t indicates the time (in terms of hour of the day) and p the price [£/MW]. If it is supposed, as examined for the auction mechanism, that the daily load demand is divided in vertical lots, each corresponding to a hour time interval, then t indicates a given hour of the day and D represents the aggregated load demand for that time interval. On the contrary, if it is supposed that it is the load duration curve to be divided in vertical lots, D represents the amount of real power requested for a time interval of t hours. In both cases, it is supposed that:

$$-\infty < \frac{\partial D(p,t)}{\partial p} < 0 \tag{12.2}$$

$$\frac{\partial^2 D(p,t)}{\partial p^2} \leq 0 \tag{12.3}$$

As in the previous chapter, the analysis is restricted to a given time interval, constant t, so that the demand function is a function only of the price:

$$D(p, t) = D(p) \tag{12.4}$$

The net demand facing generating company i when the other duopolist, j, has supply schedule $P_j = P_j(p)$, where P indicates the real power, is:

$$D_i(p) = D(p) - P_j(p) \tag{12.5}$$

Let the effective generating costs of supplying P be $C_i(P)$ and $C_j(P)$, for the duopolists with marginal costs $C'_i(P)$ and $C'_j(P)$ respectively. Moreover, a change of origin, so that the marginal costs are normalised, is assumed. This implies that the resulting supply functions have a minimum price that is zero.

The strategy for firm i is formally a function mapping price into a level of real power output, independent of time. Each firm submits its supply function to the grid dispatcher simultaneously, the day before, and the dispatcher then determines the spot price and each firm's supply by solving the price output pair that equates supply to demand at each moment. The equilibrium condition will be revealed then by:

$$D(p^*) = P_i(p^*) + P_j(p^*) \tag{12.6}$$

provided that such a price exists.

The aim of the analysis is to determine equilibrium supply functions that are supply functions which maximise the profit of each of the duopolists. The profit of the i th generating company can be expressed as follows:

$$\pi_i = p \cdot P_i - C(P_i) \tag{12.7}$$

In order to determine the profit maximising conditions, it is possible to derive Equation (12.7) with respect to the price to obtain:

$$\frac{\partial \pi_i}{\partial p} = P_i + p \cdot \frac{dP_i}{dp} - \frac{dC_i(P_i)}{dP_i} \cdot \frac{dP_i}{dp} = P_i + p \cdot \frac{\partial (D - P_j)}{\partial p} - C_i'(P_i) \cdot \frac{\partial (D - P_j)}{\partial p} = 0 \tag{12.8}$$

that is:

$$P_i + (p - C_i'(P_i)) \cdot \left(\frac{\partial D}{\partial p} - \frac{dP_j}{dp} \right) = 0 \tag{12.9}$$

and then:

$$\frac{dP_j}{dp} = \frac{P_i}{p - C_i'(P_i)} + \frac{\partial D}{\partial p} \tag{12.10}$$

Since it is supposed to consider a symmetric duopoly, it is possible to assume $P_i = P_j = P$, we obtain:

$$\frac{dP}{dp} = \frac{P}{p - C'(P)} + \frac{\partial D}{\partial p} \tag{12.11}$$

Equation (12.10) or (12.11) represents the first-order condition for profit maximisation. In order to take into account the second-order condition, it is necessary to evaluate the second derivative, with respect to the price, of the profit:

$$\frac{\partial^2 \pi_i}{\partial p^2} = 2\left(\frac{\partial D}{\partial p} - \frac{dP_j}{dp} \right) - C_i''(P_i) \cdot \left(\frac{\partial D}{\partial p} - \frac{dP_j}{dp} \right)^2 + (p - C_i'(P_i)) \cdot \left(\frac{\partial^2 D}{\partial p^2} - \frac{d^2 P_j}{dp^2} \right) \tag{12.12}$$

and verifying that it has a negative value for the solutions of Equation (12.10)

$$2\left(-\frac{P_i}{p-C_i'(P_i)}\right)-C_i''(P_i)\cdot\left(-\frac{P_i}{p-C_i'(P_i)}\right)^2+(p-C_i'(P_i))\cdot\left(\frac{d}{dp}\left(\frac{-P_i}{p-C_i'(P_i)}\right)\right)<0 \qquad (12.13)$$

It can be demonstrated that Equation (12.13) is always satisfied if linear marginal costs are taken into account. Nevertheless, it should be noted that for a symmetric duopoly, Equation (12.10) could be used only to determine the local slope of supply functions. It can be shown that all such trajectories pass through the origin where they have the same slope.

The behaviour of the differential equation that characterises the symmetric supply function equilibrium can be further analysed, determining an upper and a lower static limit for all possible equilibrium functions.

In particular, by imposing that *dP/dp* is strictly positive, Equation (12.11) can be rewritten as follows:

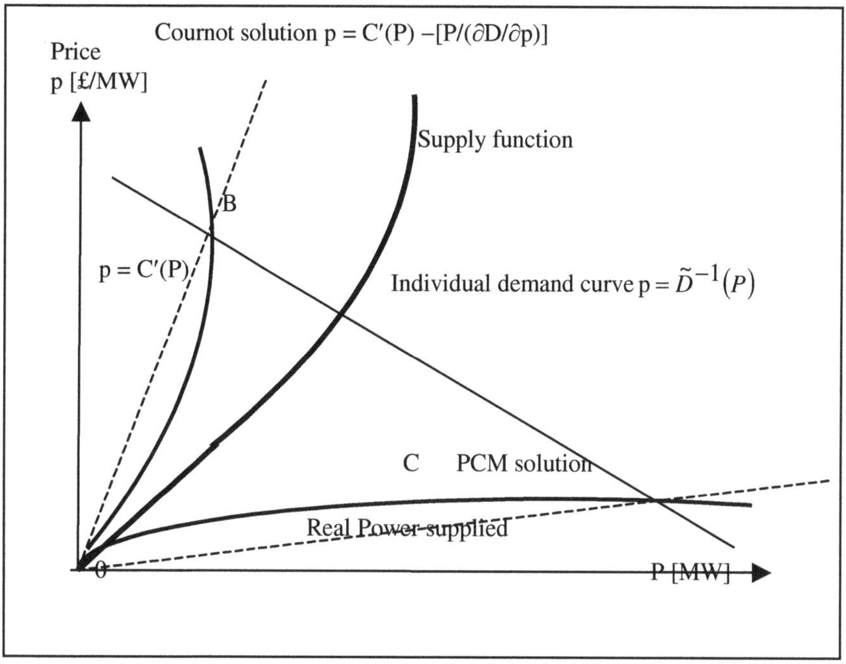

Figure 12.6. Feasible supply function equilibria

$$C'(P) < p < C'(P) - \frac{P}{\partial D / \partial p} \qquad (12.14)$$

Therefore, the duopoly supply schedule lies between these two boundaries. In particular, if both the load demand function and marginal costs have a linear expression, these curves are represented by two lines. Let us consider the first equation, $p= C'(P)$. This is the supply schedule of a perfectly competitive firm and along this curve, as shown in Figure 12.6, $dP/dp = \infty$, so $dp/dP = 0$. Any trajectory that intersects the lower boundary reaches it with horizontal slope at a point such as C. Once it has crossed the boundary it will have a negative slope, eventually reaching the ρ-axis. On the contrary, the upper boundary, represented by the following expression:

$$p = C'(P) - \frac{P}{\partial D / \partial p} \tag{12.15}$$

can be interpreted as the Cournot supply schedule.

If a trajectory reaches the upper boundary at a point such as B, its slope there will be $dP/dp = 0$, or $dp/dP = \infty$. It will cross the boundary vertically and then bend back, eventually reaching the ρ-axis.

In general, then, the duopoly supply function lies between the competitive and Cournot schedules along a trajectory such as 0A in Figure 12.6.

It should be noted that the load demand curve, illustrated in Figure 12.6, is not the aggregated demand schedule but it is represented by the individual load demand [$\rho = \tilde{D}^{-1}(P)$], that is the demand the each duopolist has to satisfy. In particular, assuming a linear aggregated demand curve, such as $p = a - b \cdot P_{tot}$, for a symmetric duopoly, $P_{tot} = 2P$, the individual demand schedule can be expressed as:

$$p = a - 2b \cdot P \tag{12.16}$$

and has a slope twice as the aggregated demand schedule.

Candidates for equilibrium supply schedules must not intersect either boundaries over the range of possible output price pairs. Under these assumptions, it is possible to define two limit supply functions, denoted by 0B and 0C in Figure 12.1, whose position varies according to changes in the load demand function BC. In fact, in [11] it has been proved that if the demand schedule can be arbitrarily high, then there is a unique solution; otherwise there maybe a connected set of equilibria bounded by an upper and a lower supply schedule.

12.2.1 Supply Constraints

The set of possible equilibria can be narrowed by taking into account the presence of capacity constraints. Let us suppose that neither firm can supply beyond a capacity limit, $P = k$. If the j th producer would sell all its available real power, $P_j = k$, the optimal response for the i th producer would be the Cournot solution, taking into account its own capacity limits:

$$\begin{cases} D = P_i + P_j \\ P_j = k;\ P_i \le k \\ \max_{p}\left(\pi_1 = p \cdot P_i - C(P_i)\right) \end{cases} \tag{12.17}$$

The solution to this problem must satisfy the Kuhn Tucker's conditions:

$$\begin{cases} P_i = -\dfrac{\partial D}{\partial p} \cdot \left(p - C'(P_i) - \mu\right) \\ \mu \cdot (k - P_i) = 0 \\ \mu \ge 0 \end{cases} \tag{12.18}$$

where μ is the Lagrangian multiplier corresponding to the shadow price of the capacity constraint. Since firm j is facing a symmetrical problem, this would represent also its optimal response. This condition is illustrated in Figure 12.7 by the point E.

It actually represents the equilibrium solution only if the load demand level is higher than or equal to D_1, so that both producers can sell all their available real power, k. If the demand level is lower, D_2, the capacity constraints will be activated and the Cournot solution will no longer represent the optimal solution for both generating companies.

As is shown in Figure 12.7, the demand schedule meets the capacity constraint at point C. The schedule 0C satisfies Equation (12.18) and represents the lowest supply function that can be in equilibrium. If one producer supplies along a lower schedule, it will reach full capacity before demand is at its maximum. The other firm will find it profitable to deviate to the Cournot. The supply function 0B, which cuts the Cournot schedule at B vertically, is also a candidate for equilibrium since it satisfies the first and second-order conditions for an optimum and does not violate the capacity constraint. It can be concluded that the effects of capacity constraints are to narrow the range of feasible equilibria and in extreme cases in which the intersection of maximum demand with Cournot supply, B, occurs at full capacity, the equilibrium will be unique.

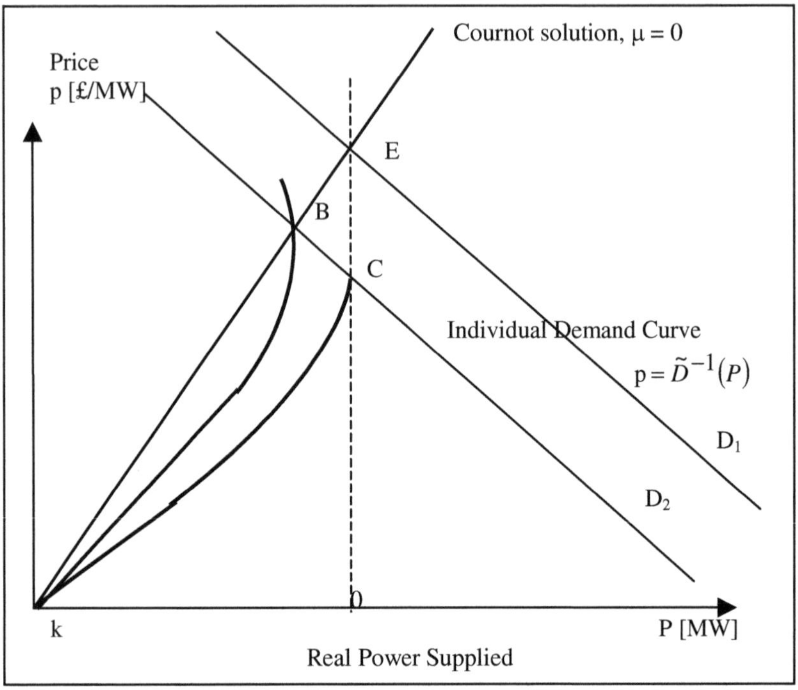

Figure 12.7. Feasible solutions with capacity constraints

12.3 Linear Strategic Supply Functions

As mentioned earlier, the great advantage of using linear supply functions to model the strategic behaviour of generating companies sharing the same electricity market, is that the resulting equilibrium condition is unique [10].

This model is based on the same assumptions made in the previous section, even if the case of an oligopoly composed of n different firms is considered.

The expression for the demand function is still the one expressed by Equation (12.1). For any equilibrium condition, then:

$$D(p^*(t),t)=\sum_{i=1}^{n} P_i(p^*(t)) \tag{12.19}$$

where p^* represents the equilibrium price. For any given time interval, the equilibrium condition is given by a set of n supply functions, one for each generating company, that maximises the producer's profit.

By following the same procedure as illustrated in the previous paragraph, the expression for the set of differential equations, enabling the profit maximising supply functions to be revealed, can be formulated as follows:

$$P_i(p)=\left[p-C_i'(P_i(p))\right]\cdot\left(-\frac{\partial D}{\partial p}+\sum_{j\neq i}\frac{dP_j}{dp}\right) \qquad i=1,2,\ldots,n \qquad (12.20)$$

Moreover, it is supposed that the aggregated demand function is a linear function of the spot price of electricity, so that:

$$\frac{\partial D}{\partial p}=constant \qquad (12.21)$$

and that marginal costs are a linear function of the real power produced:

$$\frac{d^2C_i(P_i)}{dP_i^2}=constant \qquad i=1,2,\ldots,n \quad (12.22)$$

In particular, also in this case the marginal costs are normalised, so that they are zero if no real power is produced:

$$\mathrm{C_i}(\mathrm{P_i})=\frac{1}{2}c_{2i}P_i^2,\quad \mathrm{C_i'}(\mathrm{P_i})=c_{2i}P_i \qquad i=1,2,\ldots,n \quad (12.23)$$

As already pointed out, there is a wide range of supply schedules for each generating company that satisfies the set of Equations (12.20). In this section, the solution under consideration is the linear supply function:

$$P_i(p)=\beta_i\cdot p \;\rightarrow\; p=\frac{1}{\beta_i}\cdot P_i \qquad \mathrm{i}=1,2,\ldots,\mathrm{n} \quad (12.24)$$

The supply function expression is implicitly normalised by having assumed a normalised expression for marginal costs.

By substituting Equations (12.23) and (12.24) into Equation (12.20), and dividing by the price p:

$$\beta_i=\left(1-c_{2i}\beta_i\right)\cdot\left(-\frac{\partial D}{\partial p}+\sum_{j\neq i}\beta_j\right) \qquad i=1,2,\ldots,n \quad (12.25)$$

There will be one equation of this form for each generating company acting in a strategic manner, which can be solved simultaneously to give the slopes of their supply functions, β_i. If this function is chosen by all but one of the firms, the remaining firm's best response is to follow suit, making it an equilibrium. It should be noted that even with linear marginal costs, Equation (12.20) also has many non-linear solutions. [25] proves that, both in the case of any number of symmetric firms and two asymmetric firms respectively, in the special case that the intercept of the demand could take any value between 0 and + ∞, the linear solution is the only equilibrium. The assumption of infinite demand would be inappropriate for the electricity industry, and so linear supply functions are not the unique equilibrium, but they are the most tractable.

12.4 Proposed Model

The general case of an asymmetric oligopoly, composed by n different producers, is analysed [26]. The *strategy* for the i th firm is formally a function mapping price into a level of output independent of time t. Then, each firm submits its supply function to the grid dispatcher simultaneously, the day before. Finally, the dispatcher determines the spot price and each firm's supply by solving the price output pair that matches supply and demand at any moment.

Let us assume a quadratic cost function and, consequently, linear marginal cost functions for each generating company:

$$C_i(P_i) = c_{0i} + c_{1i}P_i + \frac{1}{2}c_{2i}P_i^2 \qquad i=1,\ldots,n \quad (12.26)$$

$$C_i'(P_i) = c_{1i} + c_{2i}P_i \qquad i=1,\ldots,n \quad (12.27)$$

where P_i represents the real power output of the i th generating company.

Under these assumptions, in a perfect competitive market model, supply functions can be expressed by equalising price, p, to marginal costs incurred to produce the real power amount desired, as follows:

$$p = C_i'(P_i) = c_{1i} + c_{2i}P_i \qquad i=1,\ldots,n \quad (12.28)$$

Equation (12.28) can be rearranged as follows:

$$p = p^0_{\min_i} + \left(\frac{1}{\beta_i^0}\right) \cdot P_i \qquad i=1,\ldots,n \quad (12.29)$$

where $\rho^0_{\min_i}$ and $\frac{1}{\beta_i^0}$ are the minimum price at which the company is willing to sell and the slope of the supply function, respectively.

In order to determine the strategic supply functions in an oligopolistic market model, it is assumed that they are obtained from the perfect competitive supply functions by multiplying the minimum price level $\rho^0_{\min_i}$ and the slope $\frac{1}{\beta_i^0}$ by a coefficient k_i, k_i>0:

$$p = p_{\min_i} + \frac{1}{\beta_i} P_i \qquad i=1,\ldots,n \quad (12.30)$$

In Figure 12.8 the marginal cost bid and the strategic profit maximising bid for the i-th company are illustrated respectively.

Equation (12.30) can be rearranged in order to express the amount of real power output as function of the spot price:

$$P_i = \beta_i \cdot p - \frac{c_{1i}}{c_{2i}} \qquad i=1,\ldots,n \quad (12.31)$$

In the following it is explained how the profit maximising value of the slopes of the supply functions of each company are evaluated and the equilibrium condition of the market is determined.

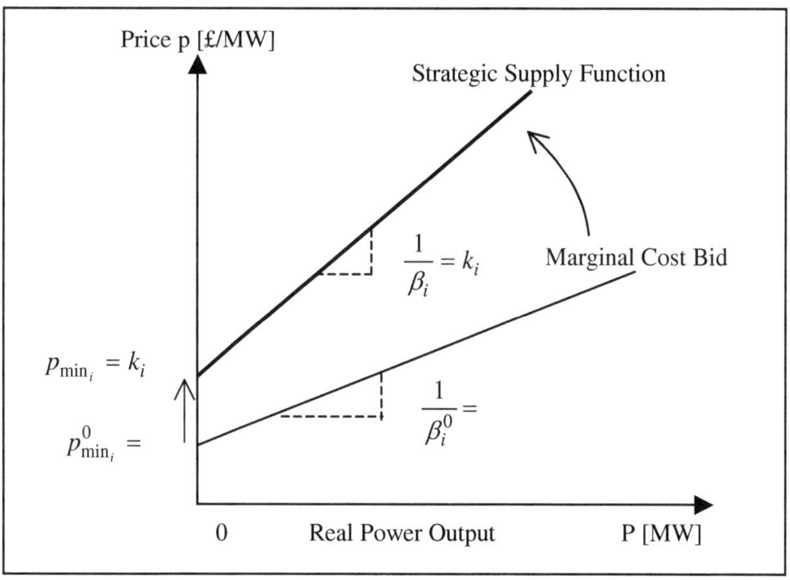

Figure 12.8. Strategic bid

12.4.1 Inverse Demand Function Evaluation

In the analysis carried out it is assumed that the total load demand, D, that the producers have to supply is a deterministic variable function of time and price:

$$D = D(p,t) \tag{12.32}$$

For any given time interval, it is assumed that the total load demand is a linear function of the price, expressed as follows:

$$D(p,\bar{t}) = \frac{a}{b} - \frac{1}{b}p \tag{12.33}$$

In this chapter the presence of transmission losses is neglected, therefore for any given time interval the total real power output produced by all the producers sharing the market has to satisfy the system load demand:

$$D(p,\bar{t}) = \sum_{i=1}^{n} P_i(p,\bar{t}) \tag{12.34}$$

By combining equations (12.33) and (12.34), it is possible to evaluate the inverse demand function, which expresses the price as a function of the total load demand and, consequently, of the total real power supplied:

$$p = a - b \cdot \sum_{i=1}^{n} P_i \tag{12.35}$$

Moreover rearranging Equation (12.34), it is possible to explicitly determine the interactions among the producers' behaviour:

$$P_i = D(p,\bar{t}) - \sum_{j \neq i} P_j \tag{12.36}$$

By combining Equations (12.31) and (12.35), the inverse demand function is evaluated in terms of strategic supply functions, as follows:

$$p = a - b \cdot \sum_{k=1}^{n} P_k = a - b \cdot \sum_{k=1}^{n} \left(\beta_k \cdot p - \frac{c_{1k}}{c_{2k}} \right) \tag{12.37}$$

and then:

$$p=\frac{a+b\cdot\sum_{k=1}^{n}\frac{c_{1k}}{c_{2k}}}{1+b\cdot\sum_{k=1}^{n}\beta_k}=\frac{a+b\cdot\sum_{k=1}^{n}\frac{c_{1k}}{c_{2k}}}{1+b\cdot\left(\beta_i+\sum_{j\neq i}\beta_j\right)} \qquad (12.38)$$

The inverse demand function, expressed in terms of supply functions, is the key point of the analysis carried out. In fact, it expresses the relation among all the producers sharing the market, showing how their behaviours are interrelated.

12.4.2. Presence of Private Contracts

An important aspect to take into account when analysing electricity market is the presence of private agreements between generating companies and customers. These private agreements may be carried out using financial derivatives like forwards contracts, futures and options contracts [7,8,9,12]. These contracts are purely financial transactions. When these contracts are set alongside physical sales, they provide insurance against excessive fluctuations in the spot price of electricity. In fact, although there are several types of contracts, they are carried out as contracts for difference. This means that if the spot price rises above the strike price, which is the price level at which the contract can be called, generators compensate retailers for the difference. But if the spot price falls below the strike price, then retailers compensate generators for the difference. Therefore, the expression for the profit for the i th producer, taking into account also the presence of bilateral contracts, can be expressed as follows:

$$\pi_i=p\cdot(P_i-X_i)+z\cdot X_i-C_i(P_i) \qquad i=1,\dots,n \qquad (12.39)$$

where X_i indicates the amount of real power traded in the bilateral contract by the i th generating company and z_i is the relative strike price. Although it is possible that every producer sets up private contract at a different strike price, in this thesis it is assumed that the strike price is the same for every contract, that is:

$$z_i=z \qquad i=1,\dots,n \qquad (12.40)$$

However, this assumption does not invalidate the results of the analysis since the strike price of each contract tends to be as close as possible to the spot price, implying that they belong to a very restricted range.

12.4.3 Final Formulation of the Model

The optimal bids submitted by each company are the result of a profit maximising process.

The first condition to determine the optimal bid is obtained by differentiating the profit with respect to the spot price:

$$\frac{\partial \pi_i}{\partial p} = P_i - X_i + p \cdot \frac{d(P_i - X_i)}{dp} - C_i'(P_i) \cdot \frac{dP_i}{dp} = 0 \qquad i= 1,\ldots,n \ (12.41)$$

Then, by substituting Equation (12.27) and (12.36):

$$P_i - X_i + (p - c_{1i} - c_{2i} P_i) \cdot \frac{d}{dp}\left(D(p) - \sum_{j \neq i} P_j \right) = 0 \qquad i = 1,\ldots,n \ (12.42)$$

Finally, taking into account Equation (12.31), which is an expression for the strategic supply functions, and Equation (12.33), we have:

$$\beta_i \cdot p - \frac{c_{1i}}{c_{2i}} - X_i + (p - c_{2i} \cdot \beta_i \cdot p) \cdot \left(-\frac{1}{b} - \sum_{j \neq i} \beta_j \right) = 0 \qquad i = 1,\ldots,n \ (12.43)$$

and then:

$$p = \frac{X_i + \dfrac{c_{1i}}{c_{2i}}}{\left[\beta_i + (1 - c_{2i} \cdot \beta_i) \cdot \left(-\dfrac{1}{b} - \sum_{j \neq i} \beta_j \right) \right]} \qquad i = 1,\ldots,n \ (12.44)$$

To guarantee that the equilibrium points determined by Equation (12.44) are profit maximising, all the solutions $[p, P_i]$ have to satisfy the second-order conditions, that is the second derivative with respect to the price has to be negative. To this purpose, by substituting in Equation (12.43), we have the following:

$$\beta_i \cdot p \frac{c_{1i}}{c_{2i}} = \frac{a}{b} - \frac{1}{b} \cdot p - \sum_{j \neq i} \beta_j \cdot p \frac{c_{1j}}{c_{2j}} \qquad i = 1,\ldots,n \ (12.45)$$

and differentiating it with respect to the price, the following set of equations is obtained:

$$-\frac{1}{b} - \sum_{j \neq i} \beta_j + \left(1 + \frac{c_{2i}}{b} + c_{2i} \sum_{j \neq i} \beta_j \right) \cdot \left(-\frac{1}{b} - \sum_{j \neq i} \beta_j \right) < 0 \qquad i = 1,\ldots,n \ (12.46).$$

Equations (12.46) are always satisfied, since coefficients c_2, b and β are positive.

By substituting the expression for the spot price, as a function of the strategic supply functions, in the set of Equations (12.44), it is possible to obtain:

$$\frac{X_i + \dfrac{c_{1i}}{c_{2i}}}{\beta_i + (1 - c_{2i}\cdot\beta_i)\cdot\left(-\dfrac{1}{b} - \sum_{j\neq i}\beta_j\right)} = \frac{a + b\sum_{k=1}^{n}\dfrac{c_{1k}}{c_{2k}}}{1 + b\cdot\left(\beta_i + \sum_{j\neq i}\beta_j\right)} \tag{12.47}$$

which can be rearranged as follows:

$$\beta_j + (1 - c_{2i}\cdot\beta_j)\cdot\left(-\frac{1}{b} - \sum_{j\neq i}\beta_j\right) = \frac{X_i + \dfrac{c_{1i}}{c_{2i}}}{a + b\sum_{k=1}^{n}\dfrac{c_{1k}}{c_{2k}}}\cdot\left[1 + b\cdot\left(\beta_i + \sum_{j\neq i}\beta_j\right)\right] \quad i=1,\ldots,n \tag{12.48}$$

Equation (12.48) represents the fundamental set of n equations in the n unknowns β_i and by solving these equations it is possible to determine the slope and the initial value of the strategic supply function of each competitor sharing the market. In fact, by solving (12.48) it is possible to calculate the coefficients β_i and since $\beta_i = \dfrac{1}{k_i \cdot c_{2i}}$, k_i and then $p_{\min_i} = k_i \cdot c_{1i}$ can be evaluated.

Equations (12.48) can be rearranged as follows:

$$\left(\beta_i\cdot\sum_{j\neq i}\beta_j\right)\cdot c_{2i} + \beta_i\left[1 + \frac{c_{2i}}{\beta_i} - b\cdot\left(\frac{X_i + \dfrac{c_{1i}}{c_{2i}}}{a + b\sum_{k=1}^{n}\dfrac{c_{1k}}{c_{2k}}}\right)\right] + \sum_{j\neq i}\beta_j\cdot\left[-1 - b\cdot\left(\frac{X_i + \dfrac{c_{1i}}{c_{2i}}}{a + b\sum_{k=1}^{n}\dfrac{c_{1k}}{c_{2k}}}\right)\right] - \frac{1}{b} - \left(\frac{X_i + \dfrac{c_{1i}}{c_{2i}}}{a + b\sum_{k=1}^{n}\dfrac{c_{1k}}{c_{2k}}}\right) \tag{12.49}$$

This is a set of non-linear equations. The solution to this set of equations is not unique. In particular, in the case of a duopoly Equations (12.49) represents two iperboles in $[\beta_1, \beta_2]$. However, if it is assumed that the strategic supply functions have a positive slope varying between the slope of the marginal cost bid *(c_{2i})* and the slope of the Cournot solution *(c_{2i} + b)*, it only determines one acceptable solution.

After solving Equation (12.49), it is possible to determine the equilibrium value of the spot price by using Equation (12.38) and then the real power output of each producer by Equation (12.31). The set of Equations (12.49) can be expressed through a vectorial syntax as follows:

$$\mathbf{F}(\mathbf{k}, \mathbf{X}, \mathbf{c}_1, \mathbf{c}_2, a, b) = \mathbf{0} \tag{12.50}$$

where:

$$\mathbf{F} = [\mathrm{F}_1, F_2, \ldots, F_n] \tag{12.51}$$

$$\mathbf{k} = [k_1, k_2, \ldots, k_n], \text{ with } k_i = 1/(\beta_i \cdot c_{2i}) \qquad i=1,\ldots,n \tag{12.52}$$

$$\mathbf{X} = [X_1, X_2, \ldots, X_n] \tag{12.53}$$

$$\mathbf{c}_1 = \lfloor c_{1_1}, c_{1_2}, \ldots, c_{1_n} \rfloor \tag{12.54}$$

$$\mathbf{c}_2 = \lfloor c_{2_1}, c_{2_2}, \ldots, c_{2_n} \rfloor \tag{12.55}$$

If it is assumed that the real power amount traded in the bilateral contracts **X**, the cost coefficients $\mathbf{c}_1$ and $\mathbf{c}_2$ and the demand parameters a and b are given, then Equation (12.50) is simplified as follows:

$$\mathbf{F}(\mathbf{k}) = \mathbf{0} \tag{12.56}$$

whose solution provides the values of the unknowns **k** which are used to determine the slopes and the minimum values of the strategic supply functions.

It should be noted that in the model developed both fixed costs c_{0i} and the strike price z do not affect the equilibrium of the market. In particular, if the fixed costs were too high or the strike price too low negative profit can be obtained, that means losses and therefore Equation (12.56) would turn into a losses minimization problem.

Moreover, capacity constraints are also taken into account. Although these constraints do not affect the strategic profit maximising bids, they influence the value of the spot price p, which is higher if some constraints are violated. To this purpose, first the equilibrium solution, in terms of strategic bids, price and real power outputs (β_i^*, p^*, P_i^*) is evaluated without taking into account the presence of constraints. Then it is checked whether the real power output of any generator exceeds its capacity constraints. If the h th generator exceeds its limit, the output is set equal to the violated limit $(P_{\max_h})$ and the new equilibrium price is evaluated as follows:

$$p_{new}^* = a - b \cdot \left(P_{\max_h} + \sum_{k \neq h} P_k \left(p_{new}^* \right) \right), \; p_{\text{new}}^* = a - b \left[P_{\max_h} + \sum_{k \neq h} \left(\beta_k \cdot p_{new}^* - \frac{c_{1k}}{c_{2k}} \right) \right] \tag{12.57}$$

$$p_{new}^{*} = \frac{a - b \cdot \left(P_{\max_h} + \sum_{k \neq h} c_{1k} / c_{2k} \right)}{1 + b \cdot \sum_{k \neq h} \beta_k}$$

It should be noted that the new equilibrium price is higher than the one resulting from solution of the unconstrained problem.

12.5 Case Studies and Results Analysis

The energy marketplace under consideration is composed of ten generating companies and each of them is assumed to own ten generators. It is assumed that all the generating units present in the system are thermal and that the upper capacity constraints are in the range [400:850] MW. Then the available capacity of a single producer varies in the range [4000:8500] MW, which means that the total average capacity available in the electricity market under consideration is around 60000 MW.

It is assumed that the cost function of the j th generator belonging to the i th generating company has a quadratic expression:

$$C_{i_j}\left(P_{i_j}\right) = c_{0i_j} + c_{1i_j} \cdot P_{i_j} + \frac{1}{2} c_{2i_j} \cdot P_{i_j}^2 \tag{12.58}$$

Moreover, under the assumption that all the generating units belonging to the i th producer have the same characteristics and that, owing to the high start-up costs of thermal units, all of them are on at the same time. The cost function of the company can be expressed as follows:

$$C_{i_{eq}}\left(P_i\right) = c_{0i_{eq}} + c_{1i_{eq}} \cdot P_i + \frac{1}{2} c_{2i_{eq}} \cdot P_i^2 \tag{12.59}$$

The coefficients $c_{0i_{eq}}$, $c_{1i_{eq}}$ *and* $c_{2i_{eq}}$ are obtained imposing the following equivalencies:

$$\begin{aligned} C_{i_{eq}}\left(P_i\right) &= 10 \cdot C_{i_j}\left(P_{i_j}\right) \\ P_i &= 10 \cdot P_{i_j} \end{aligned} \tag{12.60}$$

By substituting Equations (12.58) and (12.59) into Equation (12.60) and differentiating it twice with respect to the real power output of the single generator P_{ij}, the expression for the company cost function can be expressed as follows:

$$c_{0i_{eq}} = 10 \cdot c_{0i_j}$$
$$c_{1i_{eq}} = c_{1i_j} \qquad (12.61)$$
$$c_{2i_{eq}} = (1/10) \cdot c_{2i_j}$$

Table 12.1 gives the cost coefficients of the ten generating companies present in the market under consideration. In the last two columns of the table, the maximum capacity available for each company and the maximum amount of real power that each company decides to trade in bilateral contracts are given.

It is assumed that the total load demand of the system is a linear function of the price. Therefore the inverse demand function can be expressed as follows:

$$p = a - b \cdot D \qquad (12.62)$$

where the values of the two coefficients a and b are given below:

$a = 97$ £/MWh
$b = 0.0016$ £/MW^{2}h

These particular values are chosen since they allow an elasticity value of 0.27 to be reached at the equilibrium point.

Two different sets of simulations of the behaviour of the market have been analysed. In both the simulations, it is assumed a value of $\overline{z} = ?0.57$ for the strike price z. Moreover it is assumed that the amount of real power traded in the bilateral contracts varies with increment $X_{\max}/20$.

In the first set of simulations it is supposed that only producer 1 has set bilateral contracts. Figures 12.9 12.13, the results obtained report.

In particular, in Figure 12.9 the behaviour of the slope, β_1,of the optimal bid submitted by producer 1, of the equilibrium spot price value, of the real power output sold by producer 1 and its relative profit are illustrated as the amount of real power traded in bilateral contracts by producer1 increases.

It should be noted that the value of $\beta_1$decreases as X_1 increases. In fact, since producer 1 has a bilateral contract, it is also willing to sell more electricity than in the absence of the contract if the value of the equilibrium spot price decreases. That is because the price it will be paid is the strike price, so that a decrement in the spot price does not affect its profit.

Moreover if the strike price is lower than the spot price, the producer is highly interested in a decrement of the spot price. For this reason the spot price tends to decrease as the amount of real power traded in bilateral contracts increases. The result is an increment of the total power sold and, therefore, an increment of the market efficiency.

Table 12.1 . Generating companies parameters

Producer	$c_{0_{eq}}$ [£/h]	$c_{1_{eq}}$ [£/MWh]	$c_{2_{eq}}$ [£/MW^{2}h]	$P_{\max}$ [MW]	$X_{\max}$ [MW]
# 1	15500	24	0.0009	8500	7500
# 2	14000	24.5	0.00095	8000	6000
# 3	12500	25	0.001	7500	5500
# 4	11000	25.5	0.00105	7000	5000
# 5	9500	26	0.0011	6500	4500
# 6	8000	26.5	0.00115	6000	4000
# 7	6500	27	0.0012	5500	3500
# 8	5000	27.5	0.00125	5000	3000
# 9	3500	28	0.0013	4500	2500
# 10	2000	28.5	0.00135	4000	2000

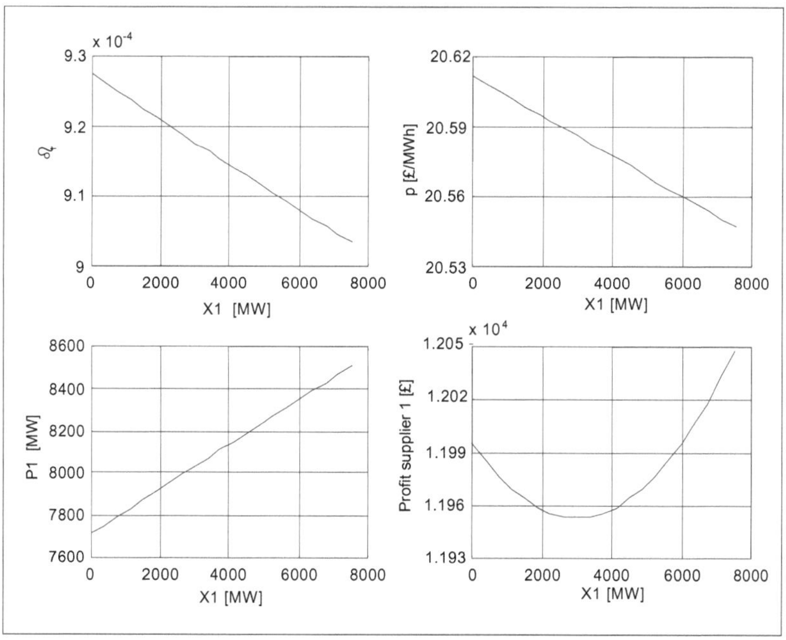

Figure 12.9. Strategic bid slope of producer 1, equilibrium price, real power output and profit of prducer 1

Figure 12.9d illustrates the behaviour of the profit as the amount of power traded in bilateral contracts increases. It should be noted that initially the profit tends to decrease as $X1$ increases, unless the spot price reaches the value $\overline{z}$ of the strike price of the contracts. This means that producer 1 at the beginning does not have any advantages from setting a contact and tries to lower its supply function to limit the disadvantages. When the spot price reaches the strike price, any further increment of $X1$ yields to an increment of producer 1's profit.

Figure 12.10 illustrates the behaviour of the slope of the supply function of all the competitors. It can be observed how the optimal bid of producer 1 is the most convenient and the one with the minimum slope.

Figure 12.11 shows that other competitors, producers 2, 3, 9 and 10, also lower their bid owing to the presence of bilateral contracts for producer 1.

The explanation is that because producer 1 decreases its supply function's slope, it will sell more power, implying a smaller residual demand for the other competitors.

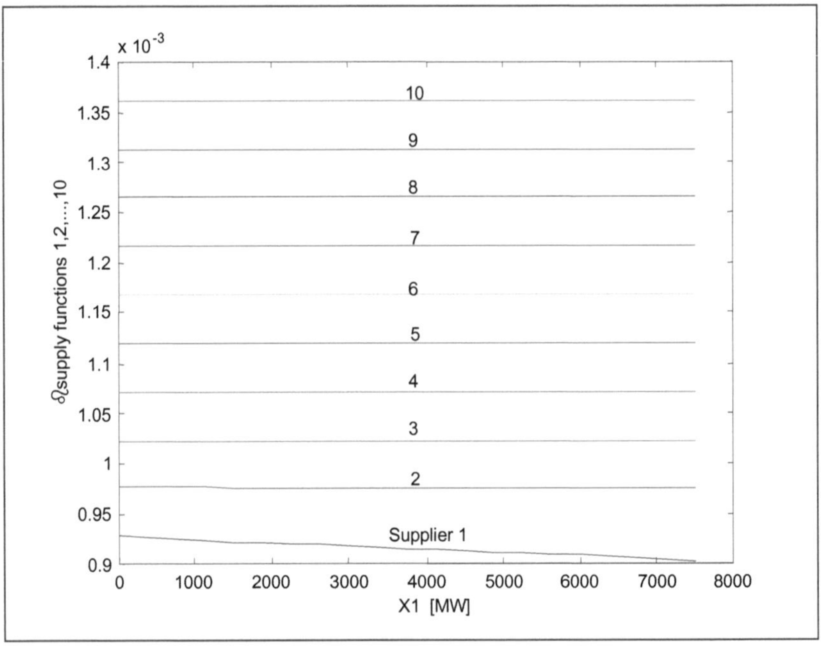

Figure 12.10. Strategic producers' bid slope

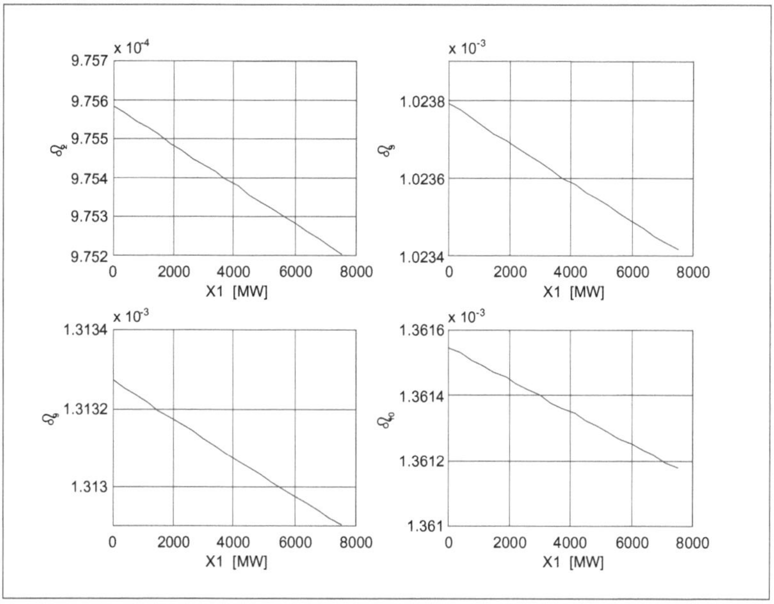

Figure 12.11. Slope of the strategic bid of producers 2, 3, 9, 10 10

To avoid this, they react and lower their supply function's slope too. In particular, in Figure 12.12 it can be noted how the real power amount sold by suppler 1 increases as the power traded in bilateral contracts increases, decreasing the output sold by the other companies.

Figure 12.13 illustrates the behaviour of the profits. By setting bilateral contracts, producer 1 not only can increase its profit but it can damage its competitors' income, lowering their profits.

In the second set of simulations, it is assumed that all the producers sharing the market have set bilateral contracts. In order to achieve a better understanding of the influence of private contracts on the behaviour of the different competitors, the results of this set of simulations are expressed in relative value, using as set values the one that would be obtained in a perfect competitive market (PCM). The results are reported in Figure 12.14.

It should be noted that the values of slopes of supply functions of the spot price and of the profits are always higher than in a PCM condition, while this does not happen for the real power output. Moreover, producers with bigger values of the slope of the supply functions, which means that they bid at a higher price than in a PCM, are the ones with lower marginal costs. This implies that they have the biggest market power. Producers with higher marginal costs bid at a higher price, looking at absolute values, but closer to their marginal costs. The reason is that producers with higher costs in order to be allowed to sell a certain amount of power, owing to the merit order criterium, are forced to bid lower, reaching the profit maximising condition. However, the effect of bilateral contracts as the traded amount of real power increases, is to lower the optimal bids closer to marginal costs offers.

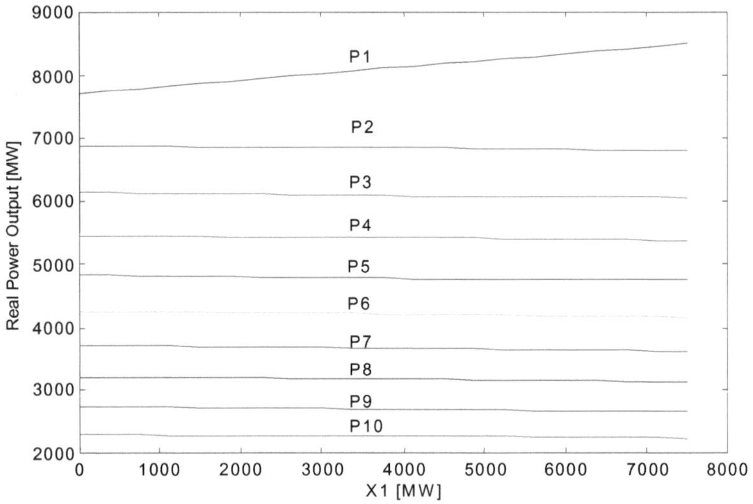

Figure 12.12. Real power output of all the producers

The amount of real power sold by each producer is not always higher than in a PCM condition. The reason is that since strategic supply functions have a bigger slope and higher value minimum price, compared to the marginal costs supply functions, the whole amount of real power sold tends to decrease, increasing the value of the spot price. In particular, it should be noted that only the power sold by bigger producers, with lower marginal costs, is less than in a PCM condition. On the contrary, producers with higher marginal costs can sell more in an oligopoly than in a perfect competitive market. Once more the presence of bilateral contracts tends to shift the whole scenario closer to the PCM condition.

Finally, by analysing the behaviour of the profits, it is revealed that producers which take the biggest advantage trading in an oligopoly, rather than in a perfect competitive market, are the mid merit generators. They are characterised by reasonable marginal costs which allow them to sell more of power than in a PCM and their profits can also be increased more than twofold. The analysis carried out shows that bilateral contracts are a very useful tool to shift an oligopolistic scenario closer to a Perfect Competitive Market condition.

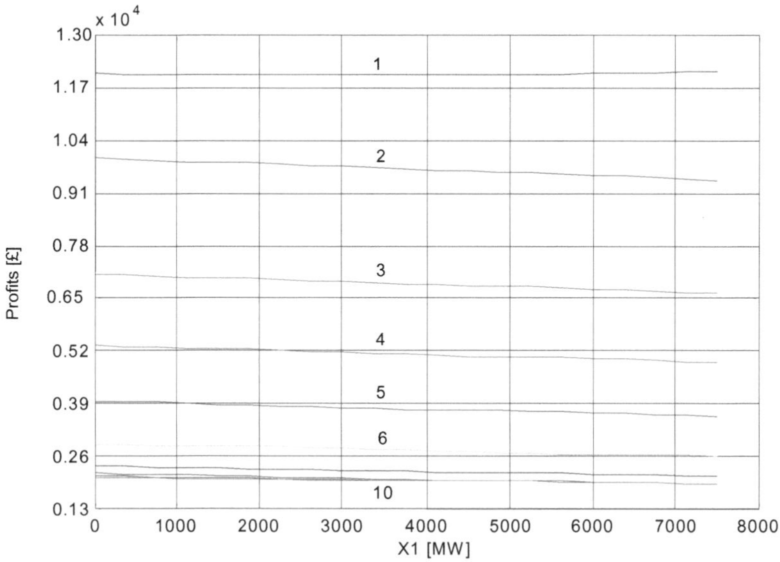

Figure 12.13. Producers' profit

12.6 Conclusions

In this chapter, a methodology to simulate the strategic behaviour of generating companies in an oligopolistic electricity market has been proposed. In this context, each firm sharing the market does not bid at marginal cost as it is in a perfect competitive market, profit-maximising strategies are carried out. In the methodology developed, the general case of an asymmetric oligopoly has been analysed and the strategic behaviour of producers investigated, assuming linear supply functions. Moreover, the proposed methodology allows the presence of private contracts, set up as Contracts for Difference (CfDs), to be taken into account, evaluating their effects on the market equilibrium conditions. The results of the analysis have been expressed in terms of market clearing price, profit-maximising value of the supply functions' slope, and hence real power output sold and profit made, of each producer sharing the market. Marginal cost bids have been evaluated and considered as desired supply function in order to determine the "market power" of producers, in terms of distance of the strategic bids from the desired marginal cost bids.

The effectiveness of the proposed model has been tested on a wholesale electricity market composed of ten different generating companies.

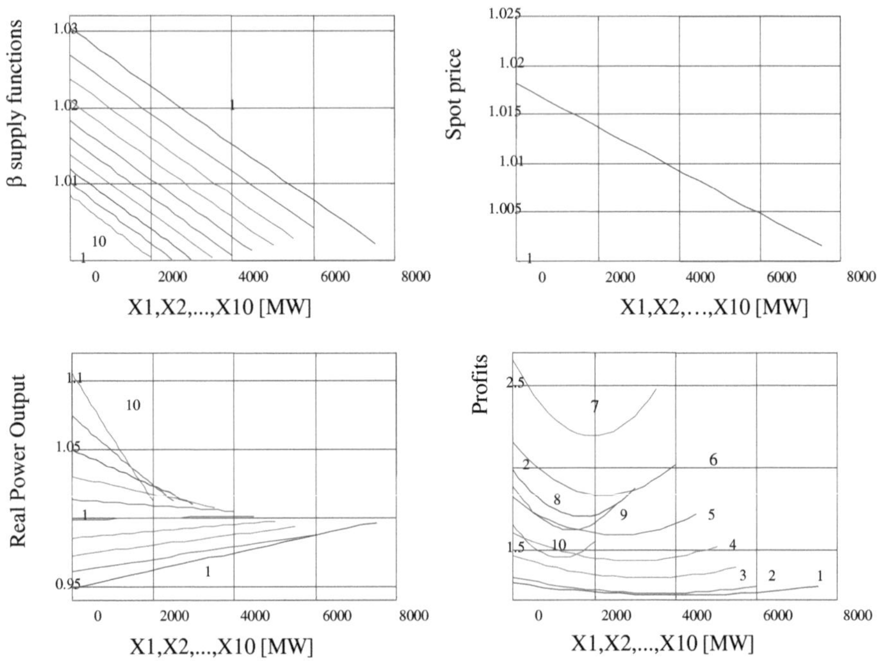

Figure 12.14. Simulation results related to a Perfect Competitive Market condition

12.7 References

[1] M. J. Arentsen, R. W. Künneke : "Economic Organization and Liberalization of the Electricity Industry", Energy Policy, Vol. 24, n. 6, pp. 541-552, 1996.

[2] "Final Order 888: Promoting Wholesale Competition Through Open Access Non-discriminatory Transmission Services by Public Utilities", Federal Energy Regulatory Commission, April 24, 1996.

[3] "Directive 96/92/EC of the European Parliament and the Council concerning Common Rules for the Internal Market in Electricity", Official Journal of the European Commission, L27 (30/1/97), pp. 20-29.

[4] C. Bier :"Network Access in the Deregulated European Electricity Market: Negotiated Third-Party Access vs. Single Buyer", CSLE Discussion paper 9906, Univ. of Saarland, June 1999.

[5] "Technical Issues, Methods and Tools in Alternative Electric Market Structures", EPRI Final Report on RP8501-1, 1992.

[6] W. Hogan, L. Ruff: "Reshaping the Electricity Industry: Competitive Market Structure and Regulatory Policy", Prepared for Wisconsin Electric Power Company, November 1994.
[7] J. S. Gans, D. Price, K. Woods: "Contracts and Electricity Pool Prices", Australian Journal of Management, Vol. 23, No. 1, pp. 83-96, June 1998.
[8] A. Powell: "Trading Forward in an Imperfect Market: the case of Electricity in Britain", The Economic Journal, Vol. 103, pp. 444-453, March 1993.
[9] N-H. M. Von Der Fher, D. Harbord: "Spot Market Competition in UK Electricity Industry", Economic Journal, 1993, Vol. 103, pp.531-46.
[10] D.M. Newbery: "Power Markets and Market Power", The Energy Journal, Vol. 16, pp. 39-66, 1995.
[11] R. W. Ferrero, J. F. Rivera, S. M. Shahidehpour: "Application of Games with Incomplete Information for Pricing Electricity in Deregulated Power Pools", IEEE Trans. on Power Systems, Vol. 13, No. 1, pp. 184-189, Feb. 1998.
[12] D. L. Post, S. S. Coppinger, G. B. Sheble': "Application of Auctions as a Pricing Mechanism for the Interchange of Electricity", 95 WM 175-0 PWRS.
[13] D. P. Mendes, D. S. Kirschen: "Assessing Pool-Based Mechanism in Competitive Electricity Markets, Proc. of "IEEE-PES 2000 Summer Meeting, Seattle", USA, July 2000, pp. 2195-2200.
[14] A. Losi, M. Russo: "A Simulation Tool for Evaluating Technical and Economical Issues in the Deregulated Electric Power Industry", Proc. of IEEE-PES 2000 Summer Meeting, Seattle, USA, July 2000, pp. 2242-2247.
[15] G. B. Shrestha, S. Kai, L. K. Goel: "An Efficient Power Pool Simulator for the Study of Competitive Market", Proc. of IEEE-PES 2000 Summer Meeting, Seattle, USA, July 2000.
[16] C. Li, A. J. Svoboda, X. Guan, H. Singh, "Revenue Bidding Strategies in Competitive Electricity Markets", IEEE Transactions on Power Systems, Vol. 14, No. 2, May 1999, pp. 492-497.
[17] Geerli, R. Yokoyama, L. Chen: "Negotiation for electricity Pricing in a Partially Deregulated Electricity Market", Proc. of IEEE-PES 2000 Summer Meeting, Seattle, USA, July 2000, pp. 2223-2228.
[18] J. Kumar, G. Sheble': "Auction Market Simulator for Price Based Operation", IEEE Transactions on Power Systems, Vol. 13, No. 1, Feb. 1998, pp. 250-255.
[19] R. J. Thomas, T. R. Scheider: "Underlying Technical Issue in Electricity Deregulation", PSERC Publication, 97-18.
[20] I. Otero -Novas, C. Meseguer, J. J. Alba: "An Iterative Procedure for Modelling Strategic Behaviour in Competitive Generation Markets", Proc. of 13th PSCC, 1999, pp. 251-257.
[21] J. W. Lamont, S. Rajan: "Strategic Bidding in an Energy Brokerage", IEEE Transactions on Power Systems, Vol. 12, No. 4, Nov. 1997, pp. 1729-1733.
[22] W. J. Elmaghraby: "Multi-unit Auction with Complementarities: Issues of Efficiency in Electricity Auctions", PSERC Publication, 98-02.
[23] A. K. David: "Dispatch Methodologies for Open Access Transmission Systems", IEEE Transactions on Power Systems, Vol. 13, No. 1, Feb. 1998, pp. 46-53.
[24] R. J. Green and D. M. Newbery: "Competition in the British Electricity Spot Market", Journal of Political Economy, 1992, Vol.100, n.51, 929-953.
[25] P. D. Klemperer and M. A. Meyer: "Supply Function Equilibria in Oligopoly under Uncertainty", Econometrica, Vol.57, n.6, Nov 1989, 1243-1227.

[26] E. Elia, A. Maiorano, Y.H. Song, M. Trovato: "Novel Methodology for Simulation Studies of the Strategic Behavior of Electricity Producers", Proc. of IEEE-PES 2000 Summer Meeting, Seattle, USA, July 2000.

13. Bidding Problems in Electricity Generation Markets

Y. He, Y.H. Song and X.F. Wang

In this chapter we discuss bidding strategies of GENCOs in electricity generation markets. First of all we give an introduction to the generation auction markets in 13.1 and then present the methodologies of bidding decision-making in 13.2. Section 13.3 and 13.4 describe two bidding models proposed for pool-based single-period market and day-ahead market respectively.

13.1 Generation Auction Markets in Electricity Markets

13.1.1 Auction Mechanism

McAfee says that an auction is a market institution with an explicit set of rules determining resource allocation and prices on the basis of bids from market participants [1]. So resource allocation and price determination are two basic functions of the auction institution, especially for products that have no standard or certain value. From this point of view, auction is an ideal pricing mechanism for electricity markets since the price of electricity depends on the supply and demand conditions at a specific moment of time.

13.1.1.1 Standard Auction Formats

Four standard formats of auction are mentioned in [1] when a unique item is to be bought or sold: English auction (also called the oral, open or ascending-bid auction); Dutch auction (or descending-bid auction); the first-price sealed-bid auction; and the second-price sealed-bids (or Vickley) auction.

In an English auction the price is successively raised until only one bidder remains. A Dutch auction is the converse of the English Auction. The auctioneer calls an initial price and then lowers the price until one bidder accepts the current price. The essential feature of the English auction and Dutch auction is that, at any point of time, each bidder knows the level of the current best bid. With the first-price sealed-bid auction and the second-price sealed-bids auction, however, potential buyers submit sealed-bids. Under the second-price sealed-bid auction a bidder submits sealed bids having been told that the highest bidder wins the item but pays a price equal not to his own bid but the second-highest bid. That is the only difference with the first-price sealed-bid auction. The basic difference between the first-price sealed-bid auction (or second-price auction) and the English auction is that, with the English auction, bidders are able to observe their rival's bids and accordingly, if they choose, revise their own bids; with the sealed-bid auction, each bidder can submit only one bid.

Electricity, as a special commodity product, cannot be treated as a unique item however. In current electricity markets, either a single-side auction (the generator side) or a double auction is used. It should be noted here that, no matter single-side auction or double auction, the seller has no information about the buyer and vice versa, i.e. generators do not know the level of current demand and customers do not know the available capacity of generators. This causes the more complication and uncertainties in bidding for both seller and buyer in electricity markets, where they face more risk than in an ordinary commodities auction market. Furthermore, electricity auction markets may have more than one commodity being bid for simultaneously, for example real-time energy, operating reserve and other ancillary service products may be bounded at the same time in an auction market. These differences need an auction mechanism that is specifically suited to electricity auction markets.

Currently nearly all the electricity markets have relied on the one-shot sealed-bid or two-step sealed-bid auction. Currently the one-shot sealed-bid auction is used less than the two-step sealed-bid auction. Compared with one-shot bidding, the two-step auction gives selected bidders one more chance to reconsider the bids and offers; it is usually the case that these selected bidders revise the bids or offers due to system transmission constraints. The market regulator releases the proposals of revision to these selected bidders who consider with due diligence costs before submitting the revised bids or offers as the second step.

Since the sealed-bid is accepted in nearly all markets, the market regulator releases the schedules for each registered facility only to the registered market participants for that registered facility. Even after the first step in a two-step sealed-bid auction, the generator has no clue about other competitors' previous bids and resulting schedules. So bidders must choose their bids without having any information about the strength of their competitors' interest in the real-time energy and operating reserve. Bidders do not know the bidding strategies of other competitors even after bidding. Such sealed-bid bidding systems are intended to restrict the gaming in electricity markets. However, what kind of bidding system is transparent for all the bidders needs further investigation and to tested in real system. The next section will discuss in some detail why multi-round bidding is not accepted in current electricity markets and what re-bidding is and how it works.

13.1.1.2 Single-round Bidding and Multi-round Bidding

Single-round bidding is employed in nearly all current electricity markets all over the world. To the authors' best knowledge, only Ontario IMO [2] and New Zealand [3] use re-bidding rules very similar to multi-round bidding and Alberta uses multi-round bidding for Power Purchase Arrangements (PPA) [4].

Apart from there, California is encouraging multi-round bidding and has started research and testing [5]. The Power Exchange (PX) in California has sponsored two separate research projects. One was done in June 1998, and the other took place in November 1998.The June research explicitly laid out the benefits from iterations and identified three costs of implementation: *(1) Feasibility Constraints.* Feasibility constraints address whether there is enough time for a sufficient number of iterations since iterations are time-consuming and the research showed that schedules would have to be very tight. *(2) Strategic Behaviour.* Strategic behaviour addresses possible gaming of the market rules. The research identified gaming opportunities involving fictitious bids that may have to be addressed before implementation. *(3) Transactions Costs.* Transaction costs relate to how much more participants would have to pay to undertake iterations profitably. Research showed that transactions costs are significant and are largely born by participants. The costs imposed by iterations include increased staff and computer resources to process the information. The November research results indicated that the costs of implementing iterations outweighed the benefits they would produce. So the iteration research in PX drew the conclusion that more testing of iterations would have provided more information about the likely impacts of implementing these rules, for instance, the impacts of strategic behaviour of suppliers and consumers.

Though multi-round bidding is still under consideration before being put into practice, the re-bidding system has been fully accepted in many electricity markets as in ROM, UK under PETA, NZEM, and IMO in Canada. The procedures and rules for re-bidding systems are varied, however. Detail is addressed in Section 13.1.2.

13.1.1.3 Simple Bids and Multi-part Bids

Bids formats are mainly of two types—simple bids and multi-part bids. Simple bids may have price/volume according to several price bands. Multi-part bids may include incremental costs, start costs and no-load costs etc. Because of the complexity of multi-part bids formats, currently the simple bids format is accepted in most electricity markets. A typical application of the multi-part bids format is in the UK under PETA and a typical simple bid format in Nord Pool. Some markets accept both, such as PJM.

The details of bid formats and bid components all over the world are discussed in Section 13.1.2 and a comparison is made in Table 13.2.

13.1.2 Existing Auction Mechanism -- Trading Arrangement and Pricing Mechanism

13.1.2.1 UK Market

Since 1990 a pool-based market has been established in UK (England and Wales) and on 27 March 2001, the New Energy Trading Arrangement (NETA) went live. We look at the UK market under the Previous Energy Trading Arrangement (PETA)
and NETA respectively.

13.1.2.2 PETA [6],[7]

1. Bids Format in PETA

The bids submitted by generators under PETA are typical multi-part bids, which include start-up price, no-load price and incremental prices as shown in Figure 13.1. Incremental prices can be dependent on the level of generation and may be given in up to three parts. Besides the offered availability and operating characteristics, which include run-up rates, run-down rates, minimum generation etc., are also needed.

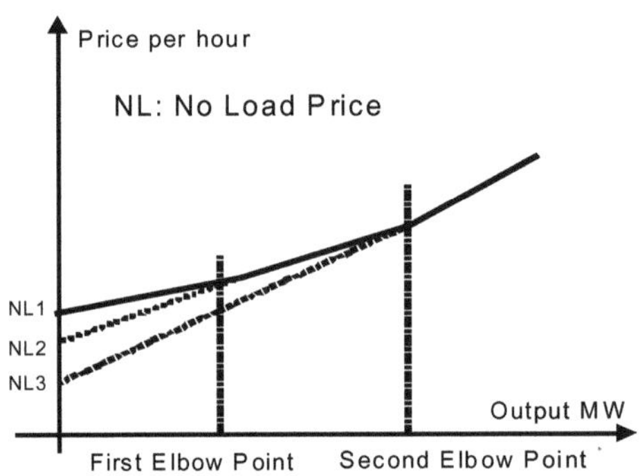

Figure 13.1. Diagram of bids offer in UK under PETA

2. Pricing Mechanism

A uniform pricing mechanism is implemented under PETA and three prices are calculated for each schedule day. The pool mechanism establishes the generation merit order to meet the forecast demand at the day-ahead stage. This schedule was called the Unconstrained Schedule. Generally, the price of the most expensive unit scheduled to meet forecast demand in each half-hour sets the price of energy, called the *System Marginal Price (SMP).* Based on SMP *Pool Purchase Price (PPP),* and *Pool Selling Price (PSP)* are calculated. PPP includes the SMP and a second element called the Capacity Payment, which equals *LOLP*(VLL - SMP),* where LOLP is the Loss of Load Probability and VLL is the Value of Lost Load. To keep daily payments through the pool balanced, an uplift element is added to

PPP to produce PSP, i.e. *PSP=PSP + UPLIFT*. These Uplift Costs reflect the difference between the cost of 'on the day' operation and the costs associated with the Unconstrained Schedule.

13.1.2.3 NETA [8][9]

Since November 1998 the main focus of work in the UK market has been on devising rules for the Balancing Mechanism and the associated settlement process. Balancing and settlement rules need to ensure efficient balancing of the system by the system operator, while encouraging generators and suppliers to contract ahead for most of their requirements in forward, futures and short-term markets. Since implementation of NETA, NGC has calculated that the Balancing Mechanism may, on average, require around less than 2 GW (5% of average demand) to be taken.

1. Balancing Mechanism

Initially, the Balancing Mechanism for a half-hour settlement period will open three and half hour before the start of the half-hour, which is called Gate Closure, seen in Figure 13.2[1]. The Balancing Mechanism will provide a basis whereby NGC, as system operator, can accept offers/bids of electricity at very short notice to solve the imbalance between forecasted system demand and actual demand.

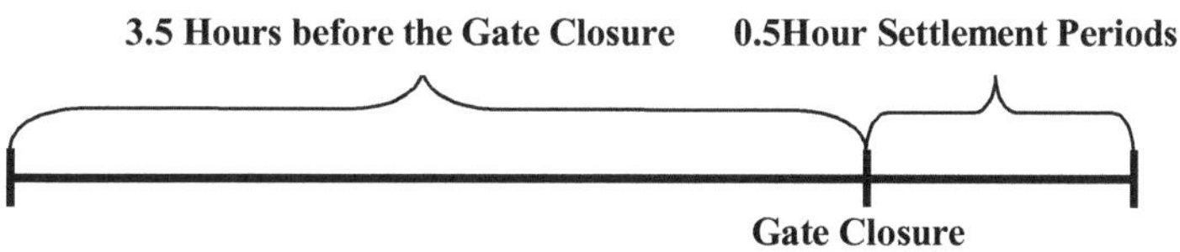

Figure 13.2. Gate closure of the balancing mechanism in UK under NETA

2. Offers and Bids Format

The submission of an offer or a bid indicates a willingness to increase or decrease the volume of active generation from the predefined level of its FPN. Offers may be delivered through increasing generation. Conversely, bids may be delivered through decreasing generation. Offers and bids will be submitted in 'Bid Offer Pairs'. Each Bid Offer Pair will consist of two prices – an offer price and a bid price, and information on the range of output (relative to the BM Unit's FPN) for which the prices apply. Figure13.3 shows an example of bid/offer pairs.

In principle, participants should be permitted to submit bid offer pairs at any time. However, for practical reasons, it is proposed that for initial implementation, bid offer pairs must be submitted by Gate Closure. Participants will effectively be able to withdraw bid-offer pairs (either partially or wholly) by reducing their Maximum Export and Import Limits (MEL/MIL). Conversely, by increasing their Maximum Export and Import Limits, they can effectively increase the available capacity of bid offer pairs.

[1] Gate Closure has been changed to 1 hour ahead since 2 July 2002.

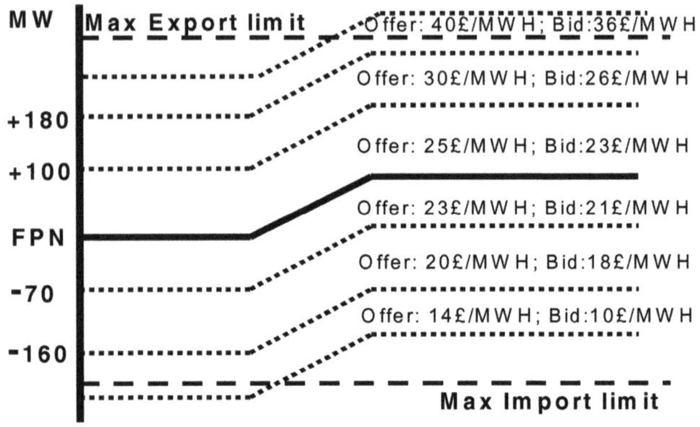

Figure 13.3. Bid/offers pairs in the UK market under NETA

13.1.2.4 California [10]

Power Exchange (PX) is an auction for the generators seeking to sell energy and for loads that are not otherwise being served by bilateral contracts. The PX is responsible for determining hourly Market Clearing Prices (MCP), which is usually the intersection of the aggregated supply curve and aggregated demand curve. The bids format accepted in California is a simple-bid format. As several crises have occurred in California since summer 1999, FERC has adopted many price remedies aimed at fixing the malfunctioning electricity market, which included ending the California Power Exchange's buy/sell requirements, penalising under-scheduling of loads, and monitoring spot prices above $150 on an interim basis etc.

13.1.2.5 PJM [11]

1. Bids format since 1st April 1999

Generating resources within the PJM control area have types of offers: cost-capped offers; historic LMP-capped offers and market-based offers. Historic LMP-capped offers use historic PJM LMPs at the generator bus where the energy is injected. This value is determined by calculating the average LMP at the generation bus during all hours over the past six months in which the resource was dispatched above minimum. For resources with less than 30 hours of qualifying economic operation within the last six months, the price cap must be calculated using the cost-cap method. Also generating resources within the PJM control area can submit generation offers that are market-based, not capped at cost. See Table 13.1 for bid components of these three types of offer.

2. Trading and pricing

ISO, as system coordinator in PJM, accepts both bilateral schedules and voluntary bids of market participants. The ISO implies an economic, security-constrained central dispatch using these schedules and bids. The Local Marginal Prices (LMPs) are used for settlement that can differ substantially across locations when the transmission system is constrained.

Table 13.1. Generator's bid components in PJM

	Cost-capped & historic LMP-capped offers	Market-based offers
Start-up	Daily; 3 types (hot, intermediate, and cold)	Optional; once every 6 months; 3 types (hot, intermediate, and cold)
No-load	Yes	Optional; once every 6 months
Incremental cost	Energy from min to max	Energy from 0 to max
Maintain minimum	Yes	No
Cooling requirement times	Basic	Optional; once every 6 months
Maximum offer	Based on cap methodology	$1000/MWh

13.1.2.6 Nord Pool [13]

1. Bids format in Nord Pool

The bids format by generators in Nord Pool is similar to the bids submitted in California, that is a simple bid format. The difference is bids in the form of continuous increasing linear curve in Nord Pool and not an increased stair-cased curve as in California. The prices in the bid and offer form are treated as breakpoints on a continuous bid and offer curve with linear interpolation between the points.

2. Price calculations: system price (ps); area price (po) and capacity fee

The participants' collective bids and offers are grouped together in an offer curve sale and in a demand curve purchase. The trade price is set according to the balance price at the meeting point between offers and demand (the equilibrium point). A system price (ps) is calculated at first based on the condition that there are no transmission restraints in the central grid. If the calculations show that the power flow between two or more bidding areas exceed capacity limits, two or more area prices are calculated (po). The capacity fee (pc) in each price area is defined as the difference between the system price and the area price (pc= ps - po).

Figure 13.4 shows the relationship among system price, area price and capacity price. If the price calculation shows that the capacity between the bidding areas is not exceeded, there will be only one price area where the area price equals the system price and the capacity fee is zero. If the power flow between two areas exceeds grid capacity, the price in the surplus area is reduced to that the low-priced area and increased in the deficit area to that the high-priced area.

The physical power flow will always go from the low-priced area to the high-priced area. This means that Nord Pool's purchases and sales of capacity between two price areas normally will provide an income that corresponds to capacity (PoH-PoL). This income is part of the income for the grid owners.

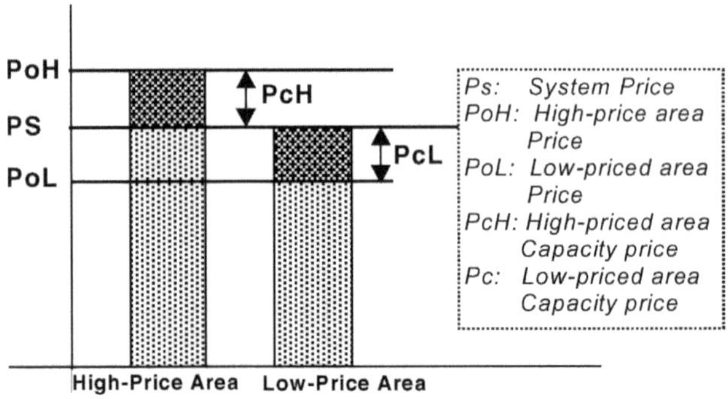

Figure 13.4. Capacity fee, system price and area prices in Nord Pool

13.1.2.7 Australia National Electricity Market (NEM) [14]

The NEM commenced operation on 13 December 1998. All the electricity output from generators is pooled, and then scheduled to meet electricity demand.

Unit energy dispatch bid/offer data is price band. The total maximum capacity registered for a dispatchable unit must be submitted in the form of ten price bands (herein called Energy Bands). Each Energy Band associates a block or quantity of electrical power output with a local price. Band Prices (in $/MW) cannot be re-bid, however Band Quantities (in MW) are re-biddable. Beside the Energy Bands there are some other bid components in NEM such as Energy Availability (in MW), Energy Ramp Rates, Daily Energy Limit (in MWh) and Fixed Loading and so on.

Daily bids must be received before 12:30pm on the day before supply is required. Re-bids are submitted after 12:30pm and can be received until approximately five minutes prior to dispatch. From all offers submitted, NEMMCO selects the generators required to produce power and at what times throughout the day based on the most cost-efficient supply solution to meet specific demand. Usually the cheapest generation unit will be selected to meet demand. When the scheduling of generators is constrained by the capacity of the interconnectors between the regions, higher price generators within the region will be called on to meet this demand. This is said the spot price of electricity between regions may be varied.

13.1.2.8 New Zealand [3]

M-co previously (EMCO) was established in 1993 by the electricity industry to develop and operate the New Zealand wholesale electricity market. NZEM prices are determined by the competitive action of buyers and sellers in the market. There are a total of 480 nodal grid exit and injection points in the New Zealand electricity system. Pricing information is collated via COMIT on the 244 nodes included within the ambit of NZEM. The nodal pricing methodology adopted by NZEM enables a price for energy, reserves and transmission losses to be established for each node. Prices are calculated for 48 half-hour trading periods every day of the year. This represents a massive inflow of information. NEZM accepts a simple bids format.

13.1.2.9 Ontario Electricity Market in Canada [2]
Two major electricity markets are operating in Canada, Ontario and Alberta. We discuss here only the Ontario market. A simple-bid format is implemented in the Ontario Market. As an incremental offer, each energy offer must contain at least two and may contain up to 20 price quantity pairs for each dispatch hour. Except for incremental offer, the generators submit ramp up/down rate as well. The rebidding is allowed in IMO, participates may modify bids and offers without restriction up to 4 hours ahead of the current pre-dispatch hour and by no more than 10% up to 2 hours ahead of current time. Within 2 hours ahead of the pre-dispatch hour no changes to bids and offers are allowed without IMO approval. The IMO releases the pre-dispatch schedules and security constraints (or load curtailment) for each registered facility only to the registered market participants for that registered facility.

13.1.2.10 Summary
Based on the above discussion, a full comparison of auction mechanism is shown in Table 13.2.

13.1.3 Market Power in Generation Auction Markets

With the start of deregulation, generation is separated from the generation transmission distribution network, which was run as a single entity before restructuring. The vertical integrated monopoly is broken up and the transmission network provides open access to any generation unit without discrimination. The generation unit provides bids/offers to market regulator/system regulator instead of its real costs. The bidding strategies, used by generation companies with the goal of maximising their own profits, show various potential possibilities to exercise market power. The market power discussed in this section mainly focuses on that associated with generation companies. So, this section is arranged as below. The first part goes a definition of market power and some methods to measure market power. Some examples of the exercise of market power are discussed in the second part. The last part discusses how to mitigate market power.

13.1.4 Getting to Know Market Power

1. Definition of Market Power
Market power is simply the power that market participants hold to manipulate the market in their own favour. Exercising market power by a generation company can be in the situation that it procures more market share or bids up market price so as to snatch more profit when the market has less effective competition or the market restrains effective competition for some reason. This must be differentiated from the situation that the market faces a scarcity problem.

2. Exercise of Market Power
The reasons for the existence of market power can be various such as market concentration, congested transmission, market structure and market rules, and so on.

Table 13.2. Comparison of bid formats and pricing mechanisms worldwide

Market	Trading period	Pricing mechanism	Price settlement	Bids format	Rebidding opportunities
UK (PETA)	Half-hour	Uniform price (SMP) +CP and Uplift	Ex-ante	Multi-part	Only availability and dynamic data allowed
UK (NETA)	Half-hour	No	No	Offers/bids pairs above or below FNPs	Only volumes, MIL/ MEL and dynamic data allowed
California	Hourly	Uniform Price(MCP)	Ex-ante	Simple Bids	Revise schedule one hour ahead of real time
PJM	Hourly	Local Marginal Price (LMP)	Ex-ante	Multi-part/ simple bids	No details
New England[12]	Hourly	Uniform Price (ECP)	Ex-ante	Multi-part	Prices not allowed
Nord Pool	Hourly	Uniform+ Zonal Price	Ex-ante	Simple Bids	NO
Australia (NEM)	Half-hour	Uniform Price + regional price	Ex-post	Simple Bids + Ramping Rates	Prices are not allowed
New Zealand (NZEM)	Half-hour	Nodal Prices	Ex-post	Simple Bids + Ramping Rates	Both prices and volumes up to 2 hours ahead
Ontario (IMO)[2]	Hourly	Uniform Price (HOEP)	Ex-ante	Simple Bids + Ramping Rates	Both prices and volumes up to 4 hours

First, market concentration can lead to market power. The general case for market concentration is when there are a few large generation companies and they have large market share.

A large company can easily manipulate energy prices that are set far from its marginal cost as illustrated in [15]. In addition, a congested transmission network can lead to market power even when the market is not concentrated. Reference [16] shows that some suppliers can take advantage of the geographic location and transmission capacity limitation to exercise the market power. Furthermore, market structure and market rules are also important causes for some kind of exercise of market power, such as what pricing mechanism is implied (uniform price or pay-as-bids), if the efficient forward or future markets are established, the demand is elastic or inelastic, etc.

The reasons for market power need further investigation and attention because all markets have more or less loopholes that market participants could exploit to exercise market power.

13.1.5 Measuring Market Power

The extent to which market power exists is of great concern for market designers and system operators. This needs tools or methods to measure it. The traditional way is to measure market concentration. New methods deal with the actual existing strategic market power in electricity markets.

1. The Classic Methods: To Measure Market Concentration

The traditional way is to measure market concentration. The two well-known methods are the Four-firm Concentration Ratio (I4) and the Herfindahl Hirshman Index (HHI). The HHI is used more widely than the Four-firm Concentration Ratio. Apart from the I4 and HHI method, the Pivotal Supplier Index (PSI) is an additional method. The PSI calculates the frequency that some quantity from a given supplier is required to serve market demand. Under such conditions, the firm is a monopoly supplier for the portion of demand that cannot be served by any other firm. The PSI is defined as:

$$PSI_{it} = \begin{cases} 1 & if \ D_t - \sum_{j \neq i} GenCap_j - MaxIMPORTS > 0 \\ 0 & if \ D_t - \sum_{j \neq i} GenCap_j - MaxIMPORTS \leq 0 \end{cases} \tag{13.1}$$

Where *GenCapj* is the capacity of firm *j*, and *MaxIMPORTS* is the aggregate import capability into the region. The *PSI* calculates the percentage of time over a given period for which a firm achieves this pivotal status by summing the number of hours during a time period *T* (for example one year) in which that firm is pivotal as Equation#(13.2)

$$PSI_t = \frac{1}{T} \sum_{t=1}^{T} PSI_{it} \tag{13.2}$$

However, I4, HHI and PSI are only indexes to measure the market concentration. The reasons for market power can be other than market concentration as we mentioned above, so the index for market concentration is a poor indicator of potential for, or existence of, market power.

So what is the best method to measure the actual market power in a dynamic and strategic electricity market? This will be discussed in the next section.

2. Dynamic Methods to Measure Actual Market Power

It is important to measure the actual market power in a strategic electricity market as such a measurement of market power takes into account the actual network structure, potential (or actual) generation competition, specified market rules and market structure. Generally there are two approaches to measure actual market power: the first is ex-post analyses existing markets where the margin between the actual market price and marginal costs of production is the index of market power. The second method is to model and simulate the specified markets and to calculate equilibria. In the second method, the margin between prices of equilibria and the marginal cost of production is treated as the actual market power index for a specified market.

Ex-post Analysis of Existing Markets

In a perfect competitive mark*et al*l participants are assumed to be price-takers. The profit-maximizing response to such a competitive market is that a competitive generation company is always willing to sell a unit of output as long as its cost of selling that unit is less than the price it receives for that unit. Its offer price will always be its marginal cost, which will be the greater of its marginal production cost and its opportunity cost of selling the power elsewhere. Based on this principle, Reference [17] uses the margin between actual market price and the marginal cost of production to measure market power in the California market from June 1998 through September 1999.

This method overcomes the shortcoming of traditional HHI or PSI method. However, it measures market power based on the true costs of generators and historical data, so it has two drawbacks, one is how to proceed if the true costs of generators are not available, and second, it is a way of ex-post assessment.

Modelling and Calculating the Equilibria of Specified Markets

Another methodology, used by [18,19] to analyse market power, is to simulate the strategic behaviour of firms in the market. The interaction of participants can be modelled in three ways: Bertrand competition, Cournot competition, and supply-function competition.

The well-implemented Cournot model in [18,19] is based on the cost and production characteristics of the actual set of generators. Only the larger firms are modelled as Cournot competitors and very small firms are modelled as price-takers, both in their own behaviours and in how they are viewed by strategic players in the market. For each demand level, the Cournot equilibrium is calculated iteratively. It is to be noted that in each iteration only the output of each Cournot player is adjusted by maximising its own profit given that the others output remain unchanged. One big drawback is it might be multi-equilibrium by this simulation and underestimation or overestimation of market power may exist because the selected Cournot players might not be the true market power exerciser. Even the small companies, which are treated in this method as price-takers might exercise market power if they are in a special geographical location and specific operational situation.

13.1.5.1 Mitigating Market Power

Reference [20] and [21] present some approaches to mitigate market power. We summarise these as follows:

1. *Administrative and organisational approaches:* These include generation divestiture and internal re-organisation.
2. *Commercial and financial tools:* The most popular implemented is bilateral contract which can be financial bilateral contract and physical bilateral contract.
3. *Market rules related approaches:* These mainly include price caps and bid caps.
4. *Market structure related approaches:* The most widely used structure related approach is demand-side bidding.

13.1.6 Uncertainties and Risk Mitigation

Previously market participants face greater risks in electricity markets now than before. Many factors make market participants do so, such as privatisation of generation, transmission network and distribution, open-access to electric network, competitively bidding in electricity markets, demand-side participation, etc. Apart from traditional risks with vertically integrated regulation, all the factors mentioned above have arisen since deregulation. The market participants (such as GENCOs, TRANCOs, DISCOs, consumers, system operator and market operator) have to consider these risks seriously. The following will discuss mainly the risks faced by GENCOs. However some risks and means of risk mitigations are suited not only to GENCOs but also to other types of participants. First we analyse the uncertainties in electricity markets and then discuss the means of risk mitigation.

13.1.6.1 Uncertainties in Electricity Markets

1. Volatile Electric Spot Price

The electric spot price is the most uncertain factor in electricity markets. As electric power cannot be stored the demand and supply of electricity must balance at any time, while maintaining electric network security. So relatively small changes in load or generation may cause large changes in prices. Rafal drew several conclusions about the volatility pattern of electricity price in [22].

- The price of electricity is far more volatile than that of other commodities normally noted for extreme volatility.
- When electric energy spot prices are measured on different time scales, it is found that electricity price volatility does not behave like most financial instruments and commodities. That means it can overestimate premiums of long-term options written on electricity by applying traditional financial formulas to short-term electricity prices.
- Electricity price volatility has its daily, weekly and seasonal characteristics.

Many factors cause specific volatility of electricity price. The basic reason, as mentioned before, is the non-storable feature of electric power compounded by the requirement for real-time balance of supply and demand. This situation can be worse when the following happens: unexpected weather changes, special events (like special galas) unexpected system failure (such as generator outage or congestion after faults occurring), generators' unreasonable bidding strategies etc. Furthermore, the analysis in [23] shows that market structure and market rules cause significant differences in the behaviour of spot prices of electricity. These include uncontrollable monopoly in the electric generation market, less elastic demand-side response to spot prices and the type of fuel used for generation etc.

2. Weather Risk
As we know, end-users in electricity markets are largely weather-dependent consumer. Such weather-dependent consumers will affect the accuracy of forecasted load demand and therefore make the energy price volatile. From this point of view, weather risk is bound up with price risk and weather uncertainty can be incorporated into the uncertainty of energy price. However, weather risk is unique. It has special attributes that set it apart from price risk and other sources of risk. Experience and theory suggest that energy prices and weather indices do not correlate well in electricity markets [24]. This makes it virtually impossible to manage weather risk with a price hedge. The effects of weather on energy consumption can be summarised as below:

- Colder than normal summers reduce electric power consumption and idle capacity, which raises the average cost of power production.
- Above average winter temperatures reduce electric power production to lower space heating requirements.
- Lower than normal precipitation upstream of hydropower facilities reduces power production.

The bidding strategies of GENCOs and IPPs take weather risk into consideration whether they transact through the pool or directly by bilateral contracts with consumers. However no research has been found describing GENCOs or IPPs building bidding strategies taking weather risk into account. However some brokers have taken weather risk seriously and provided financial derivatives to help energy sellers and energy buyers to hedge weather risk. The method they used will be address in the next section on risk mitigation.

3. Rival Bidding Strategies
Competitors' bidding strategies is an important issue and has often been addressed since deregulation. Since market participants have very limited information about other competitors, so the rival's bidding strategies are highly uncertain. The bidding strategies for GENCOs depend on energy provider type (fossil fuel power plant or hydro plant), market structure and market rules, forecasted or estimated competitor bidding strategies and so on.

13.1.6.2 Risk Mitigation
Risk mitigation means taking effective means to hedge market participants' risk. In electricity markets market participants face various uncertainties as discussed above. How to hedge the risk is a pressing topic in current research. In this section we present three basic approaches [25]: risk avoidance, decision-making under uncertainties and risk management.

1. Risk Avoidance
This has been the traditional approach adopted by engineering design and regulatory bodies. Typically it consists of screening alternative courses of action by per-

forming a risk assessment, and enforcing a threshold criterion for acceptable risk. Such criteria are set on the basis of expert opinion and policy considerations including political compromise. In the American electricity market, in order to mitigate the risk of price volatility, the highest bids that each GENCO or IPP can bid in the generation market are capped.

2. Decision-Making under Uncertainties
This approach, known as Decision Analysis, evaluates alternative actions by model the uncertain outcomes of each alternative. The method of model takes into consideration subjective and objective information including the assignment of values to potential outcomes and the decision-makers' risk preference. In the section 'Decision-making and Strategies in Generation Auction Markets' the methodologies for decision-making used in electricity markets are discussed in length.

3. Risk Management
This approach aims to control financial consequences of uncertain outcomes through economic mitigation. Such mitigation often takes the form of trading contingent claims whose financial settlement depends on realisations of uncertain state variables. Contractual arrangements such as insurance contracts are examples of such contingent claims. Using risk sharing agreements and syndication is another form of economic risk mitigation. These are mechanisms for 'spreading the risk' or more precisely the financial consequences of the risk. Hedging is a third form of economic risk mitigation in which a risk bearer reduces its exposure by creating a portfolio of ventures whose outcomes happen to be correlated so as to reduce total variability.

13.2 Decision-making and Strategies in Generation Auction Markets

13.2.1 Overview of Decision-making

In this section, we first present some basic knowledge of decision-making and its analysis methodologies. Then we focus on the decision-making problems faced by market participants in electricity markets, especially by the generation companies in electric generation markets.

13.2.1.1 Decision-making, Model and Algorithm

1. Get to Know Decision-making
Decision-making in almost all realms of human endeavour has always been a difficult task. Decision-making requires careful gathering and evaluation of facts, ascertaining relative merits of chosen alternatives and reasoning about consequences. Mathematics is concerned with manipulation of information, problem presentation and arriving at conclusion.

The decision problem can be classified into the following problems, which are based on time frame from the point of view of organization:

- Strategic Planning: Carried out by top management and the main goal is to acquire and develop productive resources. The time frame might vary from 5 to 10 years.
- Tactical Planning: Undertaken by middle management. The main goal is efficient resource utilisation and the time horizon is around 6 months to 1 year.
- Operational Control: Carried out by employees and focuses on the optimal execution of utilising resources.

Figure 13.5. Diagram of decision-making problems

Figure 13.5 shows the time frame and information flow of these three decision-making problems. Which type of problem is focused on by a decision-maker depends on the requirement of the specific problem.

2. Models and Algorithms Decision-making
To date, many decision-making problems have been well modelled, such as the shipping problem, route planning by telecom and so on. Many algorithms can be employed to solve decision-making problems as follows.

- Optimisation-based Method: The basic aim in developing an optimisation model of an operational problem is to be able to predict what the optimal solution should be, given the initial conditions of the problem.
- Network Flow: Derived from the typical decision-making problems like the travelling-salesman problem, transportation problem, and assignment problem, it has the basic features of LP problems. For such specific problems, however, some algorithms are more efficient and distinct, such as graph and tree, the Hungarian method etc.
- Games and Decision Tree: The main feature of game theory is to find a good enough, strategy considering the possibilities of rivals' behaviours. Decision tree (shown in Figure 13.6) analysis requires a knowledge of all possible and/or relevant alternative solutions. The decision tree can be interpreted as a graph or a network in which the nodes represent decisions or actions and the arcs or branches represent the strategy selections.

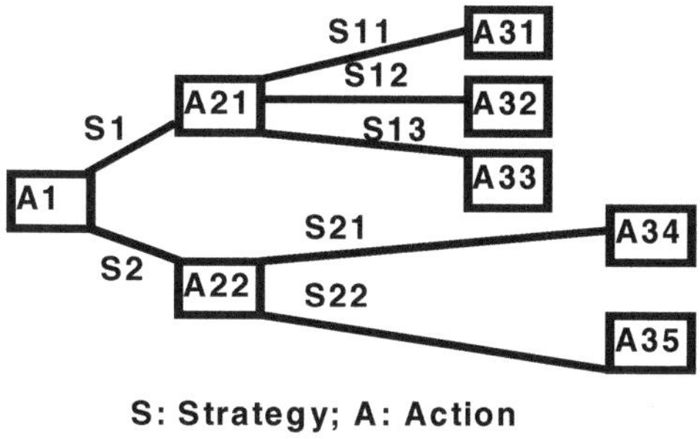

Figure 13.6. Diagram of decision tree

- Artificial Intelligence Algorithm: Neural networks and expert systems are two main tools for artificial intelligence based decision-making support system, which provide automated computation and data analysis but rely on knowledgeable users or consultants to direct system operations and interpret results. Figure 13.7 shows a simple presentation of a typical expert system. The core of the system consists of a knowledge base and inference engine.

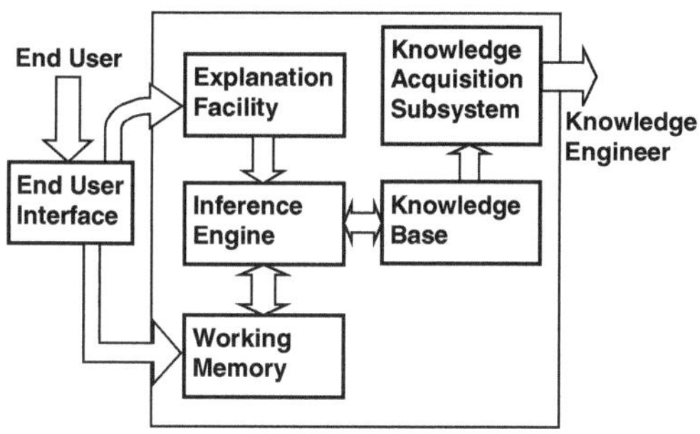

Figure 13.7. A simple presentation of a typical expert system

13.2.1.2 Decision-making in Electricity Markets

In electricity markets, each market participant faces its own decision-making problems as shown in Figure 13.8. The ISO makes transactions in the spot market (usually under are auction mechanism) by maximising the social welfare or minimising the costs. Sometimes the ISO might also play the role of market regulator, where it needs to decide transactions in future markets and forward markets. These transactions may range from real-time power, operating reserve, reactive power to transmission rights and other ancillary services. The TransCOs, which own the transmission network and provide the transmission service, decide how to make profitable transmission services, and when and where to expand transmission network etc. The DISCOs, most of which are in the role of regional electricity suppliers, make the decision and at what how price to buy/sell the electricity from the network. Customers have more and more flexibility to choose suppliers. In some market they can also contract with large power generation companies directly. They face the decision to choose suppliers and find beneficial contract prices. Brokers are emerging in some electricity markets, where they provide the various services for both customers and suppliers.

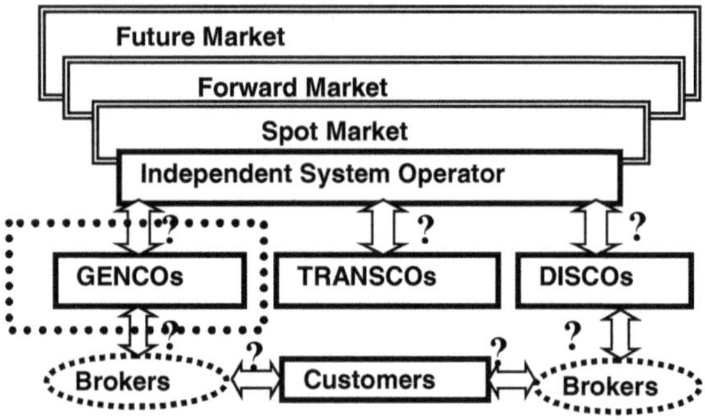

Figure 13.8. Diagram of decision-making problems in electricity markets

They might provide some types of financial contract to hedge the risk for their customers. The decisions of GENCOs (within the dotted box in Figure 13.8) will be discussed in next section and that is the focus of this chapter.

13.2.1.3 Decision-making in Electricity Generation Market

As shown in Figure 13.5, in electricity generation markets each GENCO has also three types of decision-making problems, strategic planning, tactical planning and operational control problem. The main problem for operational control of GENCOs is to make offers daily or hourly in the spot market. So this chapter will only focus on the bidding strategies of GENCOs at the operational control level. Research into the other two decision-making levels is beyond the scope of this chapter.

1. Factors Influencing the Bidding Strategies of GENCOs

In the generation auction market the generator offers generation bids (price/quantity), which reflect a way to maximise its own profit while considering other participants' reaction and the different kind of constraints and so on. Many factors depend on the strategies taken by the suppliers.

- Types of Electricity Market: Competitors choose different strategies in different electricity market environments, as in a perfect competition market and an imperfect competition market, a coordinated market and an uncoordinated market.
- Different Constraints: These various constraints include technical constraints and regulatory decisions.
- Contracts: The various types of contract among supplier, customer and dealer effect the bids of suppliers.

- Predicted Demand and Rivals' Strategies: This is one of the main factors effecting the bidding strategy. Participants may obtain the predicted demand from pool or auction centre. The rivals' strategies, however, cannot be obtained directly.
- Previous Market Prices: Like other common commodity auction markets, the generation auction market seems to have the same character that bids in the following trade days would reflect the fluctuation of previous market price.

There are other factors affecting the generation auction market than those mentioned above, especially now when the electricity market is undergoing developing and self-improving reconstruction in every country. These factors need to be quantified in detail in the models.

2. Application of Decision-making in Electric Generation Market
To date, research work shows that the following approaches have been attempted in the generator bidding strategies problem: game theory based methods, optimisation based methods, and other heuristic methods such as GA, ANN, ES based methods. In the next three sections, we discuss the applications of these methods.

13.2.2 Game Theory Applications in Generation Auction Market

13.2.2.1 Basic Concept of Game Theory and its Application in Electricity Markets

1. Brief Introduction to Game Theory
Game theory is a discipline that is used to analyse problems of conflict among interacting decision-makers. Generally game theory can be classified into two categories, cooperative and non-cooperative [26], [27]. Most work uses non-cooperative game theory in analysis of market structure, market rules and participants' behaviours in electricity markets.

2. Application of Game Theory in Electricity Markets
The well-known game theory is applied to analyse how market structures and market rules affect the optimal strategies of participants and how the market participants exercise market power potentially. Reference [28] studies how electric network structure affects market participants' behaviours. Two-node and four-node networks without AC transmission power flow constraints are used in the following three scenarios. First is perfect competition where each producer provides the bid curve equal to its marginal cost function. The second is imperfect competition where each generator chooses the supply function that maximises its profits given a fixed set of rival bids. The third is monopoly competition where there is only one owner of all generation, who chooses a supply function that maximises its overall profit. Reference [29] uses a Cournot Model to study the Nash equilibrium with two-node and three-node networks. It points out that when there is only one Nash equilibrium or the participants agree on a particular Nash equilibrium, then, and only then, equilibrium game theory predicts the game's outcome, which are often accurate. However, its two-node network with transmission constraint produces a situation in which there are multi-equilibria. It suggested a simple 'fictitious play'

algorithm based mixed strategy to find the mixed-strategy Nash equilibrium. From good cases we can see what is in participants' behaviours that are expected to predict the market's outcome and to predict the potential market power.

Apart from using game theory to study market power and market structure, game theory can also be used to study ancillary services pricing as in [30] and by GENCOs to produce bidding strategies, which is the focus in this section.

13.2.2.2 Game theory-based Bidding Strategies

1. Generator Bids Alone

The application of game theory to generator bidding strategy mainly adopts the concept of Nash equilibrium. The UC model can be implemented for generators and the OPF model can be used to solve system transmission constraints. The way to find Nash equilibrium can be summarised to the methods detailed two in Figure 13.9.

The first method is to use the dominant strategy as in [31]. The dominant strategy method needs the profit matrix to select the Nash equilibrium from. This method cannot be implemented when a large number of players are considered. The iteration method, however, can overcome this disadvantage to solve large-scale problems.

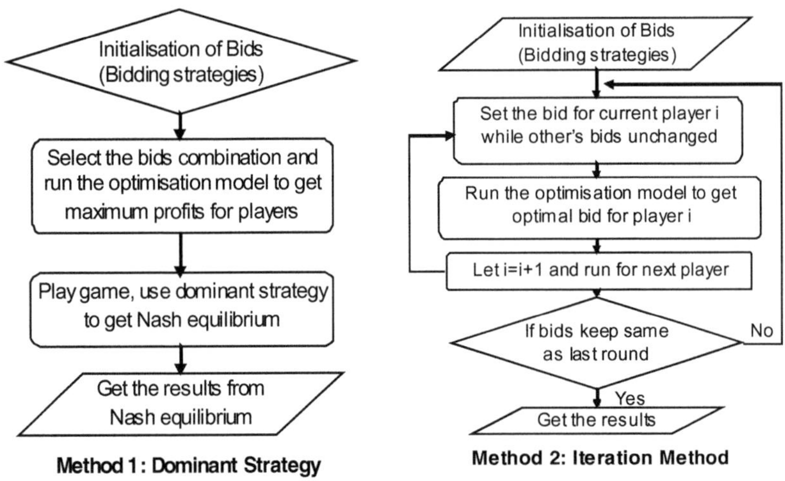

Figure 13.9. Flowcharts of two methods to find Nash equilibrium

2. Generator Bids in Coalitions

When considering coalitions, things get more complicated. The top concern is the burden of computation, since we know there are 2^{n-1} combinations of potential coalitions if n players are considered. Optimisation for these 2^{n-1} combinations requires huge computational effort and is not practical under current conditions. We hope heuristic methods can reduce the effort to an acceptable level, but unfortunately no work on this has get been done. Several papers in the literature attempt to study coalition based on a simple pricing model or network model, which has only two or three players.

The basic steps for bidding strategy decision-making in coalitions are as below:

- Information collection. This includes other players' information either by archiving from ISO or by making estimate from historical data etc.
- Determine the potential coalitions.
- Find the optimal solution for each potential coalition. This is done by maximising the group's profit (common payoff) or by maximizing the player's own profit by imposing some allocation criteria for the common payoff among the group members in the coalition.
- Finally obtain the global optimum from all of the potential coalitions.

A typical study of coalition is [32]. Ferrero in [32] focuses on the perspective of the system operator; this study, however, can also guide an individual player to select the bidding strategy.

Some research work [33], 35] models intelligent agents representing human decision-makers in electricity markets. Each agent applies the game theory and produces the priority list of possible coalitions and negotiates with other agents representing other generators.

Thus generators can bid alone or bid collusively. The best bidding strategy for the generator is the optimal choice among bidding alone and bidding collusively.

13.2.3 Optimisation-based Approaches to Making Bidding Strategies Making

13.2.3.1 Bidding Decision making by Optimisation-based Market Simulator

Since market prices are very uncertain, it will be helpful to produce bids with a market simulator built-in bidding decision-making system. This market simulator can produce market prices based on stochastic market situations, as is usually done in ISO, PX or Power Pool. In other words, such a market simulator is the simulation model of ISO, PX or Power Pool. It should be addressed that due to the highly incomplete information, the simulation is actually an estimation of the approaching market situation.

Reference [36] simulates the PX in California with central economic dispatching at the high level of bidding decision-making and a unit commitment model at the lower level. Central economic dispatching is performed in each auction round and produces tentative market prices before starting the next auction round till the MCP is stable. In each round all suppliers use the UC model to produce bidding curves by dynamic programming. The whole process is shown in Figure 13.10.

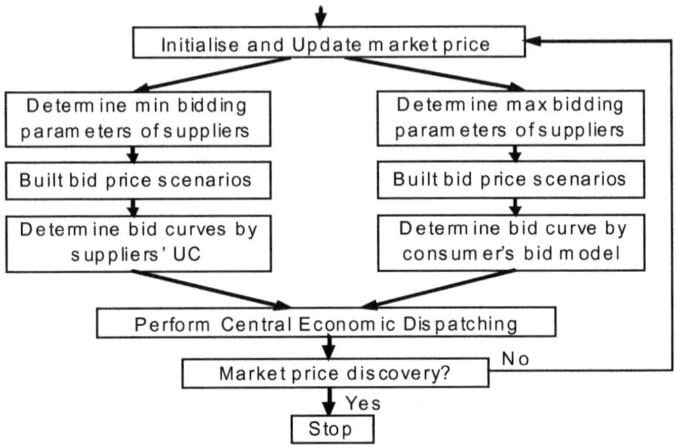

Figure 13.10. Bidding process based on market simulator in Cal_PX

The simulator used in [36] needs to perform the UC for all suppliers, and this is not practical from the perspective of individual suppliers because it has little information about the other suppliers, such as ramp-up/down rate, start-up/shut-down time and so on.

Another simulation model is used to study the bidding strategies for generators in [37] where a price-clearing process and unit commitment run in an iterative way. Similar to [36] this market simulation model needs each generator to run UC in turn. Furthermore the lower bound is set for each generator according to its cost data in [37] while the minimum bidding parameter of the generator in [36] is optimised each time after the market price is updated. The UC problem is solved by Mixed Integer Programming (MIP) in [37].

To the authors' knowledge the only market simulation model of the England and Wales Power Pool (EWPP) under PETA is done in [38]. A large-scale nonlinear program is applied to simulate the EWPP with a multi-part bid format and solved with a Lagrangian Relaxation method. With the assumption of perfect competition, i.e. the bid of any bidder has negligible effect on the system marginal and reserve price, the conclusion is drawn that regardless of generation resource, costs and constraints, a generator maximises profits by bidding to supply generation at cost and at maximum capacity.

13.2.3.2 Optimal Bidding based on Formulation of Market Prices with Generators' Behaviours Embedded

Generally, apart from forecasting, the uncertainty of market price can be solved in two ways; one is to simulate the market price based on any information the supplier has, and the second is to find the relationship between the market price and other uncertainties. Since the market price has no explicit relationship with competitors' behaviours or system demand etc., the simulation is a good choice.

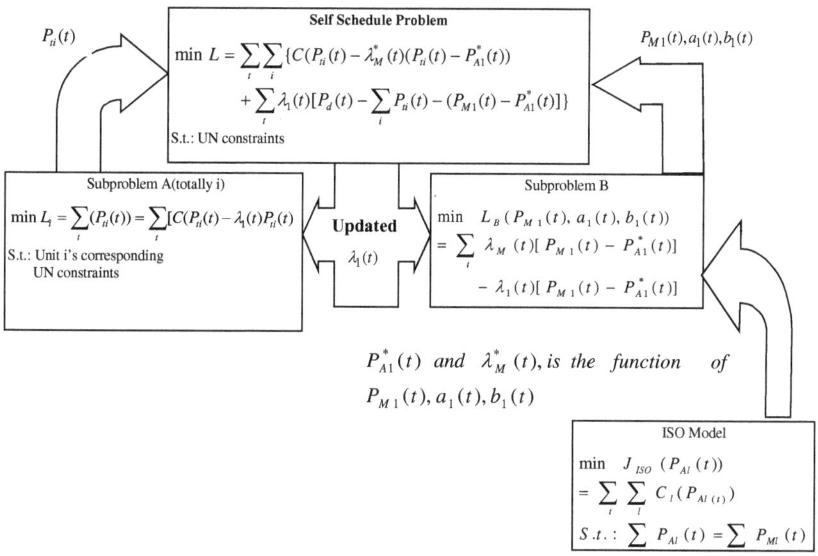

Figure 13.11. Flowchart of Lagrangian Relaxation based bidding decision-making

However, the disadvantage of simulation is the huge computation effort. If, with some simplification, competitors' behaviours can be embedded into the formula for market price, this can significantly reduce computing effort in applications. The simplification makes market prices inaccurate, however.

Figure 13.11 shows the bidding process in [39]. The notation in the figure is as in [39]. Reference [39] studies the New England power market and uses a Lagrangian relaxation method to solve the generators' self-schedule problem. The objective of generators is to maximising profit. With considerable simplification in ISO side, the formula for market prices shows it is a function of each competitor's bidding strategy. So, given the competitors' bids, the market price can be formulated as a function with respect to its own bids. The self-schedule problem is decomposed into two sub-problems and a Lagrangrian relaxation method can be applied to find the optimal bids for the generator under consideration.

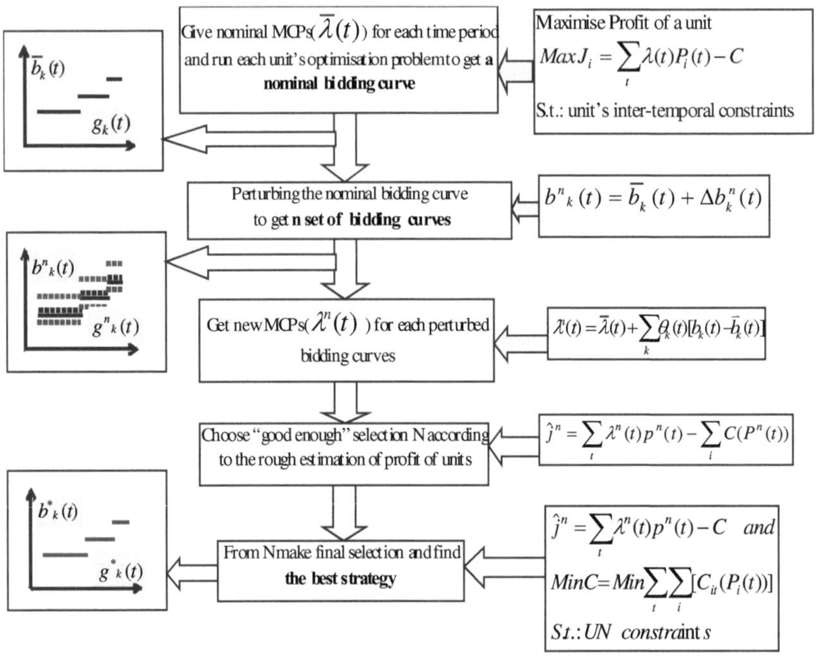

Figure 13.12. Flowchart of ordinal optimisation based bidding decision-making

In [40] the MCP in the California market is formulated as nominal market price plus a perturbation, as shown in Figure 13.12. Two things should be noted, one is that the nominal bidding curve comes from estimation based on experience and expert analysis, and second, the perturbation of market price only results from the bidding curve of the generator under consideration. In other words [40] does not take into account the other competitors' bidding behaviours. A useful contribution of [40] is the application of Ordinal Optimisation. The idea is, since it is impossible to find an optimal solution due to large uncertainties, to find a 'good enough' solution. Based on this premise, it uses rough criteria to pick out the 'good enough' subset from all the solution space and uses an accurate model to choose the final best choice.

If we assume that the market price is set by other competitors only and the rule for setting MCP in California PX is considered, then an explicit linear market price function can be derived from other competitors' bidding curves, as illustrated in Figure13.13.

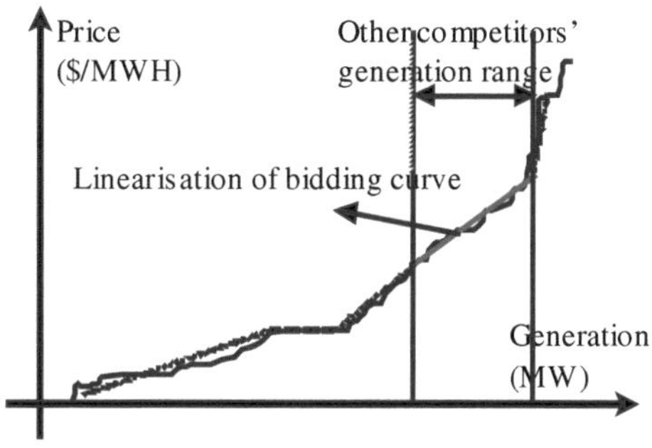

Figure 13.13. Linear simplification of competitors' bidding curves

First we estimate the other competitors' bidding curves and aggregate them into one bidding curve. Then we linearise the competitors' bidding curve at the estimated range of competitors' generation as shown in Figure13.13. The market price is then formulated with this linear curve within a specific generation range. Reference [41] uses this method to solve the self-dispatching problem for the considered generators with MIP programming. Two further factors are considered in [41]. First, this market price formulation assumes that the considered generator does not effect the market price, which is not always true in the real markets. Second, the uncertainties of competitors' bidding strategies are not taken into consideration.

13.2.3.3 Application of Markov Decision Process (MDP) in Bidding Strategies

The bidding strategies decision problem can be considered as a multi-period process. In each decision period the bidder under consideration faces a set of stochastic market situations and a set of alternative decisions (bidding strategies). The bidder chooses the most profitable decision from this set of alternative decisions for each state in each period. The bidder's optimal bidding strategy is to maximising its profit for the whole time horizon under consideration.

If we let each time period be a stage in the range from 1 to m, and let each possible market situation be a state in the range from 1 to n, then we illustrate the bidding problem as a Markov Decision Process (MDP) as in Figure13.14. Here the n and m depend on the time domain and the situation space of your study, respectively. In Figure13.14 at stage *m*, the bidder can be in any of a state *n*. In each state the bidder can choose one decision A_k from a set of alternative decision a (A_1, A_2, … …, A_K). Corresponding to a decision a_k the transition probability from a state *i* to another state *j* is given by Pr(I, J, A_k). The bidder receives a reward from each transition denoted as r(I, J, A_k). The value iteration method can be applied to find the optimal strategy for the whole time horizon. The value iteration method is similar to backward dynamic programming. This MDP approach is illustrated in [42]. The MDP is usually applied in the situation that the state space is well limited. While the states space in the bidding problem is decided by all large participants' stochastic bidding strategies, so the state space will be extremely large if no effective method is used to reduce it.

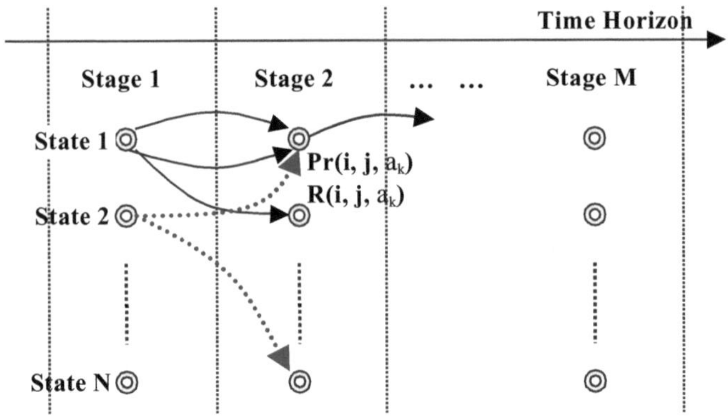

Figure 13.14. States and state transition in MDP

13.2.4 Other Methodologies for Decision-making

Apart from game theory and optimisation-based methods there are also some other methods that can be applied for the analysis of generation bidding strategies, such as expert systems, artificial intelligence approaches, probabilistic or statistic method, and control theory etc. Due to space limitation in this chapter, discussion of these methodologies is neglected.

13.3 Study of Bidding Strategies Based on Bid Sensitivities in Pool-based Spot Markets

13.3.1 Introduction

This section presents the concept of bid sensitivities and a bid sensitivities based bidding model. Bid sensitivities are derived from the Interior Point Optimal Power Flow (IPOPF) model, which are bid-price sensitivities, bid-output sensitivities, bid-profit sensitivities, and bid-line flow sensitivities. It has been found that bid sensitivities are very valuable indexes for market regulators to analyse market participants' market power and also useful information for GENCOs to analyse bidding strategies. The proposed bidding model produces bidding strategies for GENCOs in the case of both 'bid alone' and 'bid in coalition'. The competitors' bidding strategies have effects on market prices and GENCOs' outputs/bids in the proposed model.

The presented work assumes the considered generator has the basic knowledge of other competitors' bids and the bidding strategies presented are based on a multi-round bidding process. Demand-side bidding is not considered in this work. The approach presented is valid for a pool-based market without bilateral contracts (BCs).

The rest of this section is organised as follows. Section 13.3.2 derives bid sensitivities based on the IPOPF model. Section 13.3.3 shows how the generators choose the best bids based on the bids sensitivities when they bid alone, Section 13.3.4 discusses the bidding strategies when generators bid in coalition, and Section 13.3.5 provides a case study that illustrates results using a IEEE 30-bus system, and finally gives conclusions.

13.3.2 Analysis of Bid Sensitivities Based on the IPOPF Model

The following notation is used in this section:

i Index of generation units, from 1 to n

m	Index of the generation unit under consideration
$\boldsymbol{K}$	Vector of bids, K_m represents the bid offered by unit m
$\boldsymbol{S}$	Vector of scheduled outputs of generators
T	Vector of transformer tap ratio
V	Vector of node voltage
θ	Vector of phase angle
λ	Vector of spot price
C,B,A	Vector of cost coefficients of generators
$\boldsymbol{X}$	Vector of controllable and state variables except S
Bf	Benefit of consumers
g	Set of equality constraints including power flow balance etc.
h	Set of inequality constraints, such as generators' output limits; voltage limits; transformer tap ratio limits; transmission line capacity constraints and so on.

Assuming the bids $\boldsymbol{K}$ offered by units are known (this will be discussed in Section 13.3.3), the OPF problem can then be formulated as Equation (13.3) where the generator i bids K_i instead of the cost coefficient C_i, but keeps B_i and A_i unchanged. The overall objective function of the market operator is to maximise the total social welfare.

$$Max\ f_1(S_i) = Bf - \sum_{i=1}^{n}\left(K_i S_i^2 + B_i S_i + A_i\right)$$

$$\text{Subject to:} \quad \begin{array}{l} \mathbf{g}(\mathbf{X},\mathbf{S}) = 0 \\ \mathbf{h}(\mathbf{X},\mathbf{S}) \geq 0 \end{array} \qquad (13.3)$$

The constraints are similar to those in Reference [43]. The Primal Dual Interior Point Method is used to solve this OPF model and the augmented Lagrangian function can be formed as Equation(13.4)

$$L = f_1(\mathbf{z}) - \boldsymbol{\lambda}^T \mathbf{g}(\mathbf{z}) + \boldsymbol{\pi}_l^T(\mathbf{h}(\mathbf{z}) - \mathbf{s}_l - \mathbf{h}_l) + \boldsymbol{\pi}_u^T(\mathbf{h}(\mathbf{z}) + \mathbf{s}_u - \mathbf{h}_u) - \boldsymbol{\mu}(\ln \mathbf{s}_l + \ln \mathbf{s}_u) \tag{13.4}$$

Here $\mathbf{z}$ are primal variables (**X, S**), λ, $\boldsymbol{\pi}_l$, $\boldsymbol{\pi}_u$ are dual variables and $\mathbf{s}_l$, $\mathbf{s}_u$ are slack variables. Employing first-order conditions in Equation (13.4) and using a reduced KKT system to solve it, we can derive the optimal search direction of $\mathbf{Z}$

$$\Delta\mathbf{Z} = -\mathbf{H}^{-1} \cdot \nabla\mathbf{L} \tag{13.5}$$

Here $\mathbf{Z} \supset (\mathbf{z}, \boldsymbol{\lambda})$ and the functions of $\Delta\mathbf{Z}$, $\mathbf{H}$ and $\nabla\mathbf{L}$ are:

$$\begin{cases} \mathbf{H} = \begin{bmatrix} \widetilde{\mathbf{H}}(\mathbf{z}, \boldsymbol{\lambda}) & -\mathbf{J}^T(\mathbf{z}) \\ -\mathbf{J}(\mathbf{z}) & 0 \end{bmatrix} \\ \nabla\mathbf{L} = \begin{bmatrix} \widetilde{\mathbf{t}} & -\mathbf{g}(\mathbf{z}) \end{bmatrix} \\ \Delta\mathbf{Z} = \begin{bmatrix} \Delta\mathbf{z} & \Delta\boldsymbol{\lambda} \end{bmatrix}^T \end{cases} \tag{13.6}$$

where, the reduced Hessian matrix $\widetilde{\mathbf{H}}(\mathbf{z}, \boldsymbol{\lambda})$ and $\widetilde{\mathbf{t}}$ are

$$\begin{cases} \widetilde{\mathbf{H}}(\mathbf{z}, \boldsymbol{\lambda}) = \mathbf{H}(\mathbf{z}, \boldsymbol{\lambda}) + \nabla_{\mathbf{z}}\mathbf{h}^T(-[\mathbf{s}_l]^{-1}[\boldsymbol{\pi}_l] + [\mathbf{s}_u]^{-1}[\boldsymbol{\pi}_u])\nabla_{\mathbf{z}}\mathbf{h} \\ \widetilde{\mathbf{t}} = \nabla_{\mathbf{z}}\mathbf{f}(\mathbf{z}) - \mathbf{J}^T(\mathbf{z})\boldsymbol{\lambda} \quad + \nabla_{\mathbf{z}}\mathbf{h}^T(\boldsymbol{\pi}_l + \boldsymbol{\pi}_u \\ \qquad -[\mathbf{s}_l]^{-1}([\mathbf{s}_l]\boldsymbol{\pi}_l + \boldsymbol{\mu}e) - [\mathbf{s}_u]^{-1}([\mathbf{s}_u]\boldsymbol{\pi}_u - \boldsymbol{\mu}e)) \end{cases} \tag{13.7}$$

Employing the first-order partial derivative of $\mathbf{Z}$ with respect to $\mathbf{K}$, we get:

$$\frac{\partial \mathbf{Z}}{\partial \mathbf{K}} = \mathbf{H}^{-1} \cdot \frac{\partial \mathbf{H}}{\partial \mathbf{K}} \cdot \Delta\mathbf{Z} - \mathbf{H}^{-1} \cdot \frac{\partial \nabla\mathbf{L}}{\partial \mathbf{K}} \tag{13.8}$$

Now $\Delta\mathbf{Z} = 0$ at the OPF solution, so we recast Equation (13.12) as

$$\frac{\partial \mathbf{Z}}{\partial \mathbf{K}} = -\mathbf{H}^{-1} \cdot \frac{\partial \nabla\mathbf{L}}{\partial \mathbf{K}} \tag{13.9}$$

As shown above, $\mathbf{Z}$ includes primal variable $\mathbf{z}$ and dual variable λ, i.e. $\partial\mathbf{Z}/\partial\mathbf{K} = \begin{bmatrix} \frac{\partial \mathbf{S}}{\partial \mathbf{K}} & \frac{\partial \mathbf{T}}{\partial \mathbf{K}} & \frac{\partial \mathbf{V}}{\partial \mathbf{K}} & \frac{\partial \boldsymbol{\theta}}{\partial \mathbf{K}} & \frac{\partial \boldsymbol{\lambda}}{\partial \mathbf{K}} \end{bmatrix}^T$
then we obtain the following four sets of bid-sensitivities.

1. Bid-Price Sensitivities:

$\partial\boldsymbol{\lambda}/\partial\mathbf{K}$ First-order derivatives of node price $\boldsymbol{\lambda}$ with respect to bids $\mathbf{K}$ submitted by generators

2. Bid-Output Sensitivities:

$\partial\mathbf{S}/\partial\mathbf{K}$ First-order derivatives of generators' output $\mathbf{S}$ with respect to bids $\mathbf{K}$ submitted by generators

3. Bid-Profit Sensitivities:

$\partial\mathbf{P}/\partial\mathbf{K}$ First-order derivatives of profits of generators with respect to bids $\mathbf{K}$ submitted by generators

The Bid-Profit Sensitivities can be derived from Equation (13.7)

$$\frac{\partial\mathbf{P}}{\partial\mathbf{K}} = (\boldsymbol{\lambda} - 2\mathbf{CS} - \mathbf{B})\cdot\frac{\partial\mathbf{S}}{\partial\mathbf{K}} + \mathbf{S}\cdot\frac{\partial\boldsymbol{\lambda}}{\partial\mathbf{K}} \tag{13.10}$$

4. Bid-Line Flow Sensitivities:

$\partial\mathbf{P}_l/\partial\mathbf{K}$ First-order derivatives of transmission line power flow $\mathbf{P}_l$ with respect to bids $\mathbf{K}$ submitted by generators.

$$\frac{\partial\mathbf{P}_l}{\partial\mathbf{K}} = \frac{\partial\mathbf{P}_l}{\partial\mathbf{S}}\cdot\frac{\partial\mathbf{S}}{\partial\mathbf{K}} \tag{13.11}$$

Obviously the above four types of bid sensitivities are very valuable indexes to bidding strategy decision-making both for individual suppliers and market operators in the electricity generation auction market. They can be employed by individual suppliers to investigate how to decide the optimal bids, how to choose partners for potential coalitions and how to make strategies to impair other rivals' benefits according to own bid sensitivities. Also, system regulator can use this information to prevent individual suppliers from gaming so as to mitigate market power.

13.3.3 Bidding Strategies Based on Bid Sensitivities

13.3.3.1 Description of the proposed model

The bidding model presented in this paper has the objective to maximise the individual profit for each generator in the optimisation problem.

$$Max\ f_2(S_m, \lambda_m) = \lambda_m S_m - (C_m S_m^2 + B_m S_m + A_m) \tag{13.12}$$

And subject to: OPF problem as described in Equation (13.3)

Note:

1. In the objective function of the OPF problem in model Equation (13.12), the individual suppliers submit bid **K** instead of its real cost coefficient **C**. In this paper we assume the suppliers only bid the coefficient C, so B and A are kept unchanged. The optimal solution K_m of model Equation (13.12) is the optimal bid of generator M.
2. $f_2(S_m, \lambda_m)$ is not an explicit function of variable K_m but an implicit function as S_m and λ_m are all implicit functions of variable K_m. S_m and λ_m are determined by OPF.
3. The total load and demand-benefit curves are fixed, i.e. benefit Bf is constant. In this assumption, the bidding of the demand side is neglected. Demand side bidding, however, can easily be incorporated in this model, as it has similar characteristics to generation bidding.

13.3.3.2 Optimal Bids

Given other competitors' chosen strategies $K_i\,(i=1,\dots,n; i\neq m)$, the optimal bid for generator m is then decided in Equation (13.13) and Equation (13.14) by employing first-order conditions in Equation (13.12). For simplification, subscript m is ignored in the following functions.

$$-\nabla_k f_2\big|_{K_{old}} = \nabla_{k^2}^2 f_2\big|_{K_{old}} \cdot \Delta K \tag{13.13}$$

$$K_{new} = K_{old} - \left(\frac{\partial^2 f_2}{\partial K^2}\right)^{-1}\bigg|_{K_{old}} \cdot \frac{\partial f_2}{\partial K}\bigg|_{K_{old}} \tag{13.14}$$

From Equation (13.16), we get $\partial f_2/\partial K$ and $\partial^2 f_2/\partial K^2$ respectively

$$\frac{\partial f_2}{\partial K} = S\frac{\partial \lambda}{\partial K} + (\lambda - B - 2CS)\frac{\partial S}{\partial K} \tag{13.15}$$

$$\frac{\partial^2 f_2}{\partial K^2} = S\frac{\partial^2 \lambda}{\partial K^2} + 2\frac{\partial \lambda}{\partial K}\cdot\frac{\partial S}{\partial K} - 2C\left(\frac{\partial S}{\partial K}\right)^2 + (\lambda - B - 2CS)\frac{\partial^2 S}{\partial K^2} \tag{13.16}$$

From Equation (13.15) and Equation (13.16), we need $\partial S/\partial K$, $\partial\lambda/\partial K$, $\partial^2 S/\partial K^2$ and $\partial^2\lambda/\partial K^2$ to get the optimal search direction for bid K_m. $\partial S/\partial K$, $\partial\lambda/\partial K$ are the first two types of bid sensitivities, already derived. What we need in addition are second derivatives $\partial^2 S/\partial K^2$ and $\partial^2\lambda/\partial K^2$, which can be derived from Equation (13.8)

$$\begin{aligned}\frac{\partial^2 \mathbf{Z}}{\partial \mathbf{K}^2} &= -\mathbf{H}^{-1} \cdot \frac{\partial^2 \mathbf{H}}{\partial \mathbf{K}^2} \cdot \Delta \mathbf{Z} + 2\mathbf{H}^{-1} \cdot \frac{\partial \mathbf{H}}{\partial \mathbf{K}} \cdot \mathbf{H}^{-1} \cdot \frac{\partial \nabla \mathbf{L}}{\partial \mathbf{K}} \\ &\quad + 2\mathbf{H}^{-1} \cdot \frac{\partial \mathbf{H}}{\partial \mathbf{K}} \cdot \mathbf{H}^{-1} \cdot \frac{\partial \mathbf{H}}{\partial \mathbf{K}} \cdot \Delta \mathbf{Z} - \mathbf{H}^{-1} \cdot \frac{\partial^2 \nabla \mathbf{L}}{\partial \mathbf{K}^2}\end{aligned} \tag{13.17}$$

To simplify Equation (13.17) into Equation (13.18), due to $\Delta \mathbf{Z} = 0$, $\partial^2 \nabla \mathbf{L} / \partial \mathbf{K}^2 = 0$, and $\partial^2 \mathbf{H} / \partial \mathbf{K}^2 = 0$.

$$\frac{\partial^2 \mathbf{Z}}{\partial \mathbf{K}^2} = 2\mathbf{H}^{-1} \cdot \frac{\partial \mathbf{H}}{\partial \mathbf{K}} \cdot \mathbf{H}^{-1} \cdot \frac{\partial \nabla \mathbf{L}}{\partial \mathbf{K}} = -2\mathbf{H}^{-1} \cdot \frac{\partial \mathbf{H}}{\partial \mathbf{K}} \cdot \frac{\partial \mathbf{Z}}{\partial \mathbf{K}} \tag{13.18}$$

From Equation (13.8) and Equation (13.18), $\partial \nabla L / \partial \mathbf{K}$ and $\partial \mathbf{H} / \partial \mathbf{K}$ are the values we need for $\partial^2 \mathbf{Z} / \partial \mathbf{K}^2$ and $\partial \mathbf{Z} / \partial \mathbf{K}$. Once we have $\partial^2 \mathbf{Z} / \partial \mathbf{K}^2$ and $\partial \mathbf{Z} / \partial \mathbf{K}$, then $\partial f_2 / \partial K$ and $\partial^2 f_2 / \partial K^2$ are easily calculated. As K_m only appears in the objective function in model Equation (13.12), so K_m would appear only in $\partial L / \partial S_m$ (i.e. $\nabla_{S_m} L$) in the gradient ∇L.

$$\partial L / \partial S_m = 2K_m S_m + B_m \tag{13.19}$$

So $\partial \nabla_{S_m} L / \partial K_m$ and $\partial \nabla L / \partial \mathbf{K}$ are:

$$\partial \nabla_{S_m} L / \partial K_m = 2S_m \tag{13.20}$$

$$\partial \nabla L / \partial \mathbf{K} = [0 \ \ldots \ 0 \ \ 2S_m \ \ 0 \ \ldots \ 0]^T \tag{13.21}$$

Similarly, as only the derivative $\partial^2 L / \partial S^2{}_m$ (i.e. $\nabla^2_{S^2_m} L$) has K_m, that is:

$$\nabla^2_{S^2_m} L = 2K_m \tag{13.22}$$

So the $\partial \mathbf{H} / \partial K_m$ is super sparse with only one non-zero entry in the diagonal entry of $\nabla_{K_m} (\partial^2 L / \partial S^2_m)$,

$$\partial \mathbf{H} / \partial K_m = \begin{bmatrix} 0 & \cdots & 0 & 0 \\ 0 & 2 & \cdots & 0 \\ \vdots & \vdots & \vdots & \vdots \\ 0 & 0 & \cdots & 0 \end{bmatrix} \tag{13.23}$$

We therefore have enough information for Equation (13.12) to calculate the optimal bids K_m.

13.3.3.3 Nash Equilibrium Process

If we consider generator m, other competitors' bids $K_i (i = 1,...,n; i \neq m)$ should be forecast in the first round (i.e. $r = 1$). In each bidding round, all the units are calculated in order. Once one unit finds its optimal bids given the other competitors' bidding strategies, this optimal solution updates the current set of bidding strategies and the next unit processes based on this updated set of bidding strategies. The bidding round stops after all units finish this calculation one by one. In each bidding round, each unit adjusts its own bid so as to maximise its own profit. This process stops when no unit wishes to change its bid, i.e. Nash Equilibrium is found. The detailed bidding process is shown in Figure13.15.

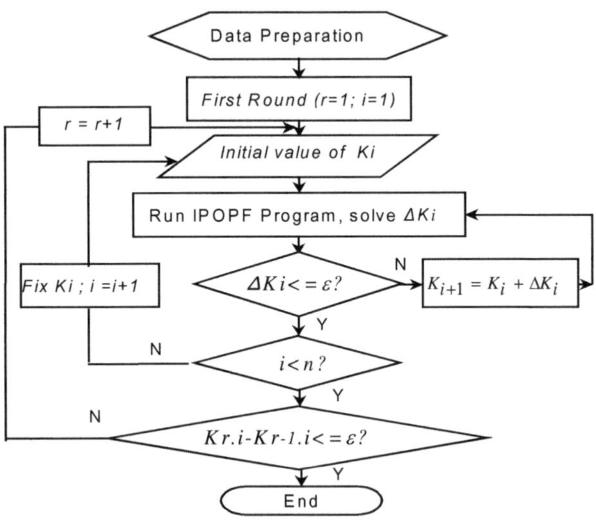

Figure 13.15. Process Bidding decision

13.3.4 Bidding Strategies when Considering Coalitions

13.3.4.1 Combinations of Potential Coalitions

Now we consider the combinations of potential coalitions from the point of view of generator *m*. Any combinations in which generator *m* is not in coalition (or say generator *m* bids individually) are ignored from the collection of combinations. Also the grand coalition is excluded since that is not allowed in the electricity market. Consider that there are *M* generators $(M \geq 3)$ and the number of combinations of potential coalitions is *R*. In the space of *M* generators, each combination of potential coalitions is denoted as C_r $(r = 1, \cdots, R)$ and the collection of all the combinations is denoted as ***C***, i.e. $\mathbf{C} = (C_1, C_2,, \cdots, C_R)$. One example of the collection ***C*** of 4-generator systems is shown in Table 13.3 where generator no.1 is under consideration. For each combination C_r in collection ***C***, there are *W* subgroups denoted as $\phi_1, \phi_2, \cdots, \phi_W$ and in each subgroup $\phi_W (w = 1, \cdots, W)$, there are N_W generators $(1 \leq N_w < M)$, i.e. N_W generators collude in this subgroup if $N_w \geq 2$ or one generator games individually (form its own subgroup with only itself) if $N_w = 1$. In the example 4-generator, the number of total combinations is 9. In the combinations of example {(1-2); (3-4)} there are two subgroups which are $\phi_1 = (1-2)$, $\phi_2 = (3-4)$.

Table 13.3. The example of c for 4-generators

Gen.1's choices for coalition	Other generators' corresponding combinations
(1-2)	{3, 4}; {(3-4)}
(1-3)	{2, 4}; {(2-4)}
(1-4)	{2, 3}; {(2-3)}
(1-2-3)	{(4)}
(1-2-4)	{(3)}
(1-3-4)	{(2)}

13.3.4.2 The Bidding Process Considering Coalitions

When considering coalitions we need to apply the cooperative game in subgroups, but the non-cooperative game among subgroups for each potential combination. With this form of gaming, the following steps are needed to find the best bidding strategies for the generator *m*.

Step 1: To find the optimal bidding strategies for generator *m* when no coalition is considered, i.e. when *M* generators bid individually. This is solved by the method presented in 13.3.3. The bidding strategies without coalition are basic strategies for generator *m*.

Step 2: To find the Nash Equilibrium in each individual combination C_r. The method will be addressed in depth later. The solutions for generator m at Nash Equilibrium for combination C_r are the bidding strategies of generator m corresponding to combination C_r.

Step 3: To pick out all the efficient combinations. We define one combination as an efficient combination if the profit of a subgroup in this combination is more than the total profit of the generators in this subgroup when they bid individually. That means the generators in a coalition subgroup have no incentive to joint this coalition if their total profit within coalition is no more than the total profit they can gain in a totally non-cooperative game (i.e. no any coalition). Here we assume the profit in a coalition subgroup is split in equal share between or among the members in the subgroup.

Step 4: To form a priority list according to the profit that generator m potentially earns and to bargain with members on the priority list. The bargaining is not included in this work.

The core work in the bidding process mentioned above is step 2: to find the Nash Equilibrium in each combination C_r. As we solve the bidding problem without coalition, bid sensitivities are applied again.

13.3.4.3 Optimal Bids of Subgroups

For each combination C_r, there are W subgroups denoted as ϕ_w $(w=1,\cdots,W)$. The optimal bidding problem for each subgroup ϕ_w can then be modelled as below:

$$\begin{gathered} Max\ f_{\phi_w}(S_{1,\cdots,}S_{N_w},\lambda_1,\cdots,\lambda_{N_w})=\sum_{n=1}^{N_w}\left[\lambda_n S_n-(C_n S_n^2+B_n S_n+A_n)\right] \\ \begin{cases} S_n \Leftrightarrow S_n(K_1,\cdots,K_i,\cdots,K_j,\cdots,K_{N_w}) \\ \lambda_n \Leftrightarrow \lambda_n(K_1,\cdots,K_i,\cdots,K_j,\cdots,K_{N_w}) \end{cases} \end{gathered} \tag{13.24}$$

Equation (13.24) is subject to OPF problem as described with Equation (13.25)

$$\begin{cases} Max \quad S = Bf - \sum_{i=1,i\subset\phi_w}^{N_w}(K_i S_i^2+B_i S_i+A_i) - \sum_{k=1,k\not\subset\phi_w}^{M}\left(K_k S_k^2+B_k S_k+A_k\right) \\ Subject\ to: \quad \mathbf{g}(\mathbf{X},\mathbf{S})=0 \\ \qquad \mathbf{h}(\mathbf{X},\mathbf{S})\geq 0 \end{cases} \tag{13.25}$$

It should be noted here again that the objective function $f_{\phi_w}(S_{1,\cdots,}S_{N_w},\lambda_1,\cdots,\lambda_{N_w})$ is not an explicit function of variables $K\in(K_1,\cdots,K_{N_w})$, but an implicit function as S_n and λ_n are all implicit functions of variables $K\in(K_1,\cdots,K_{N_w})$. S_n and λ_n are determined by OPF problem.

The objective function of Equatione(13.28) is to maximising the total profit of the generators in subgroup ϕ_w while satisfying the OPF problem. The Nash Equilibrium corresponding to coalition C_r can be found in an iterative way by solving Equation (13.24) Equation (13.25) for subgroups from ϕ_1 to ϕ_w. This process is carried out using Equation(13.16)

$$Nash\ Equilibrium\ \{\max f_{\phi_1};\ \max f_{\phi_2};\ \cdots;\ \max f_{\phi_w}\} \tag{13.26}$$

In order to get optimal bids K, $K \in (K_1, \cdots, K_{N_w})$ of the members in subgroup ϕ_w, we apply Equation (13.24) with the first order KKT condition and we get:

$$\begin{cases} \partial f_{\phi_w} / \partial K_1 = 0 \\ \cdots \\ \partial f_{\phi_w} / \partial K_i = 0 \\ \cdots \\ \partial f_{\phi_w} / \partial K_j = 0 \\ \cdots \\ \partial f_{\phi_w} / \partial K_{N_w} = 0 \end{cases} \tag{13.27}$$

For simplification the ϕ_w is ignored in the following derivation and we use vector $\mathbf{K}$ instead, then Equation (13.27) can be recast as below:

$$\begin{cases} \partial f / \partial K_1 = 0 \\ \cdots \\ \partial f / \partial K_i = 0 \\ \cdots \\ \partial f / \partial K_j = 0 \\ \cdots \\ \partial f / \partial K_{N_w} = 0 \end{cases} \Rightarrow \begin{cases} f_{\mathbf{K}}' = 0 \\ \mathbf{K} = (K_1, K_2, \cdots, K_{Nw}) \end{cases} \tag{13.28}$$

To solve Equation (13.28) we get

$$\begin{bmatrix} -f'_{K_1} \\ \cdots \\ -f'_{K_i} \\ \cdots \\ -f'_{K_j} \\ \cdots \\ -f'_{K_{Nw}} \end{bmatrix} = \begin{bmatrix} f''_{K_1^2} & f''_{K_1K_2} & & \cdots & & f''_{K_1K_{Nw}} \\ \cdots & \cdots & & \cdots & & \cdots \\ f''_{K_iK_1} & f''_{K_iK_2} & \cdots & f''_{K_iK_j} & \cdots & f''_{K_iK_{Nw}} \\ \cdots & \cdots & & \cdots & & \cdots \\ f''_{K_jK_1} & f''_{K_jK_2} & \cdots & f''_{K_jK_i} & \cdots & f''_{K_jK_{Nw}} \\ \cdots & \cdots & & \cdots & & \cdots \\ f''_{K_{Nw}K_1} & f''_{K_{Nw}K_2} & & \cdots & & f''_{K_{Nw}K_{Nw}} \end{bmatrix} \cdot \begin{bmatrix} \Delta K_1 \\ \cdots \\ \Delta K_i \\ \cdots \\ \Delta K_j \\ \cdots \\ \Delta K_{N_w} \end{bmatrix} \tag{13.29}$$

As before we use vector $\mathbf{K}$ instead and recast Equation (13.29) as Equation (13.30)

$$-f'_{\mathbf{K}} = f''_{\mathbf{K}} \cdot \Delta\mathbf{K} \tag{13.30}$$

To solve Equation (13.33) or Equation (13.34) we need f'_{K_i} , $f''_{K_i^2}$, and $f''_{K_iK_j}$, that can be derived by implying Equation (13.28) with first-order derivative and second-order derivative with respect to K_i. The results are shown in Equation (13.31)

$$\begin{cases} f'_{K_i} = \sum_{n=1}^{N_W} \left[S_n \cdot \frac{\partial \lambda_n}{\partial K_i} + (\lambda_n - 2C_nS_n - B_n) \cdot \frac{\partial S_n}{\partial K_i} \right] \\ f''_{K_i^2} = \sum_{n=1}^{N_W} \left[\begin{array}{l} S_n \cdot \frac{\partial^2 \lambda_n}{\partial K_i^2} + 2\frac{\partial \lambda_n}{\partial K_i} \cdot \frac{\partial S_n}{\partial K_i} - 2C_n \cdot (\frac{\partial S_n}{\partial K_i})^2 \\ + (\lambda_n - 2C_nS_n - B_n) \cdot \frac{\partial^2 S_n}{\partial K_i^2} \end{array} \right] \\ f''_{K_iK_j} = \sum_{n=1}^{N_W} \left[\begin{array}{l} S_n \cdot \frac{\partial^2 \lambda_n}{\partial K_i \partial K_j} + \frac{\partial \lambda_n}{\partial K_i} \cdot \frac{\partial S_n}{\partial K_j} + \frac{\partial \lambda_n}{\partial K_j} \cdot \frac{\partial S_n}{\partial K_i} - 2C_n \cdot \frac{\partial S_n}{\partial K_i} \cdot \frac{\partial S_n}{\partial K_j} \\ + (\lambda_n - 2C_nS_n - B_n) \cdot \frac{\partial^2 S_n}{\partial K_i \partial K_j} \end{array} \right] \\ f''_{K_jK_i} = f''_{K_iK_j} \end{cases} \tag{13.31}$$

So far we need three groups of partial derivatives to get the optimal $\mathbf{K}$. There are: (1) $\partial\boldsymbol{\lambda}/\partial K_i, \partial\mathbf{S}/\partial K_i$; (2)$\partial^2\boldsymbol{\lambda}/\partial K_i^2, \partial^2\mathbf{S}/\partial K_i^2$, and (3)$\partial^2\boldsymbol{\lambda}/\partial K_i \partial K_j$, $\partial^2\mathbf{S}/\partial K_j \partial K_i$. Once these three groups of partial derivatives are obtained Equation (13.28) Equation (13.31) can be used to find the optimal bids for the generators (from 1 to N_w) in subgroup ϕ_w. These three groups of partial derivatives can be obtained from the OPF problem based on bid sensitivities, among which the first two groups of partial equations have already been derived in 13.3.3. What we need is the second order mixed partial derivatives of vector $\boldsymbol{\lambda},\mathbf{S}$ with respect to K_i, K_j.

To derive $\partial^2\boldsymbol{\lambda}/\partial K_i\partial K_j$ and $\partial^2\mathbf{S}/\partial K_j\partial K_i$ we rewrite the equation of the first-order partial derivatives here:

$$\frac{\partial \mathbf{Z}}{\partial K_i} = -\mathbf{H}^{-1}\cdot\frac{\partial \mathbf{H}}{\partial K_i}\cdot\Delta\mathbf{Z} - \mathbf{H}^{-1}\cdot\frac{\partial \nabla\mathbf{L}}{\partial K_i} \tag{13.32}$$

To imply the second-order partial derivatives to Equation (13.32) with respect to K_j, we obtain:

$$\begin{aligned}\frac{\partial^2 \mathbf{Z}}{\partial K_i\partial K_j} &= \mathbf{H}^{-1}\cdot\frac{\partial \mathbf{H}}{\partial K_j}\cdot\mathbf{H}^{-1}\cdot\frac{\partial \mathbf{H}}{\partial K_i}\cdot\Delta\mathbf{Z} - \mathbf{H}^{-1}\cdot\frac{\partial^2 \mathbf{H}}{\partial K_i\partial K_j}\cdot\Delta\mathbf{Z}\\ &\quad -\mathbf{H}^{-1}\cdot\frac{\partial \mathbf{H}}{\partial K_i}\cdot\frac{\partial \mathbf{Z}}{\partial K_j} + \mathbf{H}^{-1}\cdot\frac{\partial \mathbf{H}}{\partial K_j}\cdot\mathbf{H}^{-1}\cdot\frac{\partial \nabla\mathbf{L}}{\partial K_i} - \mathbf{H}^{-1}\cdot\frac{\partial^2 \nabla\mathbf{L}}{\partial K_i\partial K_j}\end{aligned} \tag{13.33}$$

According to Equation (13.32), we have Equation (13.34)

$$\frac{\partial \mathbf{Z}}{\partial K_j} = -\mathbf{H}^{-1}\cdot\frac{\partial \mathbf{H}}{\partial K_j}\cdot\Delta\mathbf{Z} - \mathbf{H}^{-1}\cdot\frac{\partial \nabla\mathbf{L}}{\partial K_j} \tag{13.38}$$

So replace $\partial\mathbf{Z}/\partial K_j$ with Equation (13.34) in Equation (13.33) we get

$$\begin{aligned}\frac{\partial^2 \mathbf{Z}}{\partial K_i\partial K_j} &= \mathbf{H}^{-1}\cdot\frac{\partial \mathbf{H}}{\partial K_j}\cdot\mathbf{H}^{-1}\cdot\frac{\partial \mathbf{H}}{\partial K_i}\cdot\Delta\mathbf{Z} - \mathbf{H}^{-1}\cdot\frac{\partial^2 \mathbf{H}}{\partial K_i\partial K_j}\cdot\Delta\mathbf{Z}\\ &\quad +\mathbf{H}^{-1}\cdot\frac{\partial \mathbf{H}}{\partial K_i}\cdot\mathbf{H}^{-1}\cdot\frac{\partial \mathbf{H}}{\partial K_j}\cdot\Delta\mathbf{Z} + \mathbf{H}^{-1}\cdot\frac{\partial \mathbf{H}}{\partial K_i}\cdot\mathbf{H}^{-1}\cdot\frac{\partial \nabla\mathbf{L}}{\partial K_j}\\ &\quad +\mathbf{H}^{-1}\cdot\frac{\partial \mathbf{H}}{\partial K_j}\cdot\mathbf{H}^{-1}\cdot\frac{\partial \nabla\mathbf{L}}{\partial K_i} - \mathbf{H}^{-1}\cdot\frac{\partial^2 \nabla\mathbf{L}}{\partial K_i\partial K_j}\end{aligned} \tag{13.35}$$

Furthermore, because $\dfrac{\partial^2\nabla\mathbf{L}}{\partial K_i\partial K_j} = 0$ and at the optimal solution of OPF problem $\nabla\mathbf{Z} = 0$, so Equation(13.35) can be recast as

$$\frac{\partial^2 \mathbf{Z}}{\partial K_i\partial K_j} = \mathbf{H}^{-1}\cdot\frac{\partial \mathbf{H}}{\partial K_i}\cdot\mathbf{H}^{-1}\cdot\frac{\partial \nabla\mathbf{L}}{\partial K_j} + \mathbf{H}^{-1}\cdot\frac{\partial \mathbf{H}}{\partial K_j}\cdot\mathbf{H}^{-1}\cdot\frac{\partial \nabla\mathbf{L}}{\partial K_i} \tag{13.36}$$

In Section 13.3.3 we have

$$\frac{\partial \mathbf{Z}}{\partial K_i} = -\mathbf{H}^{-1} \cdot \frac{\partial \nabla \mathbf{L}}{\partial K_i}$$

and replace Equation (13.36) with it, we get Equation (13.37)

$$\frac{\partial^2 \mathbf{Z}}{\partial K_i \partial K_j} = -\mathbf{H}^{-1} \cdot \frac{\partial \mathbf{H}}{\partial K_i} \cdot \frac{\partial \mathbf{Z}}{\partial K_j} - \mathbf{H}^{-1} \cdot \frac{\partial \mathbf{H}}{\partial K_j} \cdot \frac{\partial \mathbf{Z}}{\partial K_i} \tag{13.37}$$

Now all the first-order partial derivatives and second order partial derivatives are calculated and are shown in Equation (13.38).

$$\begin{cases} \dfrac{\partial \mathbf{Z}}{\partial K_i} = -\mathbf{H}^{-1} \cdot \dfrac{\partial \nabla \mathbf{L}}{\partial K_i} \\ \dfrac{\partial^2 \mathbf{Z}}{\partial K_i^2} = -2\mathbf{H}^{-1} \cdot \dfrac{\partial \mathbf{H}}{\partial K_i} \cdot \dfrac{\partial \mathbf{Z}}{\partial K_i} \\ \dfrac{\partial^2 \mathbf{Z}}{\partial K_i \partial K_j} = -\mathbf{H}^{-1} \cdot \dfrac{\partial \mathbf{H}}{\partial K_i} \cdot \dfrac{\partial \mathbf{Z}}{\partial K_j} - \mathbf{H}^{-1} \cdot \dfrac{\partial \mathbf{H}}{\partial K_j} \cdot \dfrac{\partial \mathbf{Z}}{\partial K_i} \end{cases} \tag{13.38}$$

We have enough information to calculate the optimal bids $K, K \in (K_1, \cdots, K_{Nw})$ for the members in subgroup ϕ_w corresponding to combination C_r. We solve Equation (13.24) Equation (13.25) in an iterative way for each subgroup then we can get Nash Equilibrium corresponding to combination C_r.

13.3.5 Case Studies and Conclusions

The above-mentioned bidding processes are illustrated for the IEEE-30 bus system. The cost coefficients of six generators are listed in Table 13.4.

Table 13.4. Cost coefficients of generators

Gens	G1	G2	G3	G4	G5	G6
A($)	100	100	100	100	100	100
B($/MW)	2	1.75	1	3.25	3	3
C($/MW^2)	0.0375	0.175	0.625	0.0834	0.25	0.25

13.3.5.1 Bid Sensitivities

We list the Bid-Price sensitivities and Bid-Output sensitivities in Table 13.5 and Table 13.6 respectively when 6 units bid at their costs. In Table 13.5, the biggest and smallest sensitivities of each generator are highlighted, for example, the bid of unit1 influences node 5's nodal price most and node 8's price least. Here 1, 2, 5, 8, 11, 13 are generator nodes and the others are load nodes. Also we can see very clearly which generator has the strongest and weakest influence on the system nodal price, for example, in this case unit 1 has the strongest and unit 3 has the weakest as shown in the table. Table 13.6 shows the bid-output sensitivities when all the units bid at their costs. The numbers in this table show that if one generator increases its bids, the outputs of its own and other generators change. For example, if unit 1 bids up, while its own output decreases by 2141MW, unit 4 will benefit most (by increasing its output by 871.4MW). This index is particularly valuable when considering the gaming problem or coalition problem when units in different nodes are from the same generation company.

Table 13.5. Bid-price sensitivities

$\partial\lambda i/\partial Ki$	K1	K2	K3	K4	K5	K6
$\lambda 1$	155.9	7.532	0.703	28.71	3.328	3.286
$\lambda 2$	153.1	8.351	0.779	31.05	3.598	3.533
$\lambda 5$	157.6	8.593	1.062	34.39	3.978	3.856
$\lambda 8$	145.3	7.732	0.777	36.84	3.919	3.736
$\lambda 11$	147.2	7.826	0.785	34.24	4.762	3.732
$\lambda 13$	146.7	7.754	0.768	32.93	3.765	4.820
$\lambda 3$	151.9	7.853	0.764	32.38	3.763	3.779
$\lambda 6$	149.2	7.943	0.797	34.9	4.029	3.842
$\lambda 7$	153.3	8.245	0.912	35.04	4.049	3.884

Table 13.6. Bid-output sensitivities

$\partial Si/\partial Ki$	K1	K2	K3	K4	K5	K6
S1	-2141	100.4	9.374	382.8	44.37	43.82
S2	437.5	-183.7	2.227	88.72	10.28	10.09
S3	126.1	6.874	-17.98	27.51	3.182	3.085
S4	871.4	46.35	4.657	-612.6	23.49	22.40
S5	294.4	15.65	1.570	68.47	-85.88	7.464
S6	293.3	15.51	1.535	65.85	7.530	-84.92

In Table 13.7 and Table 13.8 we list the bid-profit sensitivities and bid-line flow sensitivities respectively when six units bid at their costs. All the bids-profit sensitivities are positive due to the bidding point at their costs. We can also get some negatives if they bid in other values. In Table 13.8 the most and least influenced lines as a result of adjusting bids are listed. These data show clearly how much to adjust the bids so as to increase or decrease the power flow of transmission lines.

It can also be noted that we could get different bid-sensitivities when these units bid in different values, i.e. bid-sensitivities are strongly coupled with units' current bids.

Table 13.7. Bid-profit sensitivities

∂Pi/∂Ki	K1	K2	K3	K4	K5	K6
P1	24668	1192	111.3	4544	526.7	520.1
P2	5559	303.7	28.29	1128	130.7	128.3
P3	1854	101.1	12.45	404.7	46.81	45.38
P4	10104	537.7	54.04	2558	272.5	259.8
P5	3512	186.7	18.73	817.0	113.2	89.05
P6	3468	183.3	18.16	778.5	89.00	114.0

Table 13.8. Bid-line flow sensitivities

∂PL/∂Ki	K1	K2	K3	K4	K5	K6
L4-6	-617.0	-17.54	2.638	264.7	19.55	-17.06
L6-7	130.1	12.56	9.777	-87.89	-9.559	-7.354
L25-26	-0.004	4E-04	8E-05	0.008	-0.001	-0.002
L29-30	-0.004	3E-04	1E-04	0.009	-0.001	-0.002
L1-2	-1514	98.94	7.524	240.5	27.05	24.42
L11-9	286.1	8.242	-5.925	61.58	-93.35	-5E-10
L10-9	-54.21	15.49	18.13	9.789	51.93	5E-09
L12-13	-377.7	-100.4	-86.31	-151.4	-92.32	7E-10
L28-27	-25.73	-6.948	-5.792	-22.59	-1.192	0

13.3.5.2 Bidding Process and Optimal Bids without Coalition

As detailed in Figure13.15 the bidding-decision processes in an iterative way, about five bidding rounds are needed to reach the Nash Equilibrium in IEEE 30-bus system. Figure13.16 shows the bid results after each bidding round when they bid at their costs in the initial state. In this equilibrium, the bid for each unit is optimal given that the others choose these equilibrium strategies. The final bidding strategies for six units are (0.0877, 0.2144, 0.6722, 0.1315, 0.2947, 0.2955).

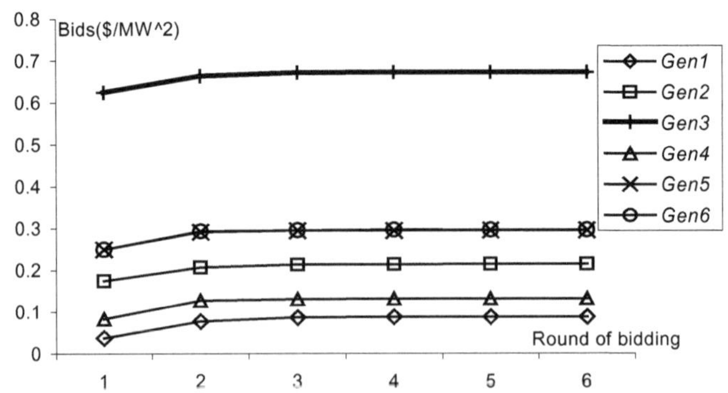

Figure 13.16. Curves of bids in IEEE 30-bus system

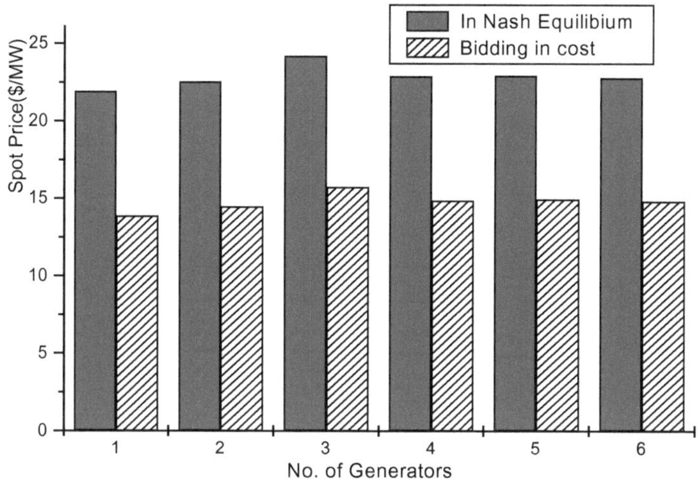

Figure 13.17. Comparison of outputs before and after convergence

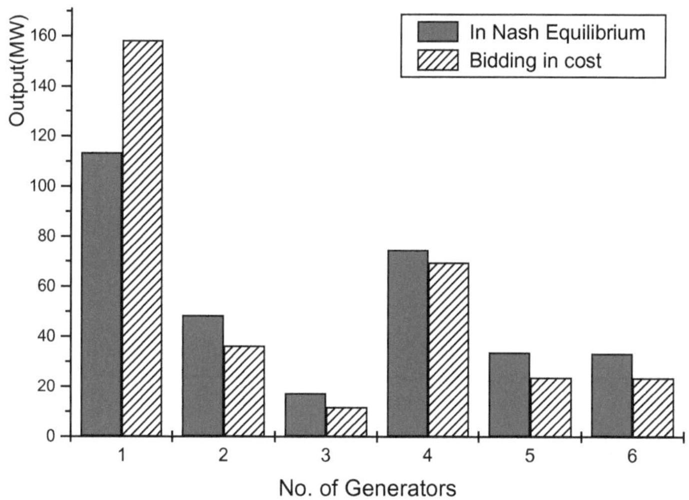

Figure 13.18. Comparison of nodal prices before and after convergence

Comparing the output and spot price of each unit when they reach equilibrium with those when they bid at their costs, we can see that the output of unit1 increases from 113.4MW to 158.2MW, but others decrease more or less, as shown in

Figure (13.17). The spot price in each generator's node goes up as its finial bid is more than its cost, as shown in Figure (13.18).

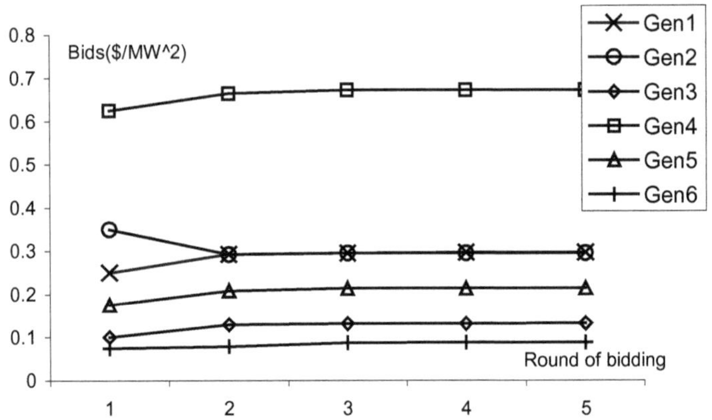

Figure 13.19. Curve of bids when initial bids not at their costs

It should be noted that the units' initial bids do not influence the final bids (when in Nash Equilibrium) as the cost functions are assumed to be convex, but affect the number of iterations needed. This is illustrated in Figure 13.19, where the units' initial bids are not their costs, but from (0.075, 0.175, 0.625, 0.1, 0.25, 0.293). In this case the finial bids in equilibrium are the same as in Figure 13.16.

13.3.5.3 Results When Bidding under Coalition

When coalition is taken into account, for simplification the collection of combinations $\boldsymbol{C}$ is reduced with the assumption that the generator m only forms a coalition with one other member and the other generators bid individually against this coalition. In our 6-generator case, generator no.1 is under coalition so the collection $\boldsymbol{C}$ comprises five potential combinations which are {(1-2), (3), (4), (5), (6)}, {(1-3), (2), (4), (5), (6)}, {(1-4), (2), (3), (5), (6)}, {(1-5), (2), (3), (4), (6)}, and {(1-6), (2), (3), (4), (5)}.

Table 13.9. Comparison of the profits of subgroups

Subgroups	(1—2)	(1—3)	(1—4)	(1—5)	(1—6)
Profit without coali-	2167	1785	2573	1960	1953
Profit under coalition	2666	1895	3418	2223	2246
Extra profit	499	110	845	263	293

Profits of the coalition subgroup for each combination are listed in Table 13.9. Table 13.9 shows that the profits of the coalition subgroup is higher than the total profit of the members in this subgroup when they bid individually. For example the profit of generator no.1 and generator no.2 increases to 2666$ if they form a coalition compared with 2167$ when they bid individually. The highest extra profit is 499$ when generator no.1 form a coalition with generator no.4 and the lowest extra profit is 110$ when it form a coalition with generator no.3. These results are consistent with the implications of the results of bid sensitivities, which suggest that apart from generator no.1, generator no.4 is the most dominant generator in the power market and generator no.3 is the weakest. So we can draw the conclusion that in the general situation (or say under perfect competition) the generator can make more profit if it joins the coalition group with more dominate members in the electricity markets. This general situation does not include cases which usually happen under imperfect competition, such as some group of generators using some 'vulnerable' transmission lines for gaming under congestion. This kind of interaction between patterns of generation ownership and network constraints could lead to market behaviour that differs from some general conclusions, as illustrated in [16] and [44].

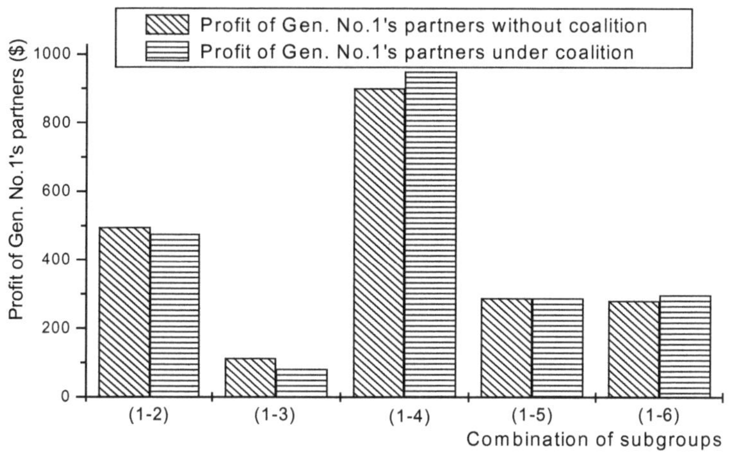

Figure 13.20. Comparison of profits of Gen. No.1's partner

Although the profits of subgroups in a coalition is greater than those without coalition, it is not always true that every member in the subgroup gets a higher profit under coalition than when they bid individually. That means, in the subgroup under coalition some members may lose own benefit but gain group benefit from coalition. That is illustrated in Figure 13.20, where the profits of generator no.1 partner in subgroup without coalition are compared with those under coalition. For instance, generator no.3's profit drops down to 81.6$ after coalition with generator no.1 compared with 114$ when it bids individually, although the total profit of generator no.1 and generator no.3 increases to 1895$ from 1785$, as shown in Table 13.9.

If we split the extra profit shown in Table 13.9 equally between the members of the coalition subgroup, we get Figure 13.21 which illustrates the split of the extra profits of generator no.1 under coalition for each combination of coalitions. The profit of generator no.1 without coalition is 1672.4\$. It obtains the highest profit of 2094.9\$ when it combines with generator no.4, and 1727.4\$ when it groups with generator no.3. So the priority list for generator no.1 is (1-4), (1-2), (1-6), (1-5), (1-3).

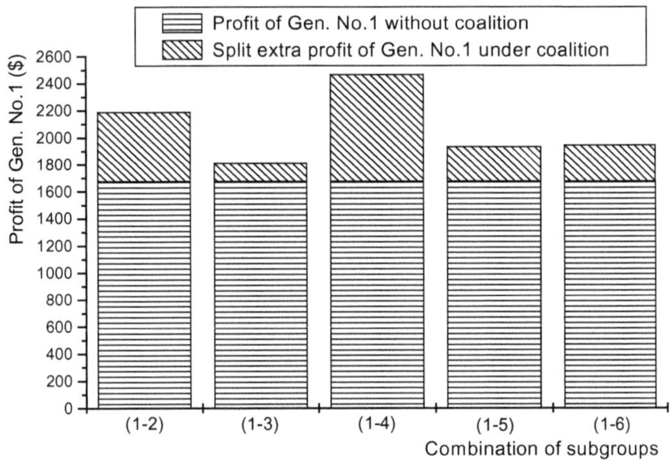

Figure 13.21. Comparison of profits of Gen. No.1

13.3.5.4 Conclusions

1. Without demand-side bidding, the results presented suggest that a liberalised market will result in reduction in social welfare, even without any market power problems. Equation (13.7) gives the optimal dispatch, which maximises the welfare. The optimal bids, i.e. the solution of Equation (13.7), result in a different dispatch with different quantities produced and prices, as shown in Figure 13.17 and Figure13.18. Hence the societal welfare would be reduced and the consumers overcharged.

2. Bid sensitivities are valuable information for individual suppliers. They are used to determine the optimal bids taking into account both maximising individual profit and system security as illustrated in this paper. Furthermore, they are also important indexes or information for individual suppliers if they are allowed to make coalitions with other units (or companies). From this point of view these sensitivities are also very useful for the market regulator to investigate potential coalitions or market power.

3. The results of coalition show that such a general potential exists in the power markets that some of the most dominant generators (indicated by bid sensitivities) might make coalitions to gain more benefit from the market. The total

benefit of the coalition group is increased over that without coalition, although some members in this coalition may gain less individual benefit.

4. The individual suppliers would submit their bids with the assumption that other rivals bid in a rational way. Further work will investigate how to bid when some players do not bid in a rational way, i.e. how to game and exercise market power. Furthermore some intra-temporal constraints like unit commitment constraints should be taken into account for the day-ahead market.

13.4 Integrated Bidding Strategies with Optimal Response to the Probabilistic Local Marginal Prices

13.4.1 Introduction

In this section we study the bidding strategies of GENCOs in a pool-based day-ahead market. The proposed bidding strategies produced by each GENCO in this section take into account cost-recovery, physical constraints, and market price fluctuation resulting from other rivals' bidding strategies. An OPF-based market price simulator is employed to produce Local Marginal Prices (LMPs) and system demands at each generation node with the corresponding probabilities reflecting rival competitors' bidding strategies. As the best response to the LMPs and system demands at its own generation node, the GENCO produces incremental step-cased price-output bidding curves with the corresponding probabilities using the market-oriented unit commitment model. These offers are called integrated offers due to the full consideration of the GENCO's cost-recovery and its own physical constraints such as ramping rates and up-time/down-time etc. Using the theory of Multiple Criteria Decision-making (MCDM), the offer having the best compromise among its payoff, market share and probability to win is selected as the final bid of the GENCO.

The rest of this section is arranged as follows. First, the description of the proposed GENCO's Bidding Decision System (GencoBDS) is given in Section 13.4.2 and then we introduce three modules of GencoBDS in Section 13.4.3. Section 13.4.4 describes a test case and gives conclusions.

13.4.2 The Proposed GencoBDS

The proposed bidding model, named GencoBDS, is based on market-oriented unit commitment and an OPF-based market price simulator, the structure is shown in Figure 13.22. In this model, single-round bidding with simple bids and discriminatory pricing (Local Marginal Price) are accepted as the basic bidding rules. The objective in this paper is for individual generation companies to produce optimal offers in the day-ahead electricity market where the physical constraints of generators and the bidding strategies of rival competitors are taken into account.

Commercial constraints and potential coalitions related to the day-ahead market are not considered in this paper. Customer bidding is not included in the simulator model, but can be introduced in a similar way to the bidding strategies of rival competitors.

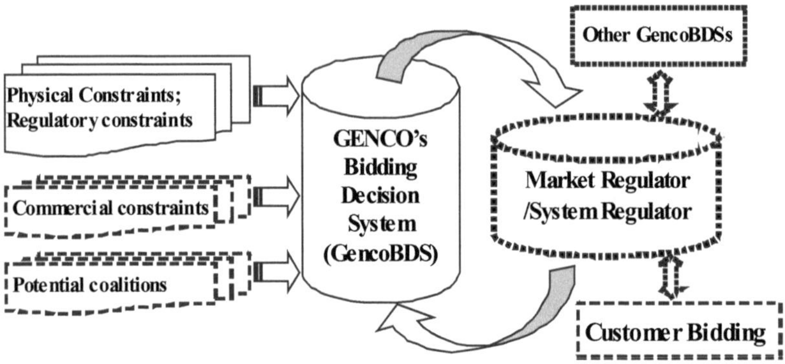

Figure 13.22. Market structure of GENCO's bidding decision system (GencoBDS)

The proposed GencoBDS in Figure 13.22 consists of three main modules: LMP simulator, self-scheduling UC model and MCDM system. The relationship of these three modules is shown in Figure 13.23. Details of each module and in-depth decision-making process will be presented in Section 13.4.3.

The main features of this presented model are:

1. The offers produced by GENCOs have taken into account cost recovery (the start-up cost, shut-down cost and no-load cost) and rival competitors' bidding strategies. It should be noted here that the highly integrated bidding strategies should consider commercial constraints and potential coalitions with partners as well.

2. Local marginal prices are produced by the probabilistic LMP simulator based on an OPF model which takes into account rivals' bidding strategies and security constraints. By incorporation of the two-segment normal distribution with probability density functions of rival offers, the LMP simulator produces stochastic LMPs for the market-oriented UC model.

3. The OPF-based LMP simulator reflects the truth that the LMPs are influenced by the offers of GENCOs and system security constraints etc. Considering that GENCOs have limited information of other rival competitors and the whole network system at hand, some simplification of the LMP simulator can be suggested as long as the simulator can provide useful system information (such as LMPs and system demands at each generation node).

4. The MCDM method is used to help find the global optimal decision. Not only payoff of the GENCO but also market share are incorporated as multi-criteria in the bidding decision-making process.

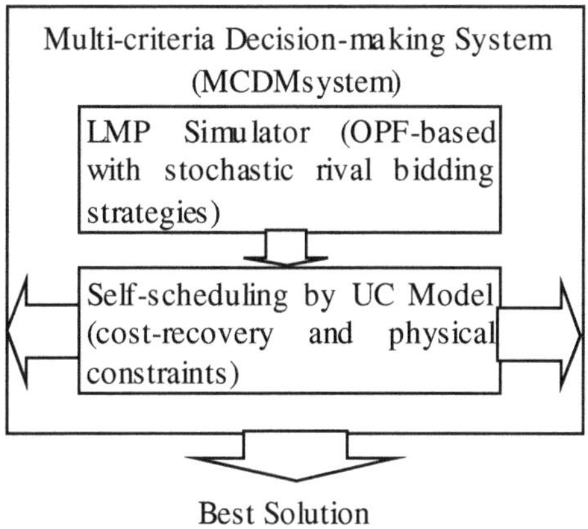

Figure 13.23. The three main modules in GencoBDS

13.4.3 Three Main Modules of GencoBDS

13.4.3.1 Security-constrained Probabilistic LMP Simulation Model

Currently two main pricing systems are implied in a deregulated power market. One is uniform pricing and the other is discriminatory pricing. The PJM energy market changed the pricing system from uniform pricing to discriminatory pricing in 1998. It should be addressed that the LMP is calculated based on actual system operating condition; the uniform price, however, is usually the highest accepted offer. The LMPs might be different from node to node and include information on system security constraints. This paper uses OPF-based probabilistic LMP simulator to simulate the LMPs with the estimation of rivals' bidding strategies and forecasted system loads and congestion information posted from the ISO. These simulated LMPs will precede feedback to the market-oriented unit commitment model so that

the GENCOs can produce offers reflecting their own marginal generation costs.

The objective function of LMP simulation model is to minimise the total purchase cost, i.e. the ISO spends the least amount to purchase generation from the GENCOs.

$$Min \;\; F(S_m, S_i) = (C_m S_m^2 + B_m S_m + A_m) + \sum_{i=1(i \neq m)}^{n} \left(C_i S_i^2 + B_i S_i + A_i \right)$$

$$\text{subject to:} \begin{cases} \mathbf{g}(\mathbf{z}) = 0 \\ \mathbf{h}_l \leq \mathbf{h}(\mathbf{z}) \leq \mathbf{h}_u \end{cases} \qquad (13.39)$$

Here i is the index of generation nodes, from 1 to n; m is the index of the GENCO under consideration; C, B, A are the offers coefficients of GENCOs; S_i is the scheduled output at each node; $\mathbf{z}$ is the vector of the controllable and state variables; $\mathbf{g}$ is the vector of equality constraints including power flow balance etc.; $\mathbf{h}$ is the vector of inequality constraints, such as generators' output limit; voltage limit; transformer tap ratio limit; transmission line capacity constraints and so on; $\mathbf{h}_l$, $\mathbf{h}_u$ are the lower limit and upper limit of inequality constraints respectively.

Now we introduce slack variables into the objective function Equation (13.39) to transform inequality constraints into equality constraints by incorporating them in a logarithmic barrier function so as to recast the equation as an augmented Lagrangian function

$$\begin{aligned} \min \quad L = \mathrm{F}(S_m, S_i) - \lambda^T \mathbf{g}(\mathbf{z}) + \pi_l^T (\mathbf{h}(\mathbf{z}) - \mathbf{s}_l - \mathbf{h}_l) \\ + \pi_u^T (\mathbf{h}(\mathbf{z}) + \mathbf{s}_u - \mathbf{h}_u) - \mu(\sum_{i=1}^{k} \ln s_{li} + \sum_{i=1}^{k} \ln s_{ui}) \end{aligned} \tag{13.40}$$

where $s_l \geq 0, s_u \geq 0, \pi_l < 0, \pi_u > 0, \mu > 0$. μ is barrier parameter, s_l, s_u are slack variables and π_l, π_u are Lagrangian multipliers with respect to lower limit constraints and upper limit constraints respectively which, in more detail, include voltage limits, output limits, transmission limits and so on. So we have $\pi_l \in (\pi_{ls} \quad \pi_{lv} \quad \pi_{ll} \quad \cdots\cdots)$ and $\pi_u \in (\pi_{us} \quad \pi_{uv} \quad \pi_{ul} \quad \cdots\cdots)$, where π_{us}, π_{ls} with respect to output upper limits and lower limits respectively, π_{uv}, π_{lv} with respect to voltage limits and π_{ul}, π_{ll} with respect to transmission limits. Based on Equation (13.40) the Local Marginal Price at node m is described as below:

$$\lambda_m = \left. \frac{\partial F(S_m, S_i)}{\partial S_m} \right|_* + \pi_{ls_m} + \pi_{us_m} \tag{13.41}$$

The sign * in Equation (13.41) represents the optimal solution. From Equation (13.41) we can see that the LMP takes into account system security constraint costs. It should be noted here that transmission line constraint cost is implicitly included in this function.

In Equation (13.39) of the LMP simulation model, each rival's "cost" function is a quadratic function $C_i S_i^2 + B_i S_i + A_i \quad (i = 1, \cdots, n, \; i \neq m)$. Rivals can produce different offers by changing the coefficients $C_i, B_i, or\ A_i$. So in order to simulate such stochastic offers we define the coefficient C_i, B_i, A_i as stochastic variables[2] by introducing a Gaussian distribution to describe the distribution of coefficients C_i, B_i, A_i.

[2] For simplification only coefficient C is employed as a stochastic variable in this paper.

In the practical bidding problem, the rival competitors would rather bid up a little bit higher than the expected value of $C_i, B_i, or\ A_i$ and it would not like to bid too much lower. So we use two segments of distribution functions with two different deviations instead of one segment of distribution function with only one deviation. Let $X_i = C_i, B_i, or\ A_i$, so the forecasted coefficients A_i, B_i, C_i are expectations of X_i denoted as $\mu = \mathrm{E}(X_i)$, and the tendency of the bidding strategy of rival competitors is the deviation of X_i denoted as $\delta = D(X_i)$. In the two-segment distribution function, $\delta = \mu/6$ when $\mu/2 < X_i \le \mu$ and $\delta = \mu/3$ when $\mu < X_i < 2\mu$. That is described in Equation (13.42) and its diagram shown in Figure 13.24.

$$f(X_i) = \begin{cases} \dfrac{6}{\sqrt{2\pi}\,\mu}\,\ell^{-\frac{18(X_i-\mu)^2}{\mu^2}} & (\mu/2 < X_i \le \mu) \\ \dfrac{3}{\sqrt{2\pi}\,\mu}\,\ell^{-\frac{9(X_i-\mu)^2}{2\mu^2}} & (\mu < X_i < 2\mu) \end{cases} \quad (\mu > 0; X_i = A_i, B_i, or\ C_i) \quad (13.46)$$

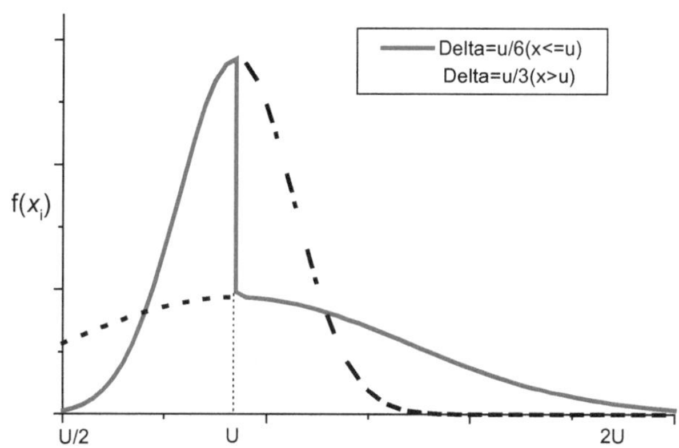

Figure 13.24. Diagram of two-segment distribution function of cost coefficients

13.4.3.2 Self-scheduling Unit Commitment Model

In the day-ahead generation market, the GENCO should produce optimal offers to ISO based on its effective forecasted spot prices, loads, and other information posted by ISO such as transmission limits and so on. In this section a detailed market-oriented unit commitment model is applied for the GENCOs to produce the optimal offers in the day-ahead electricity market. This model can give the optimal response to the forecasted day-ahead spot prices and system demands on its own generation node which is simulated in a security constrained probabilistic LMP simulator model as described above. GENCOs submit an aggregated biding curve built from n bidding curves if it has n separate generators.

The objective function is to maximise the GENCO's profit as describe in Equation (13.43). The notation is given in the Appendix.

$$Max \quad f(g_{n\cdot t}, Std_{n.t}, Dsu_{n.t}, Dsd_{n.t}) = \sum_{n=1}^{N}\sum_{t=1}^{T}(S_{n.t})$$

where:

$$S_{n.t} = \lambda_t g_{n.t} - cn_n \cdot Std_{n.t} - cv_n \cdot g_{n.t} - cu_n \cdot Dup_{n.t} - cs_n \cdot Ddn_{n.t} \tag{13.43}$$

It is subject to the following constraints.

(1) Generation Schedule Constraints:

$$S_t \geq \sum_{n=1}^{N} g_{n.t} \quad (t = 1, T) \tag{13.44}$$

(2) Unit Maximum and Minimum Output Constraints:

$$g\min_{n.t} \cdot Std_{n.t} \leq g_{n.t} \leq \overline{P}_n \cdot Std_{n.t} \quad (t = 1, T; n = 1, N) \tag{13.45}$$

(3) Ramping-up Rate Constraints:

$$g_{n.1} \leq g_{n.0} + RU_n \quad (t = 1; n = 1, N) \tag{13.46}$$

$$g_{n.t} \leq g_{n.t-1} + RU_n \quad (t = 2, T; n = 1, N) \tag{13.47}$$

(4) Ramping-down Rate Constraints:

$$g_{n.0} - g_{n.1} \leq RD_n \quad (t = 1; n = 1, N) \tag{13.48}$$

$$g_{n.t-1} - g_{n.t} \leq RD_n \quad (t = 2, T; n = 1, N) \tag{13.49}$$

(5) Minimum Time-up Constraints:
If $U_n^0 \geq Tsu_n$ or $U_n^0 = 0$ the following two constraints should be taken into consideration (the subscript n is neglected)

$$\sum_{i=k}^{k+Tup_t-1} Std_i \geq Tsu \cdot Dsu_k \quad (k = 1, \cdots, T - Tup + 1) \tag{13.50}$$

$$\prod_{i-K}^{T} Std_i \geq Dsu_k \quad (k = T - Tup + 2, \cdots, T) \tag{13.51}$$

If $0 < U_n^0 < Tsu_n$, this constraint should be added

$$Std_i = 1 \quad (i = 1, \cdots, Tsu - U^0) \tag{13.52}$$

(6): Minimum Down-time Constraints:

If $D_n^0 \geq Tsd_n$ or $D_n^0 = 0$ the following two constraints should be taken into consideration (the subscript n is neglected)

$$\sum_{i=k}^{k+Tsd_t-1} (1 - Std_i) \geq Tsd \cdot Dsd_k \quad (k = 1, \cdots, T - Tsd + 1) \tag{13.53}$$

$$\prod_{i=k}^{T} (1 - Std_i) \geq Dsd_k \quad (k = T - Tsd + 2, \cdots, T) \tag{13.54}$$

If $0 < D_n^0 < Tsd_n$, this constraint should be added

$$Std_i = 0 \quad (i = 1, \cdots, Tsd - D^0) \tag{13.55}$$

(7): Logical Constraints

$$Dsd_{n.t} + Dsu_{n.t} \leq 1 \tag{13.56}$$

$$Std_{n.t} - Std_{n.t-1} = Dsu_{n.t} - Dsd_{n.t} \qquad t \in (2, ..., T) \tag{13.57}$$

$$Std_{n.t} - Std_n^0 = Dsu_{n.t} - Dsd_{n.t} \qquad t = 1 \tag{13.58}$$

In Equation (13.48) S_t $(t = 1, T)$ is the system demand for each time period, i.e. the S_m in function Equation (13.39) of each time period. The total output of GENCO $\sum_{n=1}^{N} g_{n.t}$ after UC can be less than the system schedules S_t assuming each GENCO is not under an obligation to fully serve the system schedules. The UC model is solved using Mix Integer Programming (MIP)[45].

13.4.3.3 MCDM Method for Optimal Offers

The Multiple Criteria Decision-making (MCDM) method is accepted in the general auction problem [46]. The bidding decision-making process in the generation market is also a typical multiple criteria decision-making problem. As a GENCO, it would like to take into account the payoff for generation, the market share (a greater percentage of market share means more market power), risk-related factors etc. for bidding decision-making. This section uses the MCDM method and Expected Utility Theory (EUT) corresponding to approach 0 in the decision-making process.

As in any decision-making problem, the bidder has to choose a decision among a set of available alternatives (or situations) taking into account its preferences for the set of possible consequences. Let $\mathbf{A} = \{a_i\}$ be a set of alternatives, $i = 1,2,...,n$ and $\mathbf{P} = \{P(a_i)\}$ be a set of probabilities with respect to alternatives $\mathbf{A}$; $\mathbf{C} = \{c_k\}$ be a set of criteria which represent the factors affecting the decision in bidding strategies, $k = 1,2,...,m$ and $\mathbf{O} = \{o_k\}$ be a set of priorities with respect to criteria $\mathbf{C}$ (it is defined $\sum_{k=1}^{m} o_k = 1$); and μ_{ik} be the consequence of each alternative i with respect to each criterion k .

So the preference E_{ik} for each alternative i with respect to each criterion k is

$$\mathrm{E}_{ik} = \frac{\mu_{ik}}{\sum_{i}^{n} \mu_{ik} / n} \tag{13.59}$$

Taking into account all criteria, E_i is represented by

$$\mathrm{E}_i = \sum_{k}^{m} (\mathrm{E}_{ik} \cdot o_k) \tag{13.60}$$

and the optimal decision is

$$D = \max \{\mathrm{E}_i \cdot P(a_i)\} \tag{13.61}$$

This paper has two criteria, the payoff for generation and market share, which is the percentage of one-day's total output of GENCO with respect to the whole forecasted load demand in the day-ahead market.

13.4.3.4 Bidding Decision-making Process

The procedures for the bidding decision-making process are as below:

Step 1: To give a set of alternatives (or situations) $\mathbf{A}$ with a set of probabilities $\mathbf{P}$ resulting from the varied rivals' bidding strategies (by varying the rivals' bidding coefficient C_i).

Step 2 To get the best decision $D(a_i)$ with respect to each alternative a_i in the space of $\mathbf{A}$, the details are as follows:

(1) to give one set of basic loads for the full time band according to load forecasting results; (2) to get j sets of loads by disturbing the basic loads; (3) to run LMP simulator for each time period by employing j sets of loads to get j sets of LMPs and system demands at node k (the GENCO under consideration located at node k); (4) to run the UC model to get UC results (outputs and schedule decisions) with respect to each set of LMPs and system demands; (5) to built the ag-

gregated generation offer curve for each time band (the output with corresponding LMP).

Step 3: To calculate the payoff and market share with respect to each decision $D(a_i)$.

Step 4: To use MCDM to determine the best decision $D_{optimal}$ among the $D(a_i)$ $(i = 1,...,n)$.

13.4.4 Test Results and Conclusions

13.4.4.1 Test Results

The model and method presented in this paper are tested with the 9-bus system shown in Figure 13.25, comprising three GENCOs and three customers in the network, where GENCO no.3 is under consideration, i.e. GENCO no.1 and no.2 are rivals.

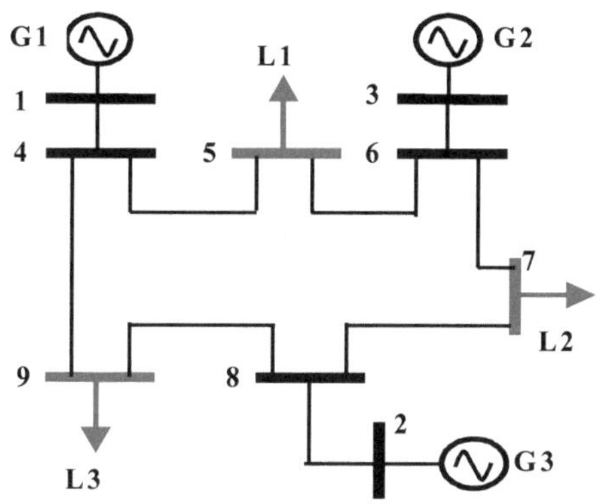

Figure 13.25. Diagram of 9-bus system

Table 13.10 shows the GENCO's characteristics and offer coefficients, where the coefficients C of rivals no.1 and no.2 are stochastic variables as given by function Equation (13.46) with $\mu_1 = 0.11$ and $\mu_2 = 0.085$. Although the method of bidding for 24 or 48 time periods during one day is accepted in most ISOs or power pools, for simplification this paper employs six time periods for illustration. The forecasted basic loads in 6 time periods are (1020,1215,1395,1635,1440,1170) MW respectively, used with disturbed loads together to assist GENCO to produce multi-section output-price bidding curves. In this case study two disturbed loads are accepted and four-section bidding curves are produced. Disturbing the basic loads by 10 percent up and 10 percent down respectively in the case study produces two sets of disturbed loads. Assume GENCO no.3 has three generators, whose parameters and initial states are shown in Table 13.11.

Table 13.10. Unit characteristics and cost coefficients

GENCO	Pmax	Pmin	A	B	C
1	570	165	1	5	Stochastic
2	750	180	2	1.2	Stochastic
3	630	160	3	1	0.1025

Table 13.11. Parameters and initial states of GENCO No.3's generators

G	Cap (MW)	Pmin (MW)	CN ($)	CV ($/MW)	CU ($)	CS ($)	RU (MW/h)	RD (MW/h)	Tsu (h)	Tsd (h)	Std_t^0	U_t^0	D_t^0
1	80	25	800	25	400	200	40	30	3	2	1	3	0
2	250	60	600	22	700	350	120	110	3	2	1	3	0
3	300	75	500	21	800	400	145	130	3	2	1	3	0

For simplification we assume the rivals keep the same bidding strategies for the whole time period, i.e. the coefficients C of rivals stay the same in different time periods. With this assumption we give the schedule results of 25 alternatives in Figure13.26 provided GENCO no.2 varies the coefficient C in the set (0.09, 0.1, 0.11, 0.12, 0.13) and GENCO no.3 varies in the set (0.065, 0.075, 0.085, 0.095, 0.105). The output in Figure 13.26 is the total output of GENCO no.3 during one day. The difference between the two curves shows that GENCO no.3 accepts outputs that are no bigger than ISO schedule results with the object of maximising its own payoff.

Figure 13.27 shows the payoff and market share of GENCO no. 3 for each alternative. Higher market share does not mean higher payoff as shown in Figure 13.26. The payoff curve follows the market share curve generally, but not always. In some cases it does not do so, for instance the market share in situation 1 is higher than in situation 2, but the payoff in situation 1 is less than in situation 2. So it is necessary to choose the market share and payoff as two different criteria in the decision-making problem. The priority for each criterion depends on experience and the decision-maker's preference. Additional relative criteria can be included in this method.

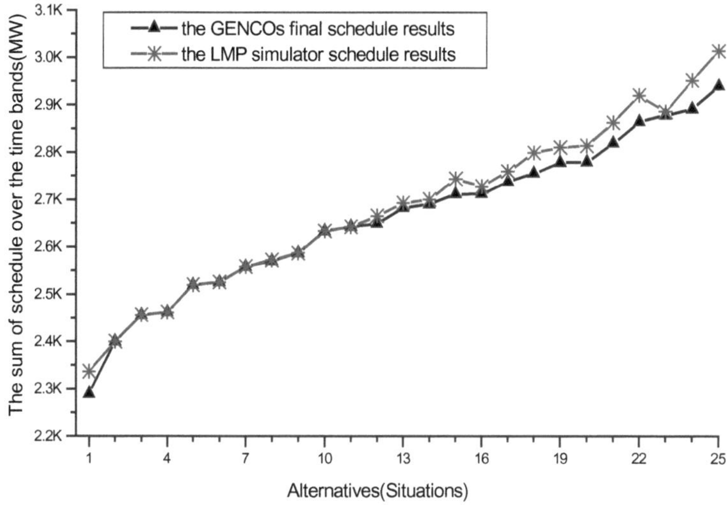

Figure 13.26. Comparison of schedule results

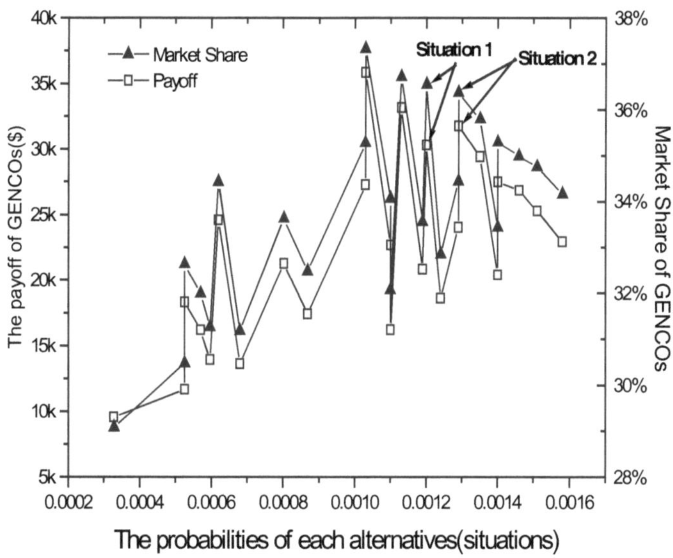

Figure 13.27. Two criteria used in bidding system

The results of preference E_i in MCDM processing are shown in Figure 13.28 where a comparison for setting different sets of priority $\mathbf{O} = \{o_k\}$ on criteria is also provided. Which set of priorities should be chosen in the decision-making problem can be decided by experience and judgement. With different sets of priorities the optimal decision might shift from one to the other. In Figure 13.28 the optimal decision shifts from decision1 to decision3 when the set of priorities is changed from set 1 to set 3.

Provided the priority of set 2 (i.e. $O_1 = 0.7$ and $O_2 = 0.3$) for GENCO no.3, the optimal bidding results of decision 2 are shown in Table 13.12. The node price when GENCO no.3 is fully scheduled is 44.72 in this case. The four-section output-price bidding curve for time period 2 is shown in Figure 13.29.

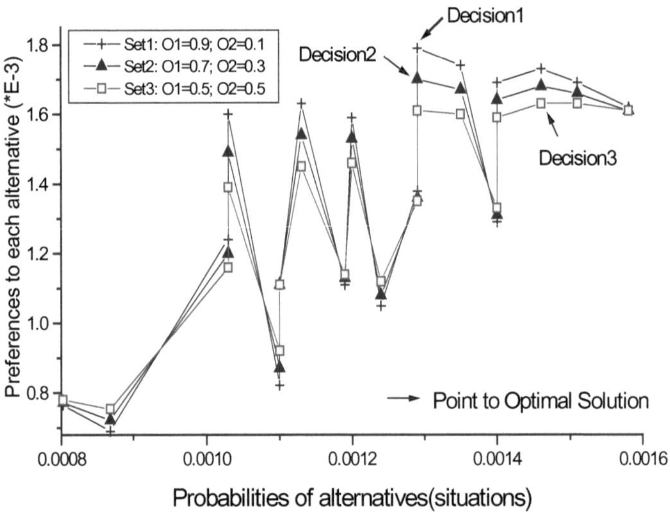

Figure 13.28. Optimal decision process with varied priorities of criteria

Table 13.12. The optimal offers submitted by GENCO no.3

Offers Time	S_1 (MW)	λ_1 (\$/MW)	S_1 (MW)	λ_2 (\$/MW)	S_1 (MW)	λ_3 (\$/MW)
T1	342.55	24.41	379.45	26.93	416.46	29.46
T2	406.35	28.77	450.66	31.79	495.14	34.83
T3	465.56	32.81	516.75	36.31	550.00	37.97
T4	545.01	38.24	550.00	42.37	550.00	48.38
T5	480.40	33.83	533.29	37.44	550.00	41.07
T6	391.66	27.77	434.25	30.67	550.00	25.42

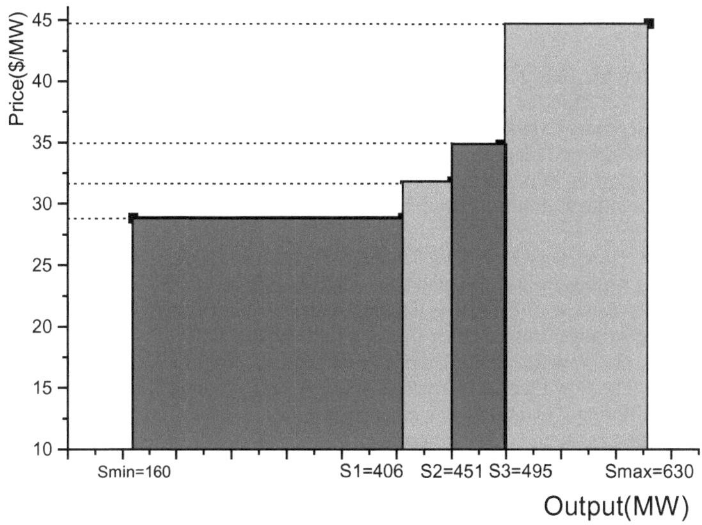

Figure 13.29. The optimal bids at time period 02

13.4.4.2 Conclusions

Given the auction rules of single round bidding and simple bids combined with the discriminatory pricing mechanism, a bidding model is built from the point of view of GENCOs. The GENCOs simulate the discriminatory prices using the LMP simulator and produce the corresponding optimal schedule results after reformed Unit Commitment. These simulated LMPs with probabilities reflect not only system security constraints costs, but also rival competitors' stochastic bidding strategies. As the bidding decision-making problem for GENCOs in the generation market might not be a decision-making problem with only one criterion, the MCDM method can be used to help each GENCO to find the global optimal decision, taking into account multiple criteria like payoff and market share are shown in this paper.

We assume that the rival competitors take the same strategy over the full time band, i.e. they keep the coefficient C the same for all time periods in this paper. Although this is not true in real systems, it is reasonable to assume that the different strategies of GENCOs will follow a basic pattern of one day's bidding curve of coefficient C, for example, the pattern with higher C at peak time and lower C at off-peak time. So the strategy set of rivals can be modelled as a set of bidding curves by moving up and moving down around the basic pattern curve. With this assumption, the solution space will be reduced to a reasonable range and can be solved in practice.

This chapter considers physical constraints on the UC model, commercial constraints will be incorporated in later work. Moreover, potential coalitions among GENCOs that locate at different nodes will be studied in the future as well.

13. References

[1]. R.Preston McAfee, John McMillan, Auctions and Bidding, Journals of Economic Literature, Vol 25, Issue 2, June 1987, pp. 699-738

[2]. The Independent Market Operator (IMO) in Ontario Canada, http://www.theimo.com/

[3]. The New Zealand Electricity Market (NZEM), http://www.transpower.co.nz/

[4]. Final Report on PPA Auction Design, 30 Dec, 1999, http://www.powerpool.ab.ca/

[5]. A report to the California Power Exchange: Iterative Bidding in the PX Market, 09 Feb, 1999

[6]. OFFER, 'Review of Electricity Trading Arrangement. Background Paper 1: Electricity Trading Arrangements in England and Wales', February 1998

[7]. OFFER, 'Review of Electricity Trading Arrangement. Background Paper 2: Electricity Trading Arrangements in Other Countries', February 1998

[8]. Ofgem, The New Electricity Trading Arrangements, Volume 1, July, 1999

[9]. Ofgem, The New Electricity Trading Arrangements, Volume 2, July, 1999

[10]. The California ISO, http://www.caiso.com/

[11]. PJM Interconnection, www.pjm.com

[12]. The ISO in New England, http://www.iso-ne.com/main.html

[13]. The Nord Pool, http://www.nordpool.no/

[14]. The NEMMCO, http://www.nemmco.com.au/publications/whitebook/introbook.htm

[15]. Yang He, Y.H.Song, Analysis of Auction Mechanism in Electricity Generation Markets by Two-Level Optimization Model, UPEC 2000, Belfast, Sep 2000

[16]. Judith B. Cardell, Carrie Cullen Hitt, William W.Hogan, etc., "Market Power and Strategy Interaction in Electricity Networks", Resource and Energy Economics 19, 1997, pp. 109-137

[17]. Severin Borenstein, James Bushnell, Frank Wolak, Diagnosing Market Power in California's Deregulated Wholesale Electricity Market, Power Working Paper, University of California Institute, Aug 2000

[18]. Severin Borenstein, James Bushnell, and Christopher R. Knittel, Market Power in electricity markets: Beyond Concentration Measures, *The Energy Journal* 20(4), Feb 1999

[19]. James Bushnell, Christopher R. Knittel, Frank Wolak, Estimating the Opportunities for Market Power in a Deregulated Wisconsin electricity markets

[20]. Harry Singh, "Market Power Mitigation in electricity markets", IEEE "Game Theory Applications in Electrical Power Systems", 1999, pp. 61-69

[21]. William W. Hogan, Electricity Market Power Mitigation, XENERGY Executive Forum, Newton, MA, May 2001

[22]. Rafal Weron, Energy Price Risk Management, Physica A 285(2000), pp. 127-134

[23]. Frank A. Wolak, Market Design and Price Behavior in Restructured Electricity Markets: An International Comparison, http://www-leland.stanford.edu/~wolak

[24]. Jack Cogen, What is Weather Risk?, PMA Online Magazine, May 1998

[25]. Shmuel Oren, Market Based Risk Mitigation: Risk Management vs. Risk Avoidance, Presented at a White House OSTP/NSF Workshop on Critical Infrastructure Interdependencies, 14-15 June, Washington DC

[26]. Graham Romp, "Game Theory-- Introduction and Applications", Oxford University Press, 1997

[27]. Harry Singh, "Introduction to Game Theory Application in electricity markets", IEEE "Game Theory Applications in Electrical Power Systems", 1999, pp. 3-6

[28]. Carolyn A. Berry etc., "Analysis Strategic Bidding Behavior in Transmission Networks", IEEE "Game Theory Applications in Electrical Power Systems", 1999, pp. 7-31

[29]. Steven Stoft, "Using Game Theory to Study Market Power in Simple Networks", IEEE "Game Theory Applications in Electrical Power Systems", 1999, pp. 33-40
[30]. X.Vieira Filho, L.G.Marzano etc, "Efficient Pricing Schemes in Competitive Environments Using Cooperative Game Theory", 13th PSCC in Trondheim, 28 June-2 July, 1999, pp. 244-250
[31]. H.Y.Yamin, S.M.Shahidehpour, "Risk Management Using Game Theory by Transmission Constrained Unit Commitment in a Deregulated Power Market", IEEE "Game Theory Applications in Electrical Power Systems", 1999, pp. 50-60
[32]. R.W.Ferrero etc., "Transaction Analysis in Deregulated Power System Using Game Theory", IEEE Trans.on PS, Vol.12, No.3, Aug.1997, pp. 1340-1344
[33]. Chris.S.K.Yeung, Ada.S.Y.Poon, Felix F.Wu, "Game Theoretical Multi-Agent Modelling of Coalition Formation for Multilateral Trades", IEEE Trans. on Power System, Vol.14, No.3, Aug.1999, pp. 929-934
[34]. V.Krishna, VC Ramesh, "Intelligent Agents for Negotiations in Market Games, Part 1: Model", IEEE Trans. on Power System, Vol.13, No.3, Aug.1998, pp. 1103-1108
[35]. V.Krishna, VC Ramesh, "Intelligent Agents for Negotiations in Market Games, Part 2: Application", IEEE Trans. on Power System, Vol.13, No.3, Aug.1998, pp. 1109-1114
[36]. Chaoan Li, Alva J.Svoboda, Xiaohong Guan, Harry Singh, "Revenue Adequate Bidding Strategies in Competitive Electricity Markets", IEEE Trans. on Power System, Vol.14, No.2, May.1999, pp. 492-497
[37]. Otero-Novas, I., Meseguer, C., Batlle, C., Alba, J.J, "A simulation model for a competitive generation market", IEEE Trans. on Power System, Vol.15, No.1, Feb.2000, pp. 250-256
[38]. G. Gross, D. J. Finlay, "Optimal Bidding Strategies in Competitive Electricity Markets", 12th PSCC, Dresden, 19-23 Aug. 1996, pp. 815-823
[39]. Daoyuan Zhang, Yajin Wang, Peter B.Luh, "Optimization Based Bidding Strategies in the Deregulated Market", IEEE Trans. on Power System, Vol.15, No.3, Aug.2000, pp. 981-986
[40]. Xiaohong Guan, Peter B.Luh, "Integrated Resource Scheduling and Bidding in the Deregulated Electricity market: New Challenges", Discrete Event Dynamics Systems: Theory and Application, V9. No4. 1999, pp. 330-341
[41]. J. Garcia, J. Roman, J. Barquin, A. Gonzalez, "Strategic Bidding in Deregulated Power Systems", 13th PSCC in Trondheim, 28 June-2 July, 1999, pp. 258-264
[42]. H. Song, C. C .Liu, J. Lawarree, R. W. Dahlgren, "Optimal electricity supply bidding by Markov decision process", IEEE Trans. on Power System, Vol.15, No.2, May.2000, pp. 618-624
[43]. K. Xie, Y. H. Song, J. Stonham, E. Yu, G. Liu.: "Decomposition Model and Interior Point Methods for Optimal Spot Pricing of Electricity in Deregulated Environments", IEEE Trans. on Power Systems., Vol. 15, No. 1, Feb. 2000, pp. 39-50
[44]. K.Seeley, .J.Lawarree, C.C.Liu: "Analysis of electricity market rules and their effects on strategic behavior in a noncongestive grid", IEEE Trans. on Power Systems, Vol.15, No.1, Feb. 2000, pp. 157-162
[45]. J.M.Arroyo, A.J.Conejo, "Optimal Response of a thermal unit to an electricity spot market", IEEE Trans. On Power System, Vol.15, No.3, August 2000, pp1098-1104
[46]. Z.A.Eldukair, "Fuzzy decision in bidding strategies", Proceedings, First International Symposium on Uncertainty Modeling and Analysis, 1990, pp. 591-594
[47] E.Gimenez-Funes, L.Godo, J.A.Rodriguez-Aguilar, P.Garcia-Calves, "Designing bidding strategies for trading agents in electronic auctions", Proceedings, International Conference on Multi Agent System, 1998, pp. 136-143

Appendix A: Notation Used in the Self-scheduling UC Model Module

n	Index of thermal unit from 1 to N
t	Index of time periods in a day from 1 to T
$S_{n.t}$	The payoff of thermal unit n at period t ($S_{n.t}$)
λ_t	Spot price of energy at period t (\$/MWH)
$\overline{P}_n$	Capacity of thermal unit n (MW)
$g\min_{n.t}$	Minimum output of thermal unit n at period t (MW)
cn_n	Fixed cost of thermal unit n (\$/P)
cv_n	Variable cost of thermal unit ^{n} (\$/MW) (the detailed variable cost function can be presented by a quadratic function)
cu_n	Start-up cost of thermal unit n (\$) (detailed modelling of start-up costs requires the use of exponential functions)
cs_n	Shut-down cost of thermal unit n (\$)
RU_n	Ramp-up limit of thermal unit n (MW/P)
RD_n	Ramp-down limit of thermal unit n (MW/P)
Tsu_n	Minimum up time of thermal unit n
Tsd_n	Minimum down time of thermal unit n
Std_n^0	Initial state of thermal unit n , 0 represents off, 1 represents on
U_n^0	Initial time periods that thermal unit n has been on
D_n^0	Initial time periods that thermal unit n has been off
$Std_{n.t}$	State variable of thermal unit n at period t , 0 decommitted, 1 committed
$Dsu_{n.t}$	Start-up variable of thermal unit n at the beginning of period t , 1 start up
$Dsd_{n.t}$	Shut-down variable of thermal unit n at the beginning of period t , 1 shut down
$g_{n.t}$	Output of energy of thermal unit n at period t

14. Transmission Services Improvement by FACTS Control

Y. Xiao, X. Wang, Y.H. Song and Y.Z. Sun

Generally speaking, transmission services should be provided in a way to support all the required functions of transmission systems reliably: there include electric power delivery to customers, possessing flexibility for changing system conditions, providing emergency support from other areas during critical situations, reducing the need for installed generating capacity and accommodating economic exchange of electric power among systems and so on. In recent decades, technological developments with rapid industrial growth have resulted in a dramatic increase in the demand for electric power. More importantly, in the last few years there has been a world-wide move towards deregulation to facilitate the development of a competitive electricity market. In effect, for transmission systems, it requires non-discriminatory open access to transmission resources, therefore, sufficient transmission capacity for transmission services supporting is a great demand. On the other hand, the set of economic, environmental and social problems faced by the utility industry, hamper the expansion and construction of power generation and transmission facilities. This situation has brought about new unprecedented challenges to electric industries all over the world. To satisfy the open access to transmission networks requirement, and to meet the demand for a substantial increase in power transfers among utilities as a major consequence of electricity market, a much more intensive utilisation of existing transmission resources is needed. Obviously, these aspects have motivated the development of strategies and methodologies to improve transmission services: there include available transfer capability (ATC) boosting, increasing the auction revenue of financial transmission rights (FTR) of the existing transmission networks, transmission congestion elimination, and system stability enhancement by flexible and rapid system control and so forth.

On the other hand, the advent of Flexible AC Transmission Systems (FACTS) technology has coincided with the major restructuring of the electric power industry. By the use of power electronics-based controllable components to control line impedance, magnitude and phase angle of nodal voltage individually and simultaneously, FACTS can provide benefits in increasing system transmission capacity and power flow control flexibility and rapidity. Therefore, it is able to play an important role in transmission services improvement. As deregulation picks up speed, meeting the demand for sufficient and improved transmission services is becoming more critical, it is imperative to investigate the capabilities and potential applications of FACTS on power networks. Taking the two important applications of FACTS control, ATC enhancement and FTR auction, as examples, this chapter addresses this issue with the intention of giving deeper insight into the ability of FACTS technology to facilitate electricity market operation and transaction.

14.1 FACTS Solutions to Power Flow Control

14.1.1 Concept of FACTS Technology

Recent decades have seen the rapid development of power electronic technology. Some power electronics based devices have already been widely used in power systems. Among them HVDC transmission and thyrister controlled static var compensator (SVC) are two good examples. The successful development of the gate turn-off technique, with the advantages of high unit power capability, high switch speed and low unit loss, provides more opportunities for power electronics based devices to be applied in power systems.

After years of supporting the development of high power electronics, in the late 1980s, the Electric Power Research Institute (EPRI) in the USA formalised the broad concept of FACTS [1]:

Alternating current transmission systems incorporating power electronic-based and other static controllers to enhance controllability and increase power transfer capability.

Generally speaking, FACTS devices are power electronics based controllers to regulate line flow and nodal voltage and, through rapid control action, mitigate dynamic disturbances. Therefore, by the use of FACTS devices, the possibility of controlling power flow without generation rescheduling or topological changes is being realised, which will provide benefits in increasing system transmission capability, control efficiency and flexibility, and ultimately transmission services improvement and system operation economy and reliability.

From the standpoint of steady-state power system operation, the transmitted power over a given line IJ is determined by the line impedance $\boldsymbol{Z}$, the magnitudes of the sending and receiving end voltages V_I and V_J, and the phase angle θ between these two voltages. Taking the simple two-machine model for example, which is illustrated in Figure 14.1, the principle is demonstrated clearly by the power flow formula:

$$P_{IJ} = \frac{V_I V_J}{\mathbf{Z}} \sin(\theta_I - \theta_J) \tag{14.1}$$

From the equation, it is concluded that using control nodal voltages, line impedance and phase angles of the related nodes, or combinations thereof, steady-state power flow can be controlled freely. According to the control principles of FACTS technology, by using appropriate thyrister-controlled and convertor-based components, FACTS devices can accomplish these functions via shunt reactive compensation, series active or reactive compensation, or specific combinations of them.

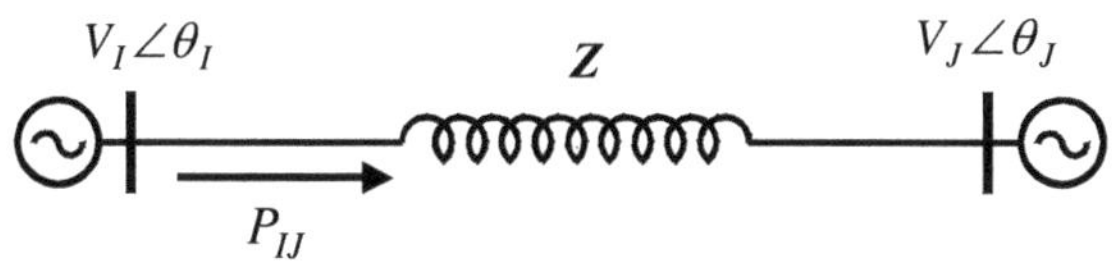

Figure 14.1. Illustration of two-machine model

14.1.2 Models of FACTS Devices

Broadly speaking, the existing steady-state models of FACTS devices can be classified into two categories: decoupled models and coupled models.

In the decoupled model, FACTS devices here usually replaced with fictitious PQ or (and) PV bus(es) [2, 3], which results in modification of the structure of the Jacobian matrix.

Generally, the coupled model consists of two major models: voltage (-current) source model (VSM) and power injection model (PIM). For different FACTS devices, the VSM has different representations. For SVC, thyristor-controlled phase-shifter (TCPS), static synchronous compensator (STATCOM), static synchronous series compensator (SSSC) and unified power flow controller (UPFC), VSM is usually formulated as series or (and) shunt inserted voltage (current) source(s) [1, 4]. As a special case, thyristor-controlled series capacitor (TCSC) can also be represented using a series inserted voltage source. The VSM is formulated according to the device's operating principle, thus can model FACTS devices in a more intuitive way. However it destroys the symmetric characteristics of the admittance matrix [5]. Moreover, trigonometric functions involved will inevitably lead to oscillation of power flow control [6]. Derived from the VSM, PIM was proposed by Han in 1982 for Phase Shifters (PS) [5]. With the conversion of inserted voltage (current) source(s) to power injections at the related buses, the PIM enables one to maintain the symmetry of admittance matrix. Because of this advantage, applications of this model are extended to nearly all FACTS devices and are used in most of the literature concerning the operation and control of FACTS-equipped power systems.

Obviously, all applications of FACTS control to transmission services improvement involve power flow control using FACTS devices. Therefore formulated PIM-based models for power flow control, including a PIM of FACTS devices for ATC enhancement and a simplified DC model for FTR auction improvement, are described in detail in the following sections.

14.1.2.1 Power Injection Model of FACTS Devices for ATC Enhancement

According to their steady-state characteristics, FACTS devices can be classified into three categories: shunt controller, series controller and unified controller, whether they are thyristor-controlled (TC) or converter-based (CB) [1]. In general, series controllers, such as TCSC, TCPS and SSSC, are characterised as active line flow control. Shunt controllers as SVC and STATCOM are used for voltage regulation. As a unified controller, UPFC can control active and reactive line flow and nodal voltage simultaneously. The classifications, main steady-state functions and the corresponding FACTS devices are summarised in Table 14.1 [7].

In PIM, since active and (or) reactive power injections represent all features of the FACTS device, it is rational to take active and (or) reactive power injections as independent control variables for power flow control. It should be noted that the power injections are only an interim result, as the original control parameters of the FACTS device need be provided in the resultant control scenario. Once the required power injections are obtained, the original control parameters can be derived easily according to the corresponding VSM of the FACTS device(s) [8].

In order to develop an effective and robust approach to power flow control with FACTS devices, first, based on the PIM, models of FACTS devices are formulated according to their classifications [7].

14.1.2.2 PIM of Shunt Controller for Voltage Control

Assuming a shunt controller connected to bus I, as shown in Figure 14.2a, there is only a reactive current source at bus I in the VSM, which is illustrated in Figure 14.2b. Therefore in the PIM, a reactive power injected directly at bus I is used to model the current source, which is represented by $Q_{II(inj)}$ in Figure 14.2c. Obviously $Q_{II(inj)} = Q_{I(inj)}$ in this model. For the shunt controller, the relationship between the control vector $\mathbf{C}_{Shunt} = \lfloor Q_{II(inj)} \rfloor$ and the voltage can be expressed as:

$$V_I = V_I(Q_{II(inj)}) \tag{14.2}$$

14.1.2.3 PIM of Series Controller for Line Flow Control

Actually, line flow control performance of FACTS devices can be understood as an additional controllable power flow through the line, which is superimposed on the 'nature' power flow. It can also be explained by the use of the parallel connected current source in the derivation of the PIM from the corresponding VSM in [5].

For a series controller installed on line L, $I-J$, it is modelled as a series inserted voltage source in the VSM, as shown in Figure 14.2b. The voltage source produces two active power injections, $P_{I(inj)}$ and $P_{J(inj)}$, to bus I and bus J respectively. Considering no loss in the device, $P_{I(inj)}$ is equal to - $P_{J(inj)}$.With the active power injection, $P_{I(inj)}$, as a control variable, i.e. $C_{Series} = \lfloor P_{I(inj)} \rfloor$, the controlled active power flow of line L, P_L, can be reformulated as power flow P_{LO} with the effect of the FACTS device, plus $P_{I(inj)}$ which flows along line L, as shown in Figure 14.2c clearly, where

$$P_{L0} = P_{L0}(V_I, V_J, \theta_I, \theta_J)$$
$$Q_{L0} = Q_{L0}(V_I, V_J, \theta_I, \theta_J) \tag{14.3}$$

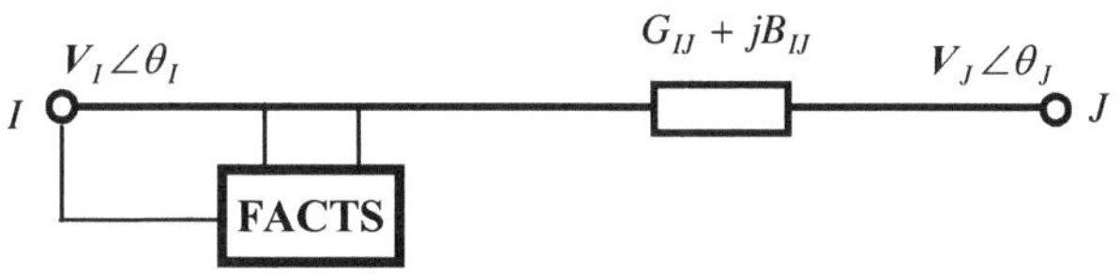

Figure 14.2a. FACTS device on line IJ

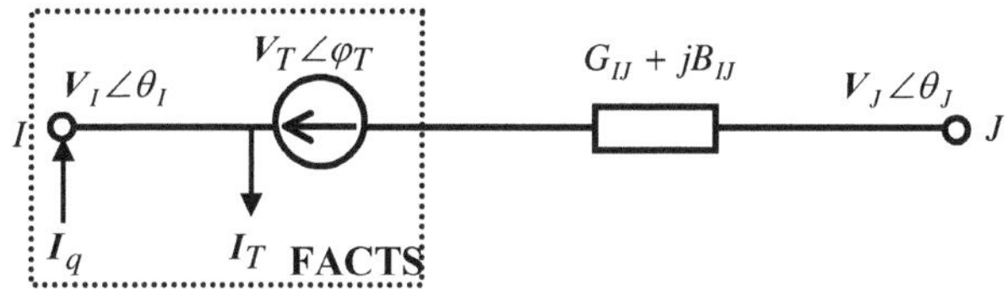

Figure 14.2b. Voltage source model of FACTS

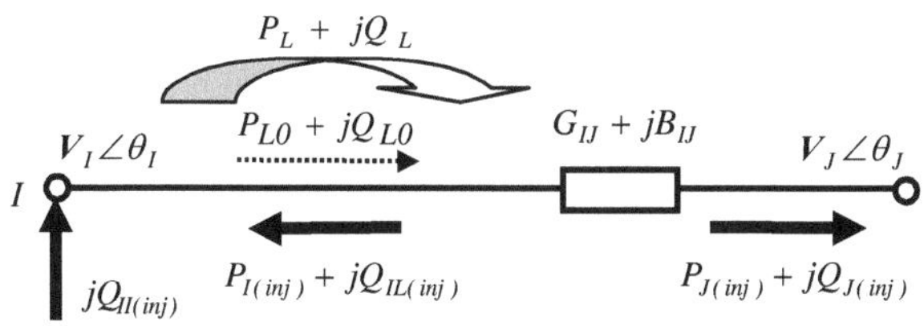

Figure 14.2c. Power injection model of FACTS devices

Since P_{LO} and Q_{LO} only pertains to the voltage and phase angle of buses I and J, which is represented by $\boldsymbol{X}_L$, for active power flow control of the series controller, the relationship between the controlled active power injection and the active line flow is expressed as:

$$P_L(\mathbf{X}_L, P_{I(inj)}) = P_{L0}(\mathbf{X}_L) - P_{I(inj)} \tag{14.4}$$

It should be noted that, different from the shunt controller, in this model, there is no reactive power injection direct to bus I, that is, $Q_{IL(inj)} = Q_{I(inj)}$. For the series controller, besides active power injection, which is taken as control variable, there are also non-control power injections involved, such as $Q_{IL(inj)}$ and $Q_{J(inj)}$. Therefore, in every iteration, the original control parameters should be derived from the obtained control variables and state variables. Then, the non-control power injections can be determined, which will be taken as known variables for the next iteration of power flow control.

14.1.2.4 PIM of the Unified Controller for Power Flow Control

With its unique combination of shunt and series compensation, as a unified controller, UPFC is able to control the active, reactive power flow and nodal voltage. For power flow control of the UPFC installed on line *IJ*, near bus I, as shown in Figure 14.2a, it is usually represented by a shunt connected reactive current source and a series inserted voltage source in the VSM of Figure 14.2b. Correspondingly, in the PIM, there are three independent power injections involved. One is $Q_{II(inj)}$ for regulating voltage, which is injected by the shunt current source directly into bus I. The series voltage source generates two independent control related power injections. One is a reactive power injection flow via line L for reactive line flow control, which is represented as $Q_{IL(inj)}$. That is, the total reactive power injection to bus I is, $Q_{I(inj)} = Q_{II(inj)} + Q_{IL(inj)}$, which is shown in Figure 14.2c clearly. The other is an active power injection along line L, $P_{I(inj)}$, for active power flow control. As in this paper, the line flow control target is set as power flow $P_L + jQ_L$ from bus I along line L, $Q_{J(inj)}$ is not taken as a control variable. The corresponding control vector is written as $C_{Unified} = \lfloor P_{I(inj)} \quad Q_{IL(inj)} \quad Q_{II(inj)} \rfloor$. As a result, for power flow control of the unified controller, the relationship between the controllable power injections and the controlled power flow can be expressed as:

$$\begin{aligned} P_L(\mathbf{X}_L, P_{I(inj)}) &= P_{L0}(\mathbf{X}_L) - P_{I(inj)} \\ Q_L(\mathbf{X}_L, Q_{IL(inj)}) &= Q_{L0}(\mathbf{X}_L) - Q_{IL(inj)} \end{aligned} \tag{14.5}$$

where $Q_{J(inj)}$ is the non-control power injection for the unified controller.

The power flow control related information of the three category FACTS devices, including control variables, available control target of the controllers and their corresponding control parameters in the VSM are described in Table 14.1. I_S is the reactive current injected by SVC or STATCOM. For TCPS, ψ is the phase shifting angle. x_c represents the controllable reactance of TCSC. V_C denotes the series inserted voltage source of SSSC. UPFC has three control parameters, magnitude V_T, phase angle φ_T of the series inserted voltage source and magnitude of the current source I_q, as I_q is in quadrature with the voltage of the connected node.

14.1.2.5 DC Model of TCSC and TCPS for FTR Auction

Since usually the FTR auction only involves active power, a DC power flow model is used for theFTR auction in this chapter. Therefore, studies of the impacts of FACTS control on FTR auction improvement are conducted on two major series FACTS devices, TCSC and TCPS, for active power flow control. The PIM presented above is simplified to model these two FACTS devices.

An equivalent circuit of a TCSC on transmission line *IJ* is shown in Figure 14.3, where the reactance of the line is x_{IJ} and the reactance of the TCSC is $-x_c$. The total susceptance between bus *I* and bus *J* is expressed as:

$$b_{IJ} = 1/(x_{IJ} - x_c) \tag{14.6}$$

Based on the DC power flow model, the active power flow along line *IJ* with the TCSC can be formulated as:

$$P_{IJ} = b_{IJ}(\theta_I - \theta_J) \tag{14.7}$$

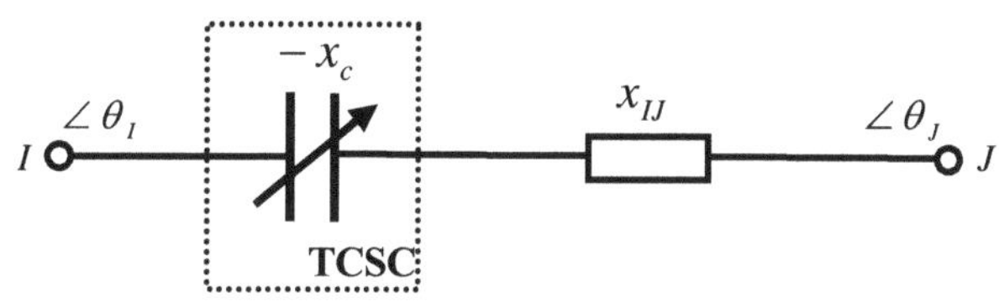

Figure 14.3. Equivalent circuit of TCSC

The equivalent circuit of a TCPS on line *IJ* is shown in Figure 14.4. Given the susceptance of the line $b_{IJ} = 1/x_{IJ}$ and the voltage shift angle ψ of the TCPS, the active power flow along the line is:

$$P_{IJ} = b_{IJ}(\theta_I - \theta_J + \psi) \tag{14.8}$$

Table 14.1. Classifications of FACTS devices and their control variables

Classification	Main steady-state functions	Control targets	Control variable(s)	Non-control power injection(s)	FACTS devices		Control parameters
Shunt Controller	Voltage regulation	V_I	$Q_{II(inj)}$	-	**TC**	SVC	I_S
					CB	STATCOM	
Series Controller	Active power flow control	P_L	$P_{I(inj)}$	$Q_{IL(inj)}$, $Q_{J(inj)}$	TC	TCPS	ψ
						TCSC	$-x_c$
					CB	SSSC	V_C
Unified Controller	Active power flow control Reactive power flow control Voltage regulation	P_L, Q_L, V_I	$P_{I(inj)}$, $Q_{IL(inj)}$, $Q_{II(inj)}$	$Q_{J(inj)}$	**CB**	UPFC	V_T, φ_T, I_q

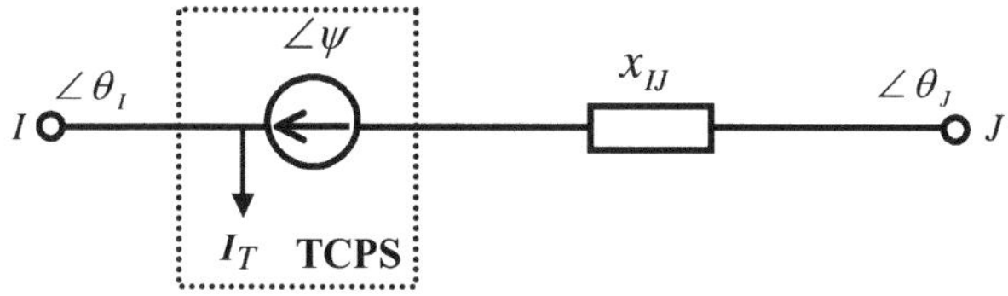

Figure 14.4. Equivalent circuit of TCPS

Formulated by replacing the device by two active power injections to both sides of the transmission line, the PIM of the series FACTS devices under the DC assumption is shown in Figure14.5. For the TCSC, the power injections can be derived as:

$$P_{I(inj)} = -P_{J(inj)} = \frac{-x_c}{x_{IJ}(x_{IJ} - x_c)}(\theta_I - \theta_J) \tag{14.9}$$

The power injections of the TCPS can be formulated as:

$$P_{I(inj)} = -P_{J(inj)} = -b_{IJ}\psi \tag{14.10}$$

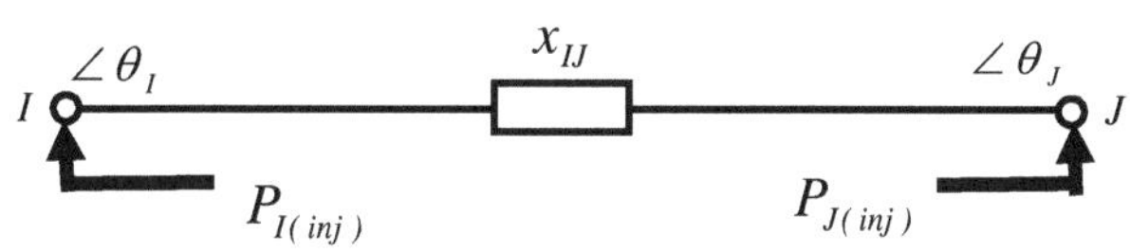

Figure 14.5. DC power injection model of TCSC and TCPS

Here it should be noted that the power injections of TCSC are functions of not only its reactance but also the difference in voltage angles between two ends of the transmission line. However, for TCPS the power injections have nothing to do with the state variables of the system.

14.2 ATC Enhancement by FACTS Control

The definition of the ATC, as well as the application of the ATC to market-based power systems has been well described and analysed in chapter 5. In practice, as economic dispatch of generation is often restrained by transmission bottlenecks, which results from the limited ATC value of some interfaces, as such, a high ATC value can lead to system operation economy. On the other hand, it is well recognised that under heavy loading situations and during emergencies, owing to the deficiency of the necessary reservation capacity, low ATC levels may mean poor reliability and security of the system. More free bilateral trades of electricity are motivated by the diversity of prices, together with the growing power export/import and free-wheeling transactions among utilities further aggravate the stressed situation of the existing transmission systems. Therefore, sufficient ATC should be guaranteed to support free market trading and maintain an economical, reliable and secure operation over a wide range of system conditions. It is crucial to seek effective solutions to boost the ATC level of the transmission networks. As a result, power suppliers will benefit from more market opportunities with less congestion and enhanced power system security; it will be more profitable for transmission owners with maximised use of existing transmission assets; and customers will also get improved services and reduced prices.

Aimed at this problem, various ATC enhancement approaches have been proposed, where adjusting the terminal voltage of generators and tap changing of the on-load tap changer (OLTC), particularly rescheduling active power of generators, are considered major control measures for ATC boosting [9, 10]. However, in an unbundled and privatized electricity market, generators usually have been scheduled ahead of the time according to the relevant optimal bidding strategies. As such, any rescheduling means worsening individual supplier's profit and attainable social welfare.

As discussed in a 1996 North American Electric Reliability Council (NERC) report - *Available Transfer Capability Definition and Determination* [11], the ability of an interconnected transmission network to reliably transfer power through prescribed interfaces may be restricted by thermal, voltage or stability limits. On the other hand, it is well recognised that, with the capability of flexible power flow control and rapid action, Flexible AC Transmission Systems (FACTS) technology has a wide spectrum of impacts on the way the transmission system is operated, in particular with respect to thermal, voltage and stability constraints. Concerning steady-state power flow, since circuits do not normally share power in proportion to their ratings, and in most situations, voltage profiles cannot be smooth, ATC values are always limited ultimately by heavily loaded circuits and (or) nodes with relatively low voltage, with the increase of system loading. As stated in [1], the FACTS concept makes it possible to use circuit reactance, voltage magnitude and angle as controls to redistribute line flow and regulate nodal voltage, thereby, mitigating the critical situation and delaying its occurrence. In addition, partly due to physical and reliability constraints, most high-voltage systems are operating far below their thermal rating [12]. By control of circuit reactance and phase angle of nodal voltages, FACTS technology enables line loading to increase flexibly, in some cases all the way up to thermal limits. Therefore, it can offer an effective and promising alternative to conventional methods for the ATC enhancement.

To resolve emerging power system problems with better transmission services support, in the late 1980s, the EPRI proposed the application of FACTS technology with two major objectives: to increase the power transfer capability of transmission systems and to control power flow over designated transmission routes [1]. However, so far there are few reports in the literature on ATC enhancement by FACTS control. A 1997 EPRI report - *FACTS Assessment Study To Increase the Arizona-California Transfer Capability* [13], demonstrated the potential solutions that FACTS technologies offer on the Arizona-California interface. The technical merits of FACTS technologies for increasing the Arizona-California transfer capability are assessed based on power flow, transient stability, and subsynchronous resonance mitigation. Two types of FACTS devices, TCSC and STATCOM, are evaluated, as well as upgrades of conventional technologies, including series capacitors and series line reactors. Economic analyses are also performed. The study indicates that use of FACTS devices was cost-effective compared to conventional methods of increasing transfer capability. It is concluded that use of TCSC and STATCOM devices could maximise the transfer capability into southern California when compared to transfers over the Arizona interface and from other power systems. The increase could be as much as 1000 MW. As stated in a California Energy Commission report [14], among all the major FACTS devices, the UPFC is the most beneficial one for increasing import capacity into San Diego Gas and Electric (SDG&E)'s service area. According to another EPRI report [15], FACTS devices can increase the capacity of individual corridors by up to 80% simply by shifting power flow from overloaded to underloaded transmission lines. In addition to that, by improving system stability through their rapid-response capability, FACTS devices in widespread use can also increase the overall capacity of a large transmission network by 20% or more. Undoubtedly, it is very important and imperative to carry out studies on exploitation of FACTS technologies to improve the ATC level of interconnected networks.

According to the 1996 NERC report [11], the ability of interconnected transmission network to reliably transfer power through prescribed interfaces may be limited by thermal limits, voltage limits and stability limits. It is well recognised that many FACTS devices can effectively modulate the dynamics of reactive power and voltage, and indirectly, improve performance of real power, such as increased transfer across wide electrical distances. The potential improvement of the transient response of a system with FACTS devices is a very important consideration. However, comprehensive ATC evaluation models that take into account dynamic aspects are still at the research and preliminary development stage. Thus, in this chapter, the application of FACTS control for ATC enhancement is only addressed from the perspective of steady-state power system operation and control. Based on a stochastic model for ATC evaluation, as presented in detail in Chapter 5, an optimal power flow (OPF)based model is proposed to evaluate the impact of FACTS control on ATC enhancement [16]. With respect to FACTS control, the ATC enhancement model is formulated to achieve maximum transfer at the specified interface. As introduced earlier, the PIM enables one to implement the control of any FACTS devices, therefore it is employed to derive the control parameters. Finally, with the IEEE 118-bus system as a test bed, case studies are conducted on all categories of FACTS devices. The results demonstrate the effectiveness of FACTS control for ATC enhancement.

This section on ATC enhancement by the use of FACTS control is organised as follows: on the basis of the ATC evaluation model described in Chapter 5, an ATC enhancement model with FACTS control is presented. A flowchart of the proposed methodology is depicted in the Implementation section. Finally, case studies for the IEEE 118-bus system are presented, with which the technical merits of FACTS technologies for increasing the ATC are demonstrated.

14.2.1 Formulated ATC Enhancement Model

The ATC enhancement model is formulated to achieve maximum power transfer by controlling the FACTS devices on interconnected lines, meanwhile, increasing all the complex loads D and generations G in current situation using a scalar loading factor λ_*, i.e.

$$\begin{aligned} &P_{D*} = \lambda_* P_D,\ Q_{D*} = \lambda_* Q_D \\ &P_{G*} = \lambda_* P_G,\ Q_{G*} = \lambda_* Q_G \end{aligned} \tag{14.11}$$

until a critical situation occurs, that is, line thermal limits or nodal voltage limits are attained.

Basic elements of the developed model are described below.

14.2.1.1 Control Variables

As elucidated in Section 14.1.2, active and reactive power injections of the FACTS device are taken as independent control variables in the ATC enhancement model.

14.2.1.2 Objective Function

The objective is to maximise the uncommitted active transfer capacity of the prescribed interface, which is represented by

$$\text{Max } F = \sum_{h=1}^{H} (P_{h*} - P_h) \tag{14.12}$$

where

H : number of tie lines across the interface, in which the active powers share the same prescribed direction;

P_h : active power flow of tie line h.

Variables with subscript * represents those at the critical equilibrium point, while variables without subscript * denote those at the current operating point.

When applying the PIM of FACTS devices to gain more insight into the objective function, Equation (14.12) can be further written as:

$$\text{Max } F(\mathbf{X}, \lambda_*, \mathbf{X}_*, \mathbf{C}_*) = \sum_{h=1}^{N_F} (P_{h*}(\lambda_*, \mathbf{X}_*, \mathbf{C}_*) - P_h(\mathbf{X})) + \sum_{h=N_F+1}^{H} (P_{h*}(\lambda_*, \mathbf{X}_*) - P_h(\mathbf{X})) \tag{14.13}$$

where

N_F : number of FACTS devices installed on the interconnected transmission lines;

X : state variables, including magnitudes and phase angles of nodal voltage, reactive output of generators performing nodal voltage control, as well as system fixed parameters.

14.2.1.3 Operating and Control Constraints

The constraints are categorised as follows:

- Equality Constraints:

As power flow equations at the current operating points are standard, they are omitted here. To ensure that the system is moving from the current equilibrium point to another one corresponding to the loading factor λ_*, the critical operating point is included into the constraints as follows:

For PQ node i which has no connection with the FACTS devices, $i = 1,..., N_1$:

$$\lambda_*(P_{Gi} - P_{Di}) - V_{i*}\sum_{j=1}^{N-1} V_{j*}(G_{ij}\cos\theta_{ij*} + B_{ij}\sin\theta_{ij*}) = 0$$

$$\lambda_*(Q_{Gi} - Q_{Di}) - V_{i*}\sum_{j=1}^{N-1} V_{j*}(G_{ij}\sin\theta_{ij*} - B_{ij}\cos\theta_{ij*}) = 0$$

For PV node i which has no connection with the FACTS devices, $i = N_1+1,...,N-1$:

$$\lambda_*(P_{Gi} - P_{Di}) - V_{i*}\sum_{j=1}^{N-1} V_{j*}(G_{ij}\cos\theta_{ij*} + B_{ij}\sin\theta_{ij*}) = 0 \tag{14.14}$$

Assuming FACTS device h installed on line $I_h - J_h$, $h = 1,..., H_F$, for node I_h:

$$\lambda_*(P_{GI_h} - P_{DI_h}) + P_{I_h(inj)*} - V_{I_h*}\sum_{j=1}^{N-1} V_{j*}(G_{I_hj}\cos\theta_{I_hj*} + B_{I_hj}\sin\theta_{I_hj*}) = 0$$

$$\lambda_*(Q_{GI_h} - Q_{DI_h}) + Q_{I_hL_h(inj)*} + Q_{I_hI_h(inj)*} - V_{I_h*}\sum_{j=1}^{N-1} V_{j*}(G_{I_hj}\sin\theta_{I_hj*} - B_{I_hj}\cos\theta_{I_hj*}) = 0$$

For node J_h:

$$\lambda_*(P_{GJ_h} - P_{DJ_h}) - P_{I_h(inj)*} - V_{J_h*}\sum_{j=1}^{N-1} V_{j*}(G_{J_hj}\cos\theta_{J_hj*} + B_{J_hj}\sin\theta_{J_hj*}) = 0 \tag{14.15}$$

$$\lambda_*(Q_{GJ_h} - Q_{DJ_h}) + Q_{J_h(inj)*} - V_{J_h*}\sum_{j=1}^{N-1} V_{j*}(G_{J_hj}\sin\theta_{J_hj*} - B_{J_hj}\cos\theta_{J_hj*}) = 0$$

where

N : number of nodes;

N_1 : number of PQ nodes;

$P_{I_h(inj)*}$: active power injection of FACTS device h to the first node I_h;

$Q_{I_hL_h(inj)*}$: reactive power injection of FACTS device h through the line L_h to the first node I_h;

$Q_{I_hI_h(inj)*}$: reactive power injection to the first node I_h directly;

$Q_{J_h(inj)*}$: reactive power injection to the second node J_h.

- Inequality Constraints:

Nodal voltage limits: (for PQ node i, $i = 1, ..., N_1$)

$$V_{i,\min} \le V_i, V_{i*} \le V_{i,\max} \tag{14.16}$$

Line thermal limits: (for line l, $l = 1,..., M$)

$$TM_l, TM_{l*} \le TM_{l,\max} \tag{14.17}$$

Generator capacity limits: (for generator i, $i = 1,..., N_G$)

$$\begin{aligned} P_{iG,\min} &\le P_{iG}, P_{iG*} \le P_{iG,\max} \\ Q_{iG,\min} &\le Q_{iG}, Q_{iG*} \le Q_{iG,\max} \end{aligned} \tag{14.18}$$

where

M : number of lines;

N_G : number of generators.

Subscript max represents the maximum limit of the variable.

To ensure the feasibility of the resultant scenario, capacity limits of FACTS devices should be taken into consideration [7]. From the implementation, it is rational to take the thermal rating of FACTS devices as limits. For the shunt controller, only the thermal limitation of the shunt transformer or converter (thyristor), T_{Shunt}, need be considered, as shown in (14.19). For the series controller, thermal limitation of the series transformer or converter (thyristor), T_{Series}, is expressed in (14.20). To be noted that, for the TCSC, in which there is a limit of controllable reactance x_c involved, the corresponding constraints on power injections can be derived from the limit. Obviously, for the unified controller, besides the two limits mentioned above, the limitation of active power transferred through transformers or converters, P_{dc}, should be included, as shown in (14.21). That is, for any FACTS device h, one has:

$$T_{Shunt} = \sqrt{(\mu_3 Q_{II(inj)})^2 + (\mu_2 P_{I(inj)})^2} \leq T_{Shunt\,,\,\max} \tag{14.19}$$

$$T_{Series} = \mu_1 \sqrt{(Q_{IL(inj)})^2 + (P_{I(inj)})^2} \leq T_{Series\,,\,\max} \tag{14.20}$$

$$P_{dc} = \mu_2 \left| P_{I(inj)} \right| \leq P_{dc\,,\,\max} \tag{14.21}$$

where for the shunt controller: $\mu_1 = \mu_2 = 0, \mu_3 = 1$,

for the series controller: $\mu_1 = 1, \mu_2 = \mu_3 = 0$,

for the unified controller: $\mu_1 = \mu_2 = \mu_3 = 1$.

The model can be rewritten as a more general optimisation model:

$$\text{Max } F(\mathbf{X}, \lambda_*, \mathbf{X}_*, \mathbf{C}_*)$$

$$\text{s.t.: } \mathbf{G}(\mathbf{X}) = 0 \tag{14.22}$$

$$\mathbf{G}_*(\lambda_*, \mathbf{X}_*, \mathbf{C}_*) = 0$$

$$\mathbf{H}_{\min} \leq \mathbf{H}(\mathbf{X}), \mathbf{H}_*(\lambda_*, \mathbf{X}_*, \mathbf{C}_*) \leq \mathbf{H}_{\max}$$

where

G : power flow equations,

H : operating constraints, including nodal voltage limits, circuit thermal limits, generator output limits and capacity limits of FACTS devices.

14.2.2 Implementation

The programme involves the development and integration of two main modules: FACTS control and an AC power flow calculation.

Compared to the Simplex method for linear programming, predictor-corrector primal-duel interior point linear programming (PCPDIPLP) is a powerful tool which leads to significant reduction in the number of iterations, especially for practical problems which usually have enormous dimensions [17]. Therefore, the method is used to solve the ATC enhancement problem. The non-linear objective and constraints must be piecewise linearised to utilise the LP algorithm to calculate the incremental values of control variables. An iterative procedure is needed to derive the control variables of FACTS devices and the loading factor for ATC maximisation.

With an increase of loading factor, system ill conditioning will be aggravated, which may result in a long period oscillation before convergence during the power flow calculation process, sometimes making the system unsolvable. As an effective method to deal with power flow calculations of ill-conditioned systems, which has been addressed in [18] intensively, the optimal multiplier Newton Raphson (OMNR) method is applied here.

The overall procedure is sketched in Figure14.6, where ε is a given threshold value for convergence.

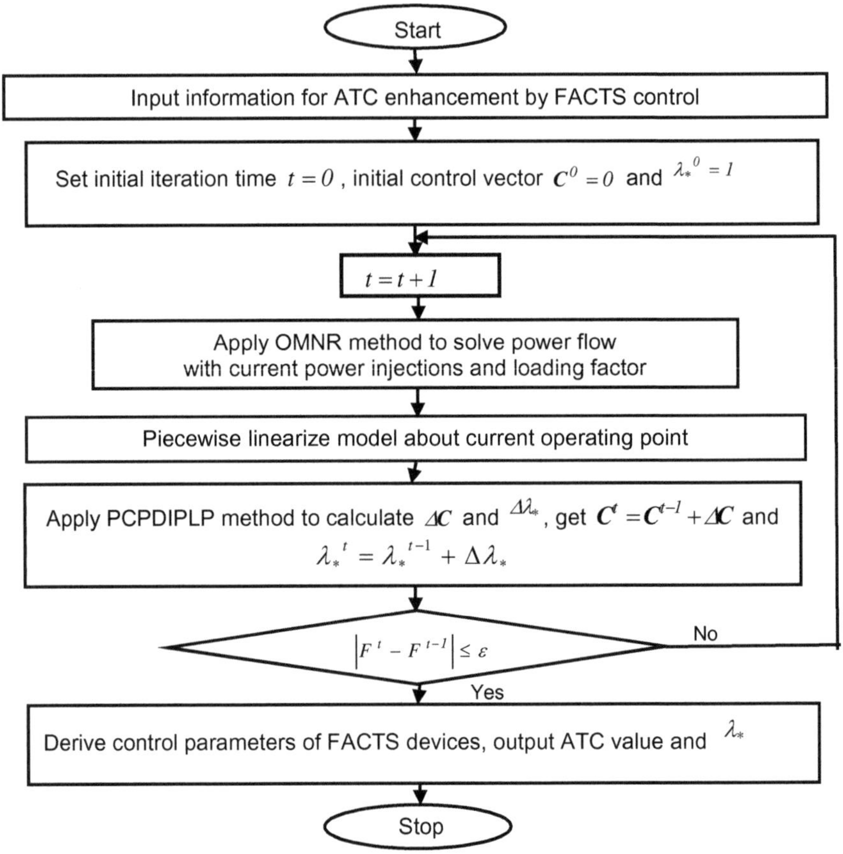

Figure 14.6. Flowchart of proposed approach

14.2.3 Case Studies

In this section, the IEEE 118-bus system [19] is employed to numerically illustrate the effectiveness of FACTS devices for ATC enhancement. According to the network structure and the power flow calculation results, the whole system is divided into two zones. The studied interface is illustrated in Figure 14.7. In this study, the per unit base is 100MVA.

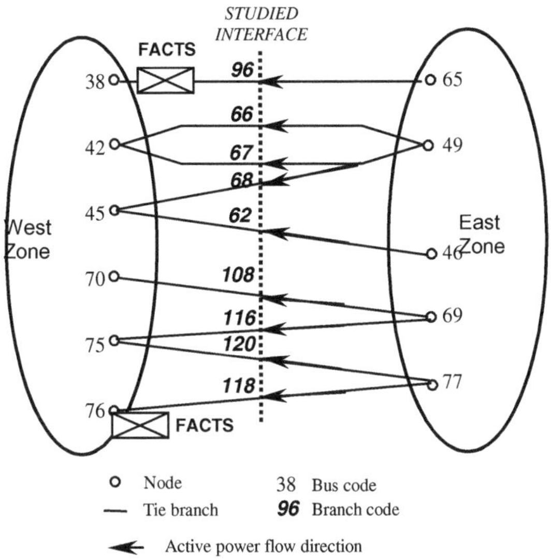

Figure 14.7. Studied interface of IEEE 118- bus system

The following criteria and assumptions are applied:

- All nodal voltages to be within the range 0.90 1.10 p.u. under normal and contingency situations.
- According to their voltage levels, line thermal limits are assumed and given in Chapter 5. For better demonstration, thermal limits of some lines, such as line 96, 108 and 116 are modified to cause unbalanced power sharing. Besides that, to achieve the full potential of FACTS control, reactive loads of nodes 20 and 118 are reduced from 0.03 p.u. and 0.15 p.u. to 0.03 p.u. and 0.00 p.u. respectively.
- As a reasonable assumption, the generation limit of each unit is set at 180% of the current output.

Power flow results for the studied interface are set out in Table 14.2, where nodes 42, 46, 49, 65 and 69 are all PV nodes. As a ratio between the apparent power of a line and the given thermal limit, the thermal burden of the tie lines is also shown in Table 14.2. It is indicated that the West-East transmission corridor is very important with a huge amount of active power transferred along it.

Table 14.2. Power flow results of studied interface in current scenario (λ_* =1.000)

Cases		Parameters of FACTS Devices (p.u.)		λ_*	ATC (MW	Iteration	CPU ti (s)
No FACTS	Case 1	-		1.365	289.97	6	2
With FACTS Devices	Case 2	SVC (node 76)	I_S =0.304	1.553	447.76	7	3
	Case 3	TCPS (line 96)	Ψ =0.157	1.612	504.31	7	3
		SVC (node 76)	I_S =0.427				
	Case 4	UPFC (line 96)	V_T =0.691, φ_T =1.094, I_q =6.156	1.783	675.96	13	8
		SVC (node 76)	I_S =1.537				

14.2.3.1 ATC Evaluation Without FACTS Device (Case 1)

First the ATC is calculated without any FACTS control as case 1. The proposed method has been coded in Fortran and run on a Pentium III 1.0GHz PC. The results are given in Table 14.3. As all studies of FACTS control are carried out on the interface, the power flow results are given in Table 14.4 in detail. To gain an intuitive understanding, thermal burden profiles of the ties and voltage profiles of the related buses in both current and critical scenarios are depicted in Figures 14.7 and 14.8 respectively. In this case, the ATC is restricted by the voltage at node 76. Obviously, reactive power compensation at this node will be an effective measure to boost the ATC.

Table 14.3. Results of ATC enhancement

Branch (I – J)	96 (38 - 65)	66, 67 (42-49)	68 (45-49)	62 (45-46)	108 (69-70)	116 (69-75)	120 (75-77)	118 (76-77)
P_{IJ} (p.u.)	-1.8158	-0.6470	-0.4967	-0.3630	1.0815	1.1013	-0.3468	-0.6111
Q_{IJ} (p.u.)	-0.5615	0.0516	-0.0174	-0.0308	0.1678	0.1710	-0.0711	-0.1844
Thermal limit (p.u.)	3.00	2.00	2.00	2.00	4.00	4.00	2.00	2.00
Thermal burden (%)	63.35	33.92	25.72	18.44	27.36	27.86	17.89	33.34

Node	38	42	45	46	49	65	69	70	75	76	77
V	0.963	0.985	0.987	1.005	1.025	1.005	1.035	0.983	0.971	0.946	1.005

Table 14.4. Power flow results of studied interface in critical scenario without FACTS (Case 1, $\lambda_* = 1.365$)

Branch (I – J)	96 (38-65)	66, 67 (42-49)	68 (45-49)	62 (45-46)	108 (69-70)	116 (69-75)	120 (75-77)	118 (76-77)
P_{IJ} (p.u.)	-2.5595	-0.9123	-0.6832	-0.5092	1.5794	1.5869	-0.4535	-0.8136
Q_{IJ} (p.u.)	-0.5175	0.1905	-0.0436	-0.1427	0.2891	0.3656	-0.0956	-0.2515
Thermal bur den (%)	87.04	46.60	34.23	26.44	40.14	40.71	23.17	42.58

Node	38	42	45	46	49	65	69	70	75	76	77
V	0.938	0.985	0.964	1.005	1.025	1.005	1.035	0.964	0.938	0.900	0.987

14.2.3.2 ATC Enhancement With Control of SVC (Case 2)

Based on the analysis above, in order to mitigate the critical situation by nodal voltage regulation, an SVC, as the most popular shunt controller, is controlled on node 76. The configuration is illustrated in Figure 14.7. The results of case 2 are given in Table 14.3. The ATC value in case 2 is higher than the original ATC in case 1 by 54.42%. This fact demonstrates the effectiveness of FACTS control on ATC enhancement.

Power flow results for the studied interface in the critical scenario of case 2 are given in Table 14.5. The corresponding tie line thermal burden profile and voltage profile of the related nodes are illustrated in Figures 14.7 and 14.8. Scanning the thermal burden in the critical scenario reveals two major facts. First it is the thermal limit of branch 96 that prevents further increase of the ATC. More importantly, the other is that the seriously unbalanced thermal burden profile means, when line 96 has reached its thermal limit, there is still plenty of space for intensive commitment of the remaining tie lines. Additionally, it is interesting to notice that the voltage of node 76 is controlled to be at its lower boundary in the critical scenario, so as to obtain the largest available active power flow on line 118. Thereby, the highest value of the ATC in case 2 can be achieved.

Table 14.5. Power flow results of studied interface in critical scenario with SVC control (Case 2, $\lambda_* = 1.553$)

Branch (I – J)	96 (38-65)	66, 67 (42-49)	68 (45-49)	62 (45-46)	108 (69-70)	116 (69-75)	120 (75-77)	118 (76-77)
P_{IJ} (p.u.)	-2.9619	-1.0581	-0.7810	-0.5886	1.8592	1.8554	-0.5038	-0.9216
Q_{IJ} (p.u.)	-0.4764	0.2818	-0.0589	-0.2057	0.3515	0.4205	-0.0890	-0.1572
Thermal burd (%)	100.00	54.75	39.16	31.18	47.30	47.56	25.58	46.74

Node	38	42	45	46	49	65	69	70	75	76	77
V	0.921	0.985	0.950	1.005	1.025	1.005	1.035	0.955	0.927	0.900	0.980

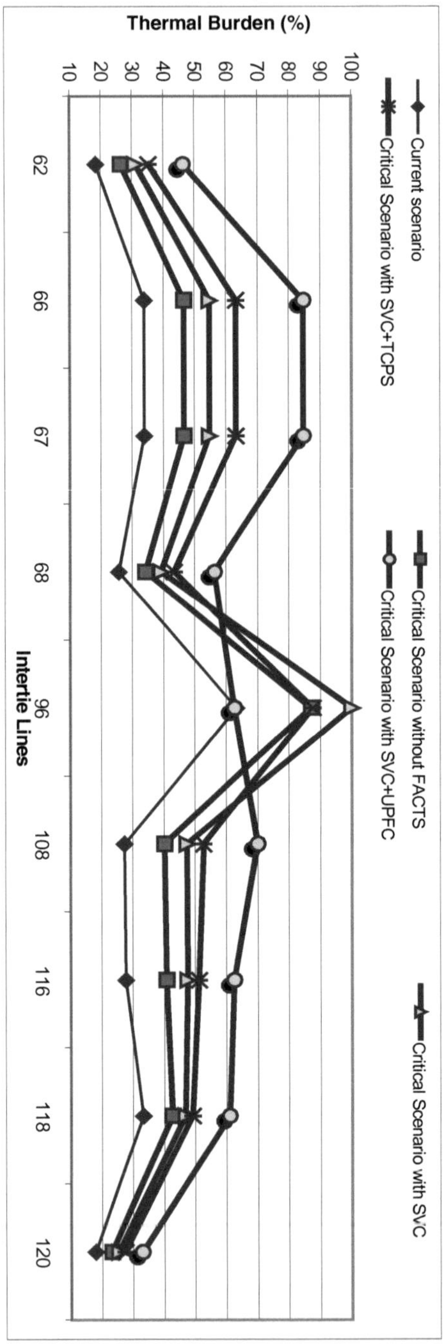

Figure 14.8. Thermal burden profiles of studied interface

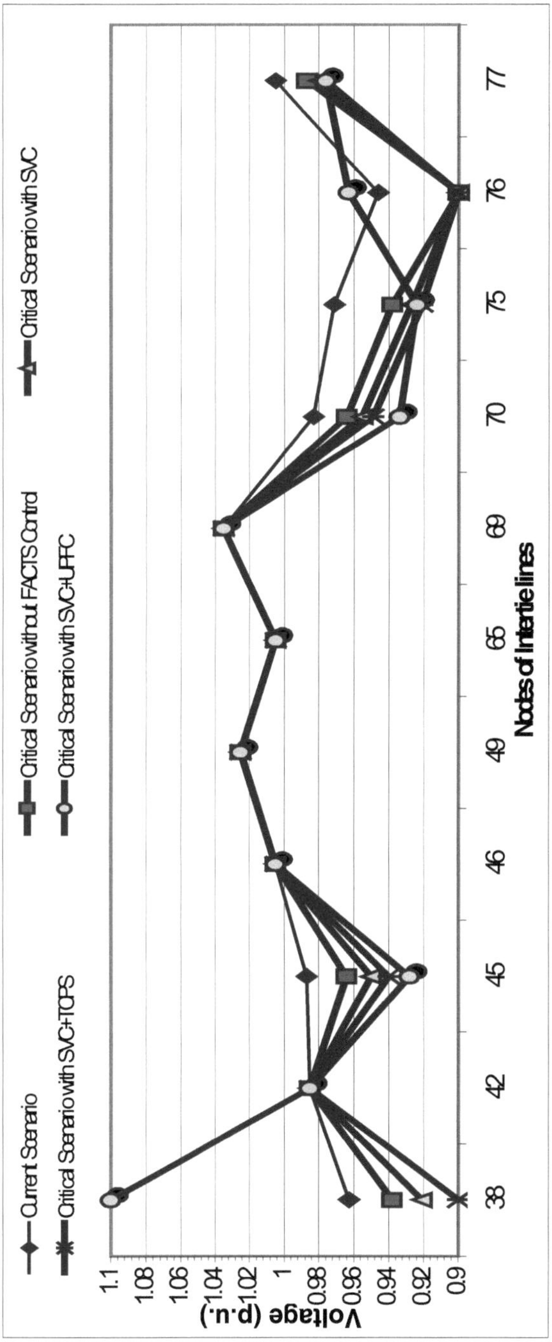

Figure 14.9. Voltage profiles of studied interface

14.2.3.3 ATC Enhancement With Control of SVC+TCPS (Case 3)

On the basis of case 2, to delay the occurrence of thermal limit violation for further ATC boosting, a viable solution is to alleviate the heavy load burden of line 96 by power flow redistribution. Aimed at this, besides the SVC on node 76 as in case 2, a Quadrature Booster (QB) type TCPS is applied on line 96. As case 3, the ATC enhancement is conducted with control of both the SVC and TCPS, where the results are given in Table 14.3. It is evident that with control of the series controller, the ATC value sees a further boost of 56.55 MW, which is about 19.50% of the original ATC level. Moreover, from the iteration numbers and CPU times of the two cases, the high efficiency of the proposed methodology is clearly seen power flow results for the interface in the critical scenario are given in Table 14.6; where the corresponding thermal burden profile and voltage profile are depicted in Figures 14.7 and 14.8 respectively.

It is to be noted that in case 1, the ATC is restricted by the voltage violation of node 76, while in case 2, further increase of transfer capability is prevented by thermal limit of line 96. The situations have been alleviated by the shunt controller and the series controller in case 3. These facts testified that, with the capability to eliminate nodal voltage and thermal limit violation, FACTS control can provide a promising alternative in ATC enhancement.

Table 14.6. Power flow results of the studied interface in the critical scenario with control of SVC and TCPS (Case 3, $\lambda_* = 1.612$)

Branch (I - J)	96 (38-65)	66, 67 (42-49)	68 (45-49)	62 (45-46)	108 (69-70)	116 (69-75)	120 (75-77)	118 (76-77)
P_{IJ} (p.u.)	-2.6210	-1.2058	-0.8660	-0.6640	2.0717	1.9988	-0.5443	-0.9758
Q_{IJ} (p.u.)	0.1465	0.3866	-0.0605	-0.2440	0.3996	0.4505	-0.0829	-0.1144
Thermal burden (%)	87.50	63.31	43.41	35.37	52.75	51.22	27.53	49.13

Node	38	42	45	46	49	65	69	70	75	76	77
V	0.900	0.985	0.940	1.00	1.025	1.005	1.035	0.949	0.921	0.900	0.976

14.2.3.4 ATC Enhancement With Control of SVC+UPFC (Case 4)

It is well known that, as the most advanced and versatile FACTS device, UPFC enables control of line flow and regulation of nodal voltage simultaneously. In order to further escalate the ATC by eliminating the critical voltage at node 38, instead of the TCPS in case 3, a UPFC is used on line 96 with the shunt part connected to node 38. Applying the SVC on node 76 and the UPFC on line 96 to control the power flow in case 4, the results are given in Table 14.3. Power flow results of the interface in the critical scenario are shown in Table 14.7. Comparing the ATC levels in case 3 and case 4, the considerable difference highlights the superior performance of the UPFC than the TCPS on ATC improvement. In this case, with the FACTS control to regulate the voltage of nodes 76 and 38, and to alleviate the heavy loading burden of line 96, the occurrence of a critical situation has been postponed from that of $\lambda_* = 1.365$ to $\lambda_* = 1.783$. Consequently, the ATC value sees a considerable increase of 385.99 MW.

It is concluded that, with the ability of flexible power flow control, FACTS devices can enhance ATC to a great degree, among them, as the most advanced FACTS devices with functions of supporting voltage and readjusting line flow simultaneously, UPFC can play an important and unique role in boosting the ATC.

It can be further calculated that, without line flow control of the UPFC to alleviate line thermal stress in the critical scenario of case 4, the thermal burden of line 96 will attain 183.80%, which demonstrates the effect of the UPFC series part on enhancement of the ATC. Figure 14.8 shows that in this case, voltages at nodes 76 and 38 have to be lifted to relatively high values, so as to prevent neighbouring nodes from violating the lower limit. In the critical scenario of case 4, the critical voltage point has been shifted from node 38 to node 74, which is 0.90 p.u., meanwhile, the heaviest thermal burden of all lines, line 41, has attained 99.94%.

Table 14.7. Power flow results of studied interface in critical scenario with control of SVC and UPFC (Case 4, $\lambda_* = 1.783$)

Branch (I - J)	96 (38-65)	66, 67 (42-49)	68 (45-49)	62 (45-46)	108 (69-70)	116 (69-75)	120 (75-77)	118 (76-77)
P_{IJ} (p.u.)	1.7206	-1.5470	-1.1243	-0.8959	2.7442	2.4682	-0.6611	-1.1614
Q_{IJ} (p.u.)	0.7534	0.6869	0.0396	-0.2274	0.5625	0.4053	-0.0166	0.3727
Thermal burden (%	62.61	84.63	56.25	46.22	70.03	62.53	33.06	60.99

Node	38	42	45	46	49	65	69	70	75	76	77
V	1.100	0.985	0.928	1.005	1.025	1.005	1.035	0.934	0.924	0.963	0.976

14.2.4 Remarks

In order to facilitate electricity market operation, sufficient transmission capability should be provided to satisfy the demand for increasing power transactions reliably. The conflict between this requirement and the restrictions on the transmission expansion in the unbundled power industry has motivated the development of methodologies to enhance the ATC of the existing transmission grid.

Based on operational limitations of the transmission system and control capabilities of FACTS technology, the technical feasibility of applying FACTS devices to boost ATC levels are analysed and identified.

From the point of view of operational planning, this section focuses on evaluation of the effects of FACTS devices on ATC enhancement. With respect to FACTS control, an ATC enhancement model is presented, which enables simulation of various FACTS controls by the use of the power injection model. With the IEEE 118-bus system as a test bed, case studies have been conducted on all categories of FACTS devices, covering shunt controllers, series controllers and unified controllers. The results demonstrated that the use of FACTS devices, particularly the UPFC which enables balance of line flow and regulation node voltage simultaneously, can enhance the ATC substantially. The considerable difference between the ATC values with and without FACTS control supports the EPRI's proposal concerning FACTS applications for ATC enhancement quantitatively. In summary, by boosting the usable power transmission capacity, FACTS technology can offer an effective and promising solution to the critical challenges posed by the present electricity market.

Finally, it is to be pointed out that the effect of FACTS devices on ATC enhancement is system dependent. For interfaces of the interconnected transmission network with relatively even load sharing and smooth voltage profile, the efficient solution has to rely on such transmission reinforcement oriented strategies as upgrading transmission lines and facilities, construction of new lines and so on.

14.3 FTR Auction Improvement by FACTS Control

The concept of FTR and the basic model of the FTR auction have been introduced in Chapter 8. The revenue from the FTR auction is determined by the transfer capability of transmission networks which may however be effected by the phenomenon of parallel flow or loop flow. It is recognised that, with flexible and effective power flow control abilities, FACTS devices could be used as an efficient measure to eliminate parallel flow and loop flow in transmission networks.

In this chapter, FACTS control is incorporated into the proposed FTR optimal auction model, with the objective of maximising revenues from use of the transmission network [20]. It also gives the market participants more opportunities to win their bids for FTRs. To make all the allocated FTRs simultaneously feasible, power flow equations and system operational limits are considered. Since the FTR auction is usually run monthly and only concerns active power, the DC power flow model is used here. Therefore, two types of series FACTS devices, TCSC and TCPS, are taken into account. The simplified PIM, which was presented earlier, is used to model these FACTS devices. The solution of this FTR optimal auction consists of the feasible sold FTRs and their prices and the derived control parameters of the FACTS devices. An 8-bus test system and a modified IEEE 30-bus system are studied to illustrate the proposed method. The comparison between cases with FACTS devices and without FACTS devices shows that proper control of FACTS devices can improve the result of FTR auction significantly.

14.3.1 Proposed Optimal FTR Auction Model

The proposed FTR auction model can be regarded as an extension of the general FTR auction model presented in Chapter 8, where two additional types of control variable are introduced:

- shifting angle of TCPS;
- reactance of TCSC.

In [21, 22] where the DC power flow model was adopted, phase shifter angle is replaced by compensation power injection at end buses while series compensation is modelled as variation in circuit reactance. The PIM of TCPS is shown to give no problem but treating circuit reactance as a control variable turns the corresponding nodal power balance constraints into non-linear equations. To ensure the problem still solvable by LP, a decomposition approach based on Benders decomposition scheme is applied in [21], while in [22] non-linear power flow control with FACTS devices was solved as a separate sub-problem.

In this chapter, a uniform PIM is adopted to represent both TCSC and TCPS to keep the problem linear and to simplify the programming. The mathematical model of the proposed method is formulated based on problem (8.X):

$$\text{Subject to: } \mathbf{B}'\boldsymbol{\theta} - \mathbf{M}\,\mathbf{P}_{FTR} - \mathbf{M}_B\mathbf{P}_B + \mathbf{M}_{(inj)}\mathbf{P}_{(inj)} = \mathbf{0} \tag{14.23}$$

$$\mathbf{P}_{FTR,\min} \le \mathbf{P}_{FTR} \le \mathbf{P}_{FTR,\max}$$

$$\mathbf{P}_{l,\min} \le \mathbf{H}\boldsymbol{\theta} \le \mathbf{P}_{l,\max}$$

$$\mathbf{P}_{(inj),\min} \le \mathbf{P}_{(inj)} \le \mathbf{P}_{(inj),\max}$$

where $\boldsymbol{P}_{FTR}$ is the matrix of winning FTR bids (MW), $\boldsymbol{b}_{FTR}$ is the matrix of bidding prices of FTR bidders, $\boldsymbol{P}_B$ is the matrix of FTR injections in the base case, $\boldsymbol{B'}$ is the linearised active power Jacobian matrix, $\boldsymbol{H}$ is the matrix of branch power flow constraint coefficients, $\boldsymbol{M}_B$ is the nodal mapping matrix of FTRs in the base case, $\boldsymbol{M}$ is the nodal mapping matrix of FTRs in the auction, subscript max and min represent the maximum and minimum limit of the variable.

Compared with problem (8.X), power injections associated with FACTS devices have been added into the nodal power balance constraints. $\boldsymbol{M}_{(inj)}$ is the connection matrix for FACTS devices, in which the elements are 1 or –1. The last set of inequality constraints is the operating limits of FACTS devices in the form of power injections. To make this model practical, constraints of FACTS internal parameters must be involved. For a TCPS, the limits of power injections can easily be derived from the limits of the phase shifter angle.

$$P_{I(inj),\min} = -b_{IJ}\psi_{\max},\; P_{I(inj),\max} = -b_{IJ}\psi_{\min} \tag{14.24}$$

But for a TCSC, the limits of power injections are more complex, as shown in (14.9). Since the power injections of the TCSC are also a function of voltage angles, which are the state linear programming variables, there are no fixed limits for power injections of the TCSC (though there are fixed limits for x_c). Therefore, its operating constraints can be formulated with two additional inequality constraints:

$$P_{I(inj)} + \frac{x_{c,\max}}{x_{IJ}(x_{IJ} - x_{c,\max})}(\theta_I - \theta_J) \geq 0 \tag{14.25}$$

and

$$P_{I(inj)} + \frac{x_{c,\min}}{x_{IJ}(x_{IJ} - x_{c,\min})}(\theta_I - \theta_J) \leq 0 \tag{14.26}$$

With the optimal solution of this problem obtained, control parameters of the TCSC and the TCPS can be derived respectively according to (14.27) and (14.28):

$$x_c^* = \frac{x_{IJ}^2 P_{I(inj)}^*}{-(\theta_I^* - \theta_J^*) + x_{IJ} P_{I(inj)}^*} \tag{14.27}$$

$$\psi^* = \frac{P_{I(inj)}^*}{-b_{IJ}} \tag{14.28}$$

The optimal control parameters of the FACTS devices obtained can be regarded as the long-term/mid-term set points of the FACTS devices. During real-time operation of the electricity market, these FACTS devices can also be controlled to relieve transmission congestion or to maintain system stability. However, as soon as the power system goes back to its normal state, the control parameters of the FACTS devices should be reset to the original operating points.

The proposed FTR optimal auction model with series FACTS devices can be actually represented by problem (14.23) plus the internal operating constraints of FACTS devices (14.24 to 14.26). This is a typical linear programming problem, which can easily be solved by Interior Point Primal-Dual Linear Programming.

14.3.2 Case Studies

To illustrate the validity of the proposed FTR optimal auction model, two test systems, an 8-bus system and a modified IEEE 30-bus system, are studied.

14.3.2.1 System I: 8-bus Test System

The 8-bus system is modified based on the first test system in [4]. The network configuration and the FTRs both in the base case and in the auction, are shown in Figure 14.10. The branch parameters and limits and the bid prices of six bidders in the FTR auction are given in Tables 1 and 4 of [4]. In addition, there is a TCPS installed on line 6 (Bus 4 to Bus 5) and a TCSC installed on line 2 (Bus 1 to Bus 4). From Figure 14.10, it can be seen that basically the system can be divided into left zone including Buses 1, 5 and 6 and right zone including the rest of the buses. The Left zone is a generation zone while the right zone is demand zone. The tie lines between these two zones are lines 1 (Bus 1 to Bus 2), 2 and 6.

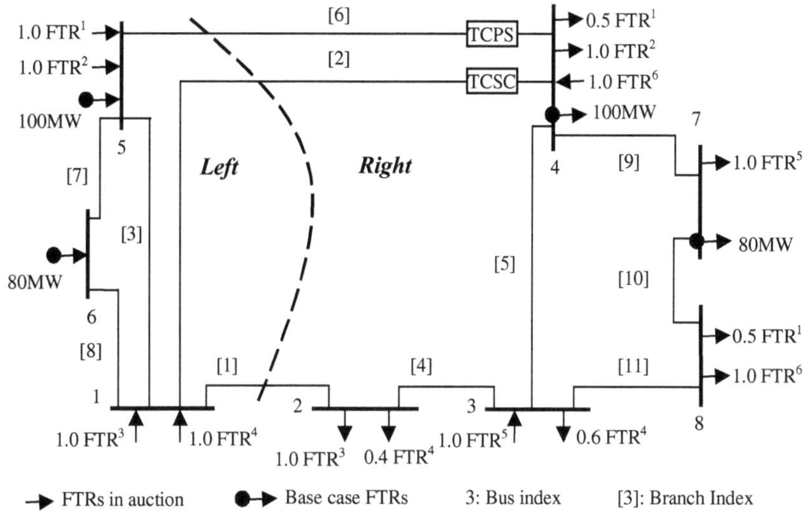

Figure 14.10. 8-bus test system

To demonstrate the impacts of different series FACTS devices on the results of the optimal FTR auction. Four cases are analysed here:

Case I: FTR auction without FACTS devices;

Case II: FTR auction with a TCSC on line 2;

Case III: FTR auction with a TCPS on line 6;

Case IV: FTR auction with both TCSC on line 2 and TCPS on line 6.

The line flows resulting from the FTR auction in four cases are shown in Table 14.8. The revenues from FTR auction in four cases are given in Table 14.9. The operating limits and the optimal control parameters of FACTS devices in Cases II, III, and IV are listed in Table 14.10, together with their corresponding power injections. In Case I, because of the bottleneck between the left zone and right zone, not all the FTR bids can be satisfied and the revenue from the FTR auction is limited. From Table 14.8, it can be noted that MW flow constraints of line 1 and line 6 are bound while line 2 still has nearly two-thirds of its transfer capacity left. To improve the transfer capability between left zone and right zone, the TCSC on line 2 is controllable in Case II while the TCPS on line 6 is controllable in Case III. The results of the both cases are better than Case I. However, the TCSC on line 2, which increases the power flow on line 2 to 296.58MW from the value of 123.69MW in Case I, has better impact on the FTR auction than the TCPS on line 6. The reason is that in Case III a new binding constraint of line flow (line 4) occurs.

Table 14.8. Line flows of auction results in four cases

Line	From Bus	To Bus	MW Flow in Case I	MW Flow in Case II	MW Flow in Case III	MW Flow in Case IV	MW Limit
1	1	2	150.00	150.00	150.00	150.00	150
2	1	4	123.69	296.58	220.63	340.00	340
3	1	5	-121.44	-236.03	-266.57	-311.98	380
4	2	3	50.85	96.64	120.00	52.60	120
5	3	4	-43.26	-82.35	30.63	-54.20	230
6	4	5	-150.00	-150.00	-150.00	-150.00	150
7	5	6	26.90	-10.35	-5.94	0.62	300
8	6	1	53.10	69.65	74.06	80.62	250
9	7	4	-258.16	-288.56	-195.94	-273.20	350
10	7	8	58.16	88.55	-4.06	73.20	340
11	8	3	-214.11	-211.44	-209.37	-226.80	240

In Case IV, both FACTS devices are controlled in the FTR auction. Obviously this is the best result of the four cases. From Table 14.8, it can be seen that all the three tie lines have reached their thermal limits in this case. In other words, the total transmission capability between left zone and right zone has been fully utilised with the control of both FACTS devices. Compared with Case II, the reactance of TCSC also reaches its maximum value but its power injection is different from Case II. The reason is that the power injection of TCSC is also a function of voltage angle difference between bus 1 and bus 4, which varies in each case. The results show that the coordination between the two FACTS devices works very well. It is due to directly embedding the power injection models of FACTS devices into the linear programming problem instead of solving the control sub-problems of FACTS devices separately.

Table 14.9. Revenues from FTR auction in four cases

Cases	Case I	Case II	Case III	Case IV
Revenues ($)	5595.24	7386.71	6167.56	7874.80

14.3.2.2 System II: 30-bus Test System

The IEEE 30-bus test system is modified to further illustrate the proposed FTR auction approach and its application to a bigger network. The bus and branch data are given in [4]. The layout of the system as well as injection and extraction points of FTRs for purchases and sales of the different bidders are shown in Figure 14.11. The data of bidders who take part in the FTR auction are given in Table 10 of [4]. The base case FTR data of the existing FTR holders who will not participate in the FTR auction are shown in Table 11 of [4].

As shown in Figure 14.11, there are three series FACTS devices installed in the system: one TCSC on line 5 (Bus 2 to Bus 5), two TCPSs on line 18 (Bus 12 to Bus 15) and line 27 (Bus 10 to Bus 21).

Table 14.10. Optimal control parameters of the FACTS devices

Cases		Optimal Value	Min Value	Max Value
Case II	x_c (TCSC)	0.02	0.00	0.02
	Power Injection at Bus 1 (MW)	-197.72	-197.72	0.00
Case III	ψ (TCPS)	0.0384	-0.10	0.10
	Power Injection at Bus 5 (MW)	-128.38	-333.33	333.33
Case IV	x_c (TCSC)	0.02	0.00	0.02
	Power Injection of TCSC at Bu (MW)	-226.67	-226.67	0.00
	ψ of the TCPS	0.0093	-0.10	0.10
	Power Injection of TCPS at Bu (MW)	-30.93	-333.33	333.33

To demonstrate the impacts of FACTS devices on the FTR auction, two cases will be studied here:

Case I: FTR auction without any FACTS device;

Case II: FTR auction with all the FACTS devices.

The auction results in the two cases, as obtained by solving the auction optimization problems (8.X) and Equation (14.20) respectively, are given in Table 14.11, which shows the distribution of FTRs in terms of bidders. Obviously, the auction result of Case II is much better than the result of Case I, since more FTRs have been traded (purchased or sold) in Case II under the control of the FACTS devices.

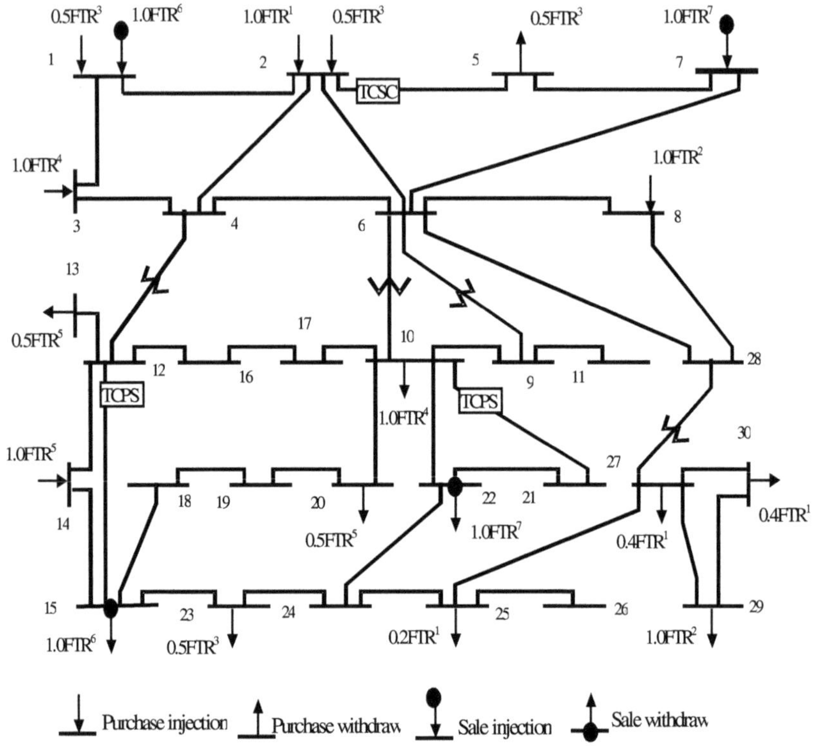

Figure 14.11. Modified IEEE 30-bus system and bidders in the FTR auction

Table 14.12 shows the branch MW flows of the FTR auction results in both cases. In the results of Case I, lines 6, 22, 29 and 37 are congested. As can be seen in Figure14.11, In the IEEE 30-bus system generation is mostly located on the up-side and demand is generally located on the downside. Consequently, the congested lines strongly limit the total transfer capability of the network. For instance, although line 6 is congested, some of its "parallel paths", such as line 4, 5, 7, and 8, still have plenty of transmission capacity left. But meanwhile bidder 4's FTR bid (from Bus 4 to Bus 10) cannot be satisfied at all in Case I. In Case II, the TCSC on line 5 becomes control variable in the FTR auction. As a result, the MW flows on lines 4, 5, 7, and 8 increase significantly while the MW flow on line 6 is still at its thermal limit. Consequently, bidder 4 wins an FTR of 85.4MW in the auction. The two TCPSs on line 18 and line 27 also improve utilisation of the network in Case II. It can be noticed that in Case II two new binding branch constraints appear. They are lines 17 and 28. The optimal control parameters of the three FACTS devices and their internal operating limits in Case II are given in Table 14.13, in the forms of both original parameters and equivalent power injections. The TCSC on line 5 has reached its operating limit.

Table 14.11. Auction results in IEEE 30-bus system

Bidder	FTR value in Case I (MW)	FTR value in Case II (MW)	FTR^{max} (MW)	FTR Type
1	27.0	45.6	100	Purchase
2	16.1	11.9	75	Purchase
3	120	120.0	120	Purchase
4	0.0	85.4	90	Purchase
5	0.0	0.0	80	Purchase
6	58.5	56.3	70	Sale
7	4.3	30	100	Sale

Table 14.12. Branch flows of FTR auction results of IEEE 30-bus system

Branch No.	From Bus	To Bus	MW Flow in Case I	MW Flow in Case II	MW Flow Limit
1	1	2	28.15	47.89	130
2	1	3	73.32	55.83	130
3	2	4	64.13	53.75	65
4	3	4	-21.68	46.19	130
5	2	5	66.00	114.74	130
6	2	6	65.00	65.00	65
7	4	6	7.74	51.28	90
8	5	7	6.00	54.74	70
15	12	4	-34.71	-48.66	65
16	12	13	-20.00	-20.00	65
17	12	14	14.62	32.00	32
18	12	15	5.24t	-14.78	32
19	12	16	0.85	17.44	32
20	14	15	-15.38	2.00	16
21	16	17	-5.15	11.44	16
22	15	18	16.00	16.00	16
23	18	19	-4.00	-4.00	16
24	19	20	-14.00	-14.00	32
25	10	20	19.00	19.00	32
26	10	17	5.15	-11.44	32
27	10	21	-12.00	-12.00	32
28	10	22	-11.03	-32.00	32
29	21	22	-32.00	-32.00	32
30	15	23	2.38	-2.49	16
36	28	27	19.42	33.97	65
37	27	29	16.00	16.00	16

Table 14.13. Optimal FACTS control parameters in Case II

		Optimal Value	Min Value	Max Value
TCSC on Line 5	x_c	0.15	0.00	0.15
	Power Injection at Bus 2 (MW)	-86.79	-86.79	0.00
TCPS on Line 1	ψ	-0.105	-0.20	0.20
	Power Injection at Bus 12 (MW)	80.62	-153.37	153.37
TCPS on Line 2	ψ	0.0314	-0.20	0.20
	Power Injection at Bus 10 (MW)	-41.96	-267.02	267.02

The nodal Market Clearing Prices (MCPs) (including both base case and auction) are given in Figure 14.12 for both cases. These prices provide valuable information for all the FTRs on how much they are worth. This information can be used for exchanging FTRs in the bilateral secondary market. Negative FTR MCPs mean that these FTRs are counter-flows, which could help eliminate network congestion.

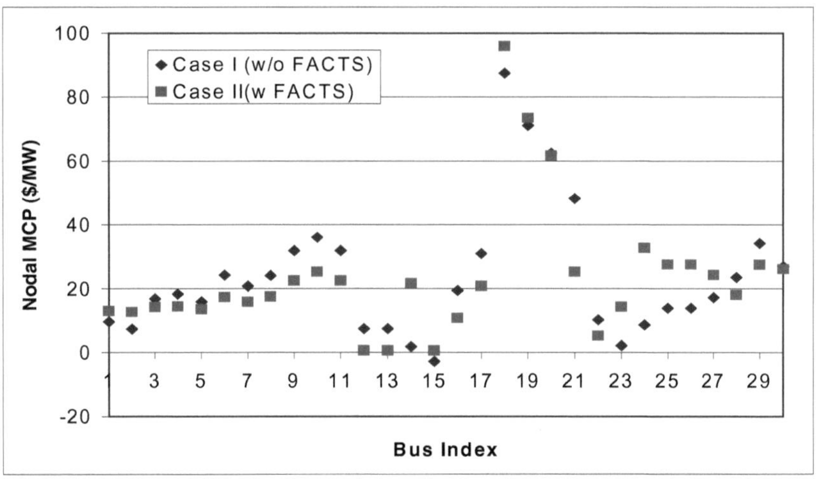

Figure 14.12. Nodal MCP results in two cases

14.3.2.3 Discussion

Compared with ATC-based curtailment, the LMP/FTR approach allows participants to buy through transmission congestion and provides a more market-based way to transmission expansion. From the test results, it can be noticed that the FTR auction is improved because the transfer capability of transmission networks is enhanced by FACTS control. But the derived optimal control parameters of FACTS devices for FTR auction would be different from the control settings for ATC enhancement. The main reason is that in ATC calculations system load increases with a certain load distribution factor while in the FTR auction the load pattern is determined by individual FTR bids, which are not necessary to match the real dispatch.

14.3.3 Remarks

Two types of series FACTS devices, TCSC and TCPS, are embedded into the FTR optimal auction problem through the PIM to allocate more existing transmission capabilities to market participants as FTRs. Some new special treatments have been applied to TCSC's internal parameter limits to keep the auction problem as one linear problem.

The proposed FTR auction model with FACTS devices is solved as one LP problem instead of dealing with the control of FACTS devices as separated subproblems. This is very important for the coordination between multiple FACTS devices in the system. An 8-bus system and the IEEE 30-bus system are analysed to illustrate this model. The test results show the differences between the auction with/without FACTS devices and the different impacts from controlling different FACTS devices at different locations. The best solution comes from appropriate coordination of FACTS devices.

14.4 References

[1] L Gyugyi (1999) "Flexible AC Transmission Systems (FACTS)", Chapter 1

[2] Y H Song & A T Johns (Edt.), IEE Power and Energy Series 30

[3] Nabavi-Niaki, M R Iravani (1996) "Steady-state and dynamic models of Unified Power Flow Controller (UPFC) for power system studies", IEEE Transactions on Power Systems, Vol. 11, No. 4 , Nov., pp. 1937 –1943

[4] D J Gotham, G T Heydt (1998) "Power flow control and power flow studies for systems with FACTS devices", IEEE Trans. on Power Systems, Vol.13, No.1, Feb., pp.60-65

[5] CIGRE TF38-01-06-Final Report (1995) Load flow control in high voltage systems using FACTS controllers, Oct.

[6] Z X Han (1982) "Phase shifter and power flow control", IEEE Trans. on PAS, Vol.101, No. 10, pp.3790-3795

[7] M Noroozian, G Andersson (1993) "Power flow control by use of controllable series components", IEEE Trans. on Power Delivery, Vol.8, No.3, July, pp.1420 –1429.

[8] Xiao Ying, Y H Song, Y Z Sun (2002) "A novel power flow control approach to power systems with embedded FACTS devices", IEEE Trans. on Power Systems, to appear

[9] Xiao Ying (2001) "Operational optimization of FACTS control for improving transmission services", Ph.D. Thesis, Brunel University, London

[10] Y Dai, J D McCalley, V Vittal (1999) "Simplification, expansion & enhancement of direct interior point algorithm for power system maximum loadability", Proceedings of the 21st International Conference on Power Industry Computer Applications, pp.170-179

[11] E De Tuglie, M Dicorato, M La Scala (1999) "A static optimization approach to assess dynamic available transfer capability", Proceedings of the 21st International Conference on Power Industry Computer Applications, , pp.238-246
[12] North American Electric Reliability Council (NERC) (1996) Available transfer capability definition and determination, June
[13] N G Hingorani, L Gyugyi (2000) "Understanding FACTS, concepts and technology of flexible ac transmission systems", The Institute of Electrical and Electronics Engineers, Inc., New York
[14] EPRI (1997) "FACTS assessment study to increase the Arizona-California transfer capability", Report TR-107934, May
[15] California Energy Commission (CEC) (1999) Flexible AC Transmission Systems benefits study, San Diego Gas and Electric, Oct.
[16] Stahlkopf Karl (1999) "Comments of the Edison Electric Institute on regional transmission organizations in Docket", No. RM99-2-000, EEI Energy Issues/News, EPRI in USA, Feb.
[17] Xiao Ying, Y H Song, Y Z Sun (2002) "Available transfer capability enhancement using FACTS devices", Accepted by IEEE Trans. on PS
[18] S Mehrotra (1992) "On the implementation of a primal-dual interior point method", SIAM Journal on Optimization, Vol.2, pp.575-601
[19] S Iwamoto, Y Tamura (1981) "A load flow calculation method for ill-conditioned power systems", IEEE Trans. on PAS, Vol.100, No.4, pp. 1736-1743
[20] IEEE 118-Node System, "Power systems test case archive", University of Washington, USA, http://www.ee.washington.edu/research/pstca/.
[21] X Wang, Y H Song, Q Lu, Y Z Sun (2002) "Optimal Allocation of Transmission Rights in A Network with FACTS Devices", IEE Proceedings on Generation, Transmission and Distribution, Vol.149, No.3, May, pp.359-366
[22] G N Taranto, L M V G Pinto, M V F Pereira (1992) "Representation of FACTS devices in power systems economic dispatch", IEEE Transactions on Power Systems, May, 7(2), pp.572-576
[23] S Y Ge, T S Chung (1999) "Optimal active power flow incorporating power flow control needs in flexible AC transmission systems", IEEE Transactions on Power Systems, May 14(2), pp. 738-744
[24] M I Alomoush, S M Shahidehpour (2000) "Generalized model for fixed transmission rights auction", Electric Power Systems Research, 54, pp.207-220

Index

MIX
Papier aus verantwortungsvollen Quellen
Paper from responsible sources
FSC® C105338

If you have any concerns about our products,
you can contact us on
ProductSafety@springernature.com

In case Publisher is established outside the EU,
the EU authorized representative is:
Springer Nature Customer Service Center GmbH
Europaplatz 3, 69115 Heidelberg, Germany

Printed by Libri Plureos GmbH
in Hamburg, Germany